(*continued on back*)

Chemical Analysis of
Polycyclic Aromatic Compounds

CHEMICAL ANALYSIS

A SERIES OF MONOGRAPHS ON
ANALYTICAL CHEMISTRY AND ITS APPLICATIONS

VOLUME 101

A WILEY-INTERSCIENCE PUBLICATION

JOHN WILEY & SONS

New York / Chichester / Brisbane / Toronto / Singapore

Chemical Analysis of Polycyclic Aromatic Compounds

Edited by

TUAN VO-DINH

Advanced Monitoring Development Group
Health and Safety Research Division
Oak Ridge National Laboratory
Oak Ridge, Tennessee

WILEY

A WILEY-INTERSCIENCE PUBLICATION

JOHN WILEY & SONS

New York / **Chichester** / **Brisbane** / **Toronto** / **Singapore**

Library of Congress Cataloging in Publication Data:

Chemical analysis of polycyclic aromatic compounds/edited by Tuan Vo-Dinh.
    p. cm.—(Chemical analysis, ISSN 0069–2883; v. 101)
    "A Wiley-Interscience publication."
    Includes bibliographies and index.
    ISBN 0-471-62889-1
    1. Polycyclic aromatic compounds—Analysis.   I. Vo-Dinh, Tuan.
II. Series.
QD335.C48   1988   88–19095
547'.6—dc 19   CIP

Printed in the United States of America

10 9 8 7 6 5 4 3 2 1

To Kim-Chi and Michelle

CONTRIBUTORS

H. J. Bowley, BP Research Centre, Sunbury-on-Thames, Middlesex, England. *Present address*: Department of Chemistry, Wake Forest University, Winston-Salem, North Carolina

R. L. M. Dobson, Proctor & Gamble Company, Cincinnati, Ohio

A. P. D'Silva, Ames Laboratory, Iowa State University, Ames, Iowa

John C. Fetzer, Chevron Research Company, Richmond, California

Zheng-Sheng Fu, Department of Chemistry, Laboratory of Analytical Micellar Chemistry, Wake Forest University, Winston-Salem, North Carolina. *Present address*: Chemistry Department, Northwestern Teachers' College, Lanzhou City, Gansu Province, People's Republic of China

D. L. Gerrard, BP Research Centre, Sunbury-on-Thames, Middlesex, England

Willie L. Hinze, Department of Chemistry, Laboratory of Analytical Micellar Chemistry, Wake-Forest University, Winston-Salem, North Carolina

Ronald A. Hites, School of Public and Environmental Affairs and Department of Chemistry, Indiana University, Bloomington, Indiana

Donald J. Kippenberger, Department of Chemistry, Laboratory of Analytical Micellar Chemistry, Wake Forest University, Winston-Salem, North Carolina. *Present address*: U.S. Army Environmental Hygiene Agency, Aberdeen Proving Grounds, Maryland

Douglas A. Lane, Environment Canada, Atmospheric Environment Service Downsview, Ontario, Canada

Gleb Mamantov, Department of Chemistry, University of Tennesse, Knoxville, Tennessee

Linda B. McGown, Department of Chemistry, Duke University, Raleigh-Durham, North Carolina

Martin D. Morris, Department of Chemistry, Laboratory of Analytical Micellar Chemistry, Wake Forest University, Winston-Salem, North

Carolina. *Present address*: U.S. Army Environmental Hygiene Agency, Aberdeen Proving Grounds, Maryland

Michael D. Morris, Department of Chemistry, University of Michigan, Ann Arbor, Michigan

Kasem Nithipatikom, Department of Chemistry, Duke University, Raleigh-Durham, North Carolina

Fawzy S. Sadek, Department of Chemistry, Laboratory of Analytical Micellar Chemistry, Wake Forest University, Winston-Salem, North Carolina. *Present address*: Department of Physical Science, Winston-Salem State University, Columbia Heights, Winston-Salem, North Carolina

Regina M. Santella, Comprehensive Cancer Center and Division of Environmental Sciences, School of Public Health, Columbia University, New York, New York

H. N. Singh, Department of Chemistry, Laboratory of Analytical Micellar Chemistry, Wake Forest University, Northwestern Teachers' College, Winston-Salem, North Carolina. *Present address*: Department of Chemistry, Aligarh Muslim University, Aligarh, India

Konstadinos Siomos, Institute of Materials Structure and Laser Physics, Technical University of Crete, Crete, Greece

Richard D. Smith, Chemical Methods and Separations Group, Chemical Sciences Department, Battelle-Pacific Northwest Laboratories, Richland, Washington

Marina Stefanidis, Comprehensive Cancer Center and Division of Environmental Sciences, School of Public Health, Columbia University, New York, New York

Phil Stout, Department of Chemistry, University of Tennessee, Knoxville, Tennessee. *Present address*: Bio-Rad Digilab Division, Cambridge, Massachusetts

Tuan Vo-Dinh, Advanced Monitoring Development Group, Health and Safety Research Division, Oak Ridge National Laboratory, Oak Ridge, Tennessee

S. J. Weeks, Ames Laboratory, Iowa State University, Ames, Iowa

Ronald W. Williams, Department of Chemistry, Laboratory of Analytical Micellar Chemistry, Wake Forest University, Winston-Salem, North Carolina. *Present address*: U.S. Army Environmental Hygiene Agency, Aberdeen Proving Grounds, Maryland

Bob W. Wright, Chemical Methods and Separations Group, Chemical Sciences Department, Battelle-Pacific Northwest Laboratories, Richland, Washington

M. Zander, Rütgerswerke AG, Castrop-Rauxel, Federal Republic of Germany

PREFACE

Polycyclic aromatic compounds (PACs) are ubiquitous environmental pollutants that represent the largest class of suspected chemical carcinogens. PACs have been a topic of great interest in a wide spectrum of research disciplines ranging from analytical chemistry to biology, toxicology, and epidemiology, as evidenced by the many monographs devoted to PACs in the last decade. In the last few years, important advances have been made in improving existing analytical methods and developing new techniques for analysis of PACs. Many of these new techniques have appeared in technical journals and symposium proceedings but have not been critically reviewed in a comprehensive monograph. The need for such a monograph is critical because much instrumentation and many analytical technologies have emerged since the early 1980s.

Probably one of the most formidable challenges in chemical analysis of PACs is the characterization of complex mixtures. PACs generally occur in "real-life" samples as complex mixtures that vary greatly in concentration of individual components. The sensitivity and specificity required of analytical techniques are critical factors for environmental assessments and human health studies because compound structure can drastically affect the biological activity of PACs.

In the quest for elucidation of PAC carcinogenicity there is a continuing evolution from investigations of the parent homocyclic systems, the polycyclic aromatic hydrocarbons (PAHs), to investigations of heterocyclic PAC systems. It is noteworthy that many PAHs undergo photochemical reactions in the atmosphere, leading to the production of heterocyclic PACs. Because PAHs cannot account for all the biological activities of many samples, this evolution has led to increased recognition of the important role of heteroatoms upon PAC genotoxicity. It is, therefore, a purpose of this monograph to deal with the broader class of PACs, which include both homocyclic and heterocyclic species.

Recently, analytical methods and instrumentation have experienced dramatic development and growth. New developments in experimental procedures and stationary phases for gas chromatography, high-performance liquid chromatography, and supercritical fluid chromatography have led to improved analysis of PACs. Luminescence spectroscopy, with its inherent sensitivity for aromatic systems, has greatly benefited from the advances in laser and detector technologies. Multicomponent analysis of PAC mixtures can be performed more effectively using phase-resolved detection. Simple and rapid techniques such as synchronous luminescence and room-temperature phosphorescence provide cost-effective screening of complex samples, whereas site-selection or line-narrowing techniques offer selective means for chemical analysis. These advances have also been complemented by new developments in micelle-mediated separation and analysis methods. New developments in mass spectroscopy and in Fourier transform and resonance-enhanced multiphoton ionization mass spectrometry have further improved the structural characterization of complex mixtures. Infrared and Fourier transform infrared spectroscopies have complemented Raman spectroscopy in both qualitative and quantitative analysis of PACs. Innovative techniques such as UV-resonance Raman and surface-enhanced Raman spectroscopies have made significant advances in improving the sensitivity of detection. A new generation of analytical techniques based on photoionization spectroscopy, photothermal spectroscopy, and immunoadsorbent techniques has also been developed and applied to PAC analysis.

Early detection, understanding, and, ultimately, prevention of PAC-related carcinogenesis are among the most important challenges facing this and future generations. The development of effective methods and instrumentation for chemical analysis should provide a critical contribution toward achieving these important goals and toward ensuring the development of ecologically viable and safe technologies. It also provides the necessary tools to support the establishment of effective strategies and rational policies to protect ecosystems as well as the health and well-being of people.

It is our hope that the information contained in this monograph will foster the critical and creative thinking needed to develop the full potential of analytical techniques for the important class of PACs and will contribute to fundamental basic knowledge while ensuring harmonious relationships among mankind, the environment, and technology.

Oak Ridge, Tennessee   TUAN VO-DINH
January 1989

ACKNOWLEDGMENTS

It is a great pleasure for me to acknowledge, with gratitude, the contributions of the authors of each chapter in this book. Their contributions provide a valuable forum for discussion and examination of the most recent advances in methodologies and instrumentation for chemical analysis of polycyclic aromatic compounds. I wish to express my deep gratitude to Dr. J. D. Winefordner for his kind encouragement and advice. I would like to thank Drs. P. J. Walsh, S. V. Kaye, and C. R. Richmond at Oak Ridge National Laboratory, as well as Drs. R. Wood, P. Duhamel, and G. Goldstein at the U.S. Department of Energy, for their continued support throughout this undertaking. The sponsorship of the Office of Health and Environmental Research, U.S. Department of Energy, under contract DE-AC05-84OR21400 with Martin Marietta Energy Systems, Inc., is gratefully acknowledged.

The completion of this work has been made possible with the encouragement, love, and inspiration of my wife Kim-Chi and my daughter Michelle.

TUAN VO-DINH

CONTENTS

Chemical Analysis of
Polycyclic Aromatic Compounds

CHAPTER

1

SIGNIFICANCE OF CHEMICAL ANALYSIS OF POLYCYCLIC AROMATIC COMPOUNDS AND RELATED BIOLOGICAL SYSTEMS

TUAN VO-DINH

Advanced Monitoring Development Group
Health and Safety Research Division
Oak Ridge National Laboratory
Oak Ridge, Tennessee

1. INTRODUCTION

We are exposed daily to an environment containing a myriad of potentially harmful substances, both natural and anthropogenic. An important class of these substances has been termed the *polycyclic aromatic compounds* (PACs). The development of effective methods and instrumentation for trace detection of PACs in the environment and for assessment of their potential hazards, ecological impacts, and human health risks is critical for the achievement of environmentally viable and safe technologies. Problem areas pertaining to identification of specific compounds or classes of compounds, i.e., analysis of complex mixtures, estimation of realistic dose regimes, and determination of biological effects, continue to create new challenges to chemical analysis of PACs.

2. POLYCYCLIC AROMATIC COMPOUNDS: FROM PAH TO PAC

For the past few centuries, PACs have been studied with great interest because many of these species have been shown to be potent carcinogens and/or mutagens. More than two centuries ago, skin cancer was observed in chimney sweeps, and later on it was observed in workers at tar factories. Frequent contact with soot and tar materials were correlated with the incidence of skin cancer. Later investigations indicated that the carcinogenic activity of these materials can be correlated with their PAC content, because PACs have been found to be carcinogenic and/or mutagenic in laboratory animal experiments. Extensive investigations led to the observation in 1930 of

1

dibenz[a,h]anthracene as the first chemical carcinogen (1) and to the isolation in 1933 of benzo[a]pyrene as a cancer-producing hydrocarbon from coal tar (2). Subsequently, numerous other polycyclic hydrocarbons were synthesized and tested for carcinogenicity (3–5). Research efforts were further devoted in the 1960s and 1970s to investigation of the parent homocyclic polyaromatic hydrocarbon (PAH) compounds. Extensive studies were performed to study *in vivo* and *in vitro* metabolism of a wide range of PAHs, including anthracene, benzo[a]anthracene, benzo[a]pyrene, dibenz[a,c]anthracene, dibenz[a,h]-anthracene, dibenzo[a,h]pyrene, dibenzo[a,i]pyrene, 7,12-dimethylbenzo[a]-anthracene, 7- and 12-methyl[a]anthracenes, 3-methylcholanthrene, and pyrene (6–15). These investigations laid the foundations for the basic knowledge on metabolic pathways of PAHs and led to subsequent studies that unveiled that the covalent binding of PAHs to DNA and protein was mediated by metabolic processes involving formation of epoxides (16, 17). The importance of the parent PAH structure was illustrated by the Pullmans' theoretical attempt to correlate the molecular reactivity of PAHs to their carcinogenicity. Pullman and Pullman (18) have associated the carcinogenic activity of these compounds with the presence of a chemically reactive phenanthrene-type double bond (K-region). They have suggested further that the primary process in cancer induction is an addition reaction of the cellular component at the K-region. The "bay region" theory of PAH carcinogenesis, which correlates carcinogenicity with ease of formation of a carbonium ion in a diol epoxide, has also been quite effective in explaining and predicting relative carcinogenicities (19). It was suggested that the importance of the bay region may be its ability to sterically hinder arene oxide detoxification processes as opposed to the usually stressed increased reactivity of diol epoxides in the region (20).

Over the last decade, the search for major genotoxic compounds has evolved from investigation of the parent PAHs (21) to the study of PAC systems (22). The emphasis evolving from the basic PAHs to heterocyclic PACs is a result of observations that homocyclic aromatic compounds cannot account for all the biological activities encountered in many environmental and biological samples. For example, it is clear that the PAH class itself is responsible for only a small fraction of the mutagenic activity of ambient airborne particulate matter. In a series of three sequential extractions of airborne particulates, the first nonpolar cyclohexane extract exhibited about one-third of the direct mutagenicity of the sum of the remaining two, namely, dichloromethane and acetone extracts (23). The study indicated that PAHs were not responsible for most of the mutagenicity of the cyclohexane extract. Another study indicated that approximately half of the mutagenic activity of some experimental samples was attributable to nitrated compounds (24).

With the evolution of interest from the parent PAHs to the more complex PACs has come the urgent need for improved or new analytical methods. Analytical techniques must be able not only to differentiate compounds with different benzenoid-ring sizes but also to identify specific substitute and/or derivative chemical groups attached to the basic structures.

An important problem area in chemical analysis is the characterization of complex mixtures. Whereas the identification and quantification of a specific compound at trace levels remain important goals, the ability to analyze complex mixtures has become a major focus. The importance of complex mixtures has arisen from the need to identify, monitor, and understand synergistic and/or antagonistic effects of multicomponent systems. One should emphasize that we have little or no unequivocal evidence that specific PACs cause cancer in humans, but we do have a wealth of evidence that many complex mixtures containing PACs do cause human and animal cancer.

Finally, another area of great importance to PAC research is the application of analytical techniques to "biological monitoring" of human exposure to PACs. This is a challenging area for research and development, and advances in analytical methodology and instrumentation are critically needed to analyze complex biochemical systems in the attempt not only to monitor the presence of PACs in the environment but also to assess the ultimate human health effects associated with PACs.

3. DEFINITION OF PACs

PACs comprise a complex class of condensed multinumbered benzenoid-ring compounds. The parent homocyclic species, which contain only carbon and hydrogen, are the familiar PAHs. In addition to the PAHs, there are thousands of substituted PAHs that could have various substitute groups, such as alkyl, amino, chloro, cyano, hydroxy, oxy, or thio groups. Also, there is a wide spectrum of PAH heterocyclic derivatives that contain one or several heteroatoms such as nitrogen, oxygen, or sulfur in the aromatic structure.

Over the past few decades, the family of PACs has been known by various names. Historically, the term "PAH" has been most widely used because the early interest was devoted to carcinogenic activities of the basic homocyclic system. Benzo[a]pyrene (BaP), which has been extensively investigated, is the most familiar PAH compound.

Nomenclature problems arise when other groups of PACs, which contain heterocyclic substituents or derivatives, are described. The use of different names by various research groups has often led to confusion. Some investigators have referred to PACs as *polycyclic organic compounds* (POCs), whereas others have preferred the term *polycyclic organic matter* (POM). The term *polynuclear aromatic* (PNA) *compounds* has also been frequently used. At present, no one term has achieved universal acceptance. To describe the class of polyaromatic compounds in a broader sense (i.e., including homocyclic as well as heterocyclic compounds) the term *polycyclic aromatic compounds* is used in this monograph.

4. OCCURRENCE AND DETECTION OF PACs

PACs are generally formed during incomplete combustion or pyrolysis of organic matter containing carbon and hydrogen. Because combustion of

Table 1. Major PAC Sources and Estimated BaP Emissions in the United States (1971–1973)[a]

Source	BaP Emissions (tons/year)	Percentage of Total Emissions
Coal refuse fires	310	34.7
Residential furnaces, coal (hand-stoked)	300	33.6
Coke production	170	19.0
Vehicle disposal (open burning)	25	2.8
Wood burning (fireplaces, etc.)	25	2.8
Mobile sources, gasoline	11	1.2
Tire degradation	11	1.2
Forest and agricultural refuse burning	11	1.2
Open refuse burning (domestic-municipal)	10	1.1
Intermediate coal furnaces	7	0.8
Enclosed incineration (apartment-municipal)	3	0.3
Other (coal steam power plants, asphalt air blowing, oil and natural gas combustion, diesel-powered vehicles)	4	0.5
Totals	894	100.0

[a] From U.S. Environmental Protection Agency (ref. 27) and from ref. 25.

organic materials is involved in countless natural processes or human activities, PACs are omnipresent and abundant pollutants in air, soil, and water. PAC formation during combustion of a wide variety of materials has been extensively investigated. Some precursors include: methane; other simple saturated and unsaturated hydrocarbons; peptides; lipids; carbohydrates; lignins; terpenes; nicotine; and leaf pigments (see ref. 25 and references therein). Formation of PACs is most favored by pyrolysis (or air-deficient combustion) at temperatures in the range of 650–900 °C (26). At present, it is not possible to list a complete inventory of all PACs produced by the most common sources of emission. So far, such an inventory is compiled for only a few sources, and it deals with a limited number of PACs. Table 1 provides a list of some frequent sources of PACs (25, 27). The various formation processes and emission sources of PACs have been previously described in detail in a number of monographs (28–35) and are therefore not described again here. The following examples are not intended to be exhaustive but, instead, is aimed to illustrate the wide variety of sources of PACs.

4.1.   Energy Sources

Among energy-related products, fossil fuels are the major sources of PACs. The primary sources of airborne PACs are associated with combustion, coal coking, and petroleum catalytic cracking. Coal and shale conversion also contribute to the production of PACs (25, 37a, 37b). Not only production but also transportation and the use of synthetic fuels and petroleum products provide emission sources for PACs. Other major sources of PACs include heat and power generation, refuse burning, and contamination by oil spills (Table 1).

4.2.   Industrial and Urban Environments

The occurrence of PACs in the workplace and industrial environments has been extensively discussed in previous works (28–30). Workers' exposure to PACs in paving and roofing operations and in the steel and silicon carbide industries has been investigated recently (36). Cost-effective luminescence techniques were used to screen PACs in industrial air samples (37c, 37d). In urban environments, an important source of PACs is diesel exhaust. The high efficiency of these engines has made them increasingly popular. Because diesel engines generally give 25% better fuel economy than gasoline engines (37), many automobile manufacturers are increasing the "diesel proportion" of their products. By 1990, it is anticipated that 15–20% of the automobiles in the United States will be powered by diesel engines (37). The exhaust emissions generally contain a soot spherule with condensed organic materials.

Various studies using chemical extraction techniques have indicated the presence of PACs in filter samples of diesel exhaust (38). For example, five specific PACs—fluoranthene, pyrene, benz[*a*]anthracene, benzo[*k*]fluoranthene, and benzo[*a*]pyrene—extracted using six solvents (acetonitrile, methanol, 2-propanol, acetone, methylene chloride, and toluene) and four chemical extraction methods were determined for diesel exhaust particulate samples (38).

Yu and Hites have studied the detailed chemical composition of soot extracts collected from diesel engines (39). The two most mutagenic fractions contained alkylated phenanthrenes, fluorenes, fluorenones, aldehydes, quinones, and other PACs. The compounds in the hexane/toluene fraction are PACs containing three to five aromatic rings. Phenanthrene and alkylated phenanthrene are the major components. Fluoranthene and pyrene, as well as their methyl homologues, were also detected. 1-Methylphenanthrene, 2-methylphenanthrene, and 9-methylphenanthrene have been found to be active in a *Salmonella typhimurium* mutagenicity bioassay (39). Their specific mutagenicities were 0.3 to 0.5 times that of BaP. 9-Methylfluorene was reported to be 0.3 times as mutagenic as BaP, and its higher homologue, 1,9-dimethylfluorene, was found to be twice as active as BaP. On the basis of these bioassay results, the alkylphenanthrenes and alkylfluorenes could be the

Table 2. PACs Found in CH_2Cl_2 Extract of Diesel Exhaust Particulates[a]

Fraction	Compound
Hexane/toluene fraction	Methylfluorenes
	Dibenzothiophene
	Phenanthrene
	C_2-fluorenes
	3-Methylphenanthrene
	2-Methylphenanthrene
	9-Methyl- and 4-methylphenanthrene
	1-Methylphenanthrene
	C_3-fluorenes
	Phenylnaphthalene
	C_2-phenanthrenes
	Fluoranthene
	C_4-fluorenes
	Pyrene
	Methylphenylnaphthalenes
	C_3-phenanthrenes
	Methylfluoranthenes and methylpyrenes
	C_2-phenylnaphthalenes

Table 2. (*contd.*)

Fraction	Compound
	C_4-phenanthrenes
	C_3-phenylnaphthalenes
	Benzo[*ghi*]fluoranthene
	Benzo[*a*]anthracene
	Chrysene or triphenylene
	Nitropyrene or nitrofluoranthene
	Benzofluoranthenes
	Benzopyrenes
	Perylene
Toluene fraction	2-Naphthaldehyde
	1-Naphthaldehyde
	6-Methyl-2-naphthaldehyde
	Biphenylcarboxaldehydes
	Fluorenone
	C_2-naphthaldehydes
	Methylbiphenylcarboxaldehydes
	Methylfluorenones
	C_3-naphthaldehydes
	2-Methylfluorenone
	C_2-biphenylcarboxaldehydes
	C_2-fluorenones
	9,10-Anthracenedione
	4*H*-Cyclopenta[*def*]phenanthrene-4-one
	Methyl-9,10-anthracenedione
	C_4-naphthaldehyde and methyl-9,10-anthracenedione
	Phenanthrenecarboxaldehydes
	2-Phenanthrenecarboxaldehyde
	C_3-fluorenones and C_4-fluorenones
	C_5-fluorenones
	Methylphenanthrenecarboxaldehydes
	C_5-fluorenones
	C_6-naphthaldehydes
	Benzanthrone/11-benzo[*a*]fluorenone
	Benzo[*b*]naphtho[2, 1-*d*]thiophen/ other isomers
	7*H*-Benz[*de*]anthrone-7-one
	Pyrenecarboxaldehyde or fluoranthenecarboxaldehyde
	Unknown

[a] From ref. 39.

major mutagens in the hexane/toluene fraction (39). The list of compounds identified in the hexane/toluene and toluene fractions of diesel exhaust particulates shown in Table 2 illustrates the wide variety of PACs in complex environmental samples. A personel monitor has been developed to measure PAC vapors in industrial environments (39a). A ranking index to characterize PACs in environmental samples has been proposed (39b).

4.3. Indoor Air

Indoor air quality is an important health factor because an individual may spend as much as 70–90% of his/her time indoors (40). PACs have been analyzed in indoor air particulate sample extracts from residential areas (37c). Moschandreas et al. have investigated the effects of woodburning on the indoor air quality in residences (41). This study of urban residences in the Boston metropolitan area revealed a potential adverse impact on indoor air quality from woodstoves and fireplaces. Indoor BaP concentrations during woodstove use averaged five times more than during nonwoodburning periods. The significance of elevated BaP during woodburning activities is clearly illustrated in Table 3, which gives examples of indoor and outdoor BaP concentrations in various residences (41). Gas facilities for heating and cooking did not appear to affect BaP levels significantly. A slight increase in indoor BaP levels was observed in the home with combined smoking and gas facilities. An analysis of the BaP concentration levels suggested that there was

Table 3. Typical 24-h Indoor and Outdoor BaP Concentrations (ng/m^3) at Various Residences in the Boston Metropolitan Area[a]

	Indoor BaP (ng/m^3)		Outdoor BaP (ng/m^3)	
Residence type	Mean	Range	Mean	Range
Gas facilities (nonsmoking)	0.4	0.1–0.6	0.5	0.1–2.2
Gas facilities (smoking)	0.9	0.4–1.8	0.4	0.1–1.4
Electric facilities (nonsmoking)	1.1	0.2–2.8	0.9	0.3–2.2
Electric facilities (woodstove no woodburning)	1.0	0.3–1.1	0.8	0.7–0.8
Woodburning	4.7	2.3–8.0	1.3	0.2–4.2

[a] From ref. 41.

little or no difference in indoor environments among the homes except when woodburning occurred.

4.4. Foods

PACs can also be found in vegetable products (41a) and in food during preparation and cooking. Crosby et al. have identified various PACs in sausage and smoked foods (42). The typical levels of several PACs in three different smoked foods are shown in Table 4.

4.5. Marine Environments

The presence of PACs has also been reported in plankton, seaweeds, filter feeding organisms, and marine sediments. PACs can enter the oceans by many routes, including petroleum spills, runoff from roads, sewage, effluents from industrial processes, and fallout from the atmosphere (43). Dunn and Stich have investigated the use of BaP in mussels in estimating PAC contamination of the marine environment (43, 44). There was a marked correlation between the presence of human activity in coastal areas and the contamination of mussels by BaP. Wharfs and docks and structures containing creosoted pilings or timbers appeared to be particularly strong sources of carcinogenic compounds (44).

4.6. Plants

Many plants contain linear furanocoumarins, or psoralens, which cause photosensitization of these plants (see ref. 45 and references therein). Plants containing linear furanocoumarins can cause photosensitization in livestock and poultry, resulting in economic losses. Celery handlers and field workers are also affected with photosensitization of the fingers, hands, and forearms. These skin rashes are referred to as *celery dermatitis* (45). Pearlman et al. have investigated theoretical models for DNA lesion caused by psoralens (46).

5. IMPORTANCE OF HETEROCYCLIC PACs

A number of reviews and monographs have previously described the occurrence, significance, and characterization of PAHs (28–34). However, most of these works deal mainly with the parent PAHs and do not place particular emphasis on heterocyclic PACs. Only nitrated PACs are the topic of a previous monograph (35). As previously discussed, the aim of this present work is to deal with the broader class of PACs, which includes not only the

Table 4. Levels of PACs Found in Three Different Smoked Foods[a]

| Food | PAC ($\mu g\,kg^{-1}$) | | | | | |
	Fluoranthene	Benzo[k]-fluoranthene	Benzo[b]-fluoranthene	Benzo[a]-pyrene	Indeno[1,2,3-cd]pyrene	Benzo[ghi]-perylene
Bacon	7.8	0.05	0.30	0.05	2.5	3.75
Kippers	2.4	0.10	0.35	0.10	2.7	4.3
Cheese	4.2	0.15	0.30	0.20	ND[b]	0.60

[a]From ref. 42.
[b]ND=not detected (less than 0.02 $\mu g\,kg^{-1}$).

10

parent PAHs but also the heterocyclic PACs. Many heterocyclic PACs are produced in conjunction with their parent PAH compounds. From both epidemiological and experimental studies, many heterocyclic PACs are highly suspect as causative agents in human cancer. The importance of heterocyclic PACs to the environment and human health is illustrated in the following examples describing the occurrence and biological effects of some specific classes of these compounds.

5.1. Nitrogen-Containing PACs

Nitrogen-containing (nitro) PACs are produced primarily as the result of incomplete combustion processes. Well-known sources of nitro-PACs include: exhaust emission from diesel engines (47); aluminum smelting effluent (48); and wood smoke and gasoline engine exhaust (49). Nitro-PACs have also been identified in fly ash, soot from woodburning stoves, urban air, and earlier forms of xerographic toner (50). Nitro-PACs are also found in tobacco and tobacco smoke (51). These compounds include: carbazoles, indoles, naphthylamines, and quinolines. As shown in Figure 1, the N-heterocyclic compounds in tobacco and tobacco smoke constitute a wide variety of different types of PACs. Among the more than 65 N-heterocyclic PACs reported in Table 5 are those with benzenoid-ring systems containing two, three, and five fused rings (Figure 1).

Another possible source of nitro-PAC production is atmospheric transformations involving reactions of gas-phase or adsorbed PACs with oxides of nitrogen and/or nitric acid and other atmospheric reactive species (52). For example, the identification of 2-nitrofluoranthene and 2-nitropyrene in ambient air (52, 53) provided evidence of atmospheric formation of nitro-PACs, since these species have been detected neither in direct emissions nor in laboratory experiments dealing with the direct interactions of PACs and oxides of nitrogens. Thus, it is believed that these two nitro-PACs must be formed in the atmosphere during transport.

Nitro-PACs have been recognized as one of the most important classes of PAC derivatives because of their biological activities (32, 35, 54). Among the nitroarenes, nitropyrenes are of particular interest because in microbial assays they have been identified as primary mutagenic components of diesel emission particulates (55). Because they do not react directly with DNA (56), their biological effects are presumably mediated through cellular conversion of the parent compound into a species that reacts readily with DNA to form adducts. Heflich et al. showed that the pathway for activation of 1-nitropyrene (1-NP) proceeds as illustrated in Figure 2 (57). The critical first step, a rate-limiting step, is enzymatic reduction to 1-nitrosopyrene. This is followed by a subsequent reduction to the corresponding hydroxylamine, a species capable

Table 5. Fused-Ring N-Heterocyclic PACs in Tobacco and Tobacco Smoke[a]

Two Fused Rings

Indoles

1,3-Dimethylindole
1,4-Dimethylindole
1,5-Dimethylindole
1,6-Dimethylindole
1,7-Dimethylindole
1,2-Dimethylindole
2,3-Dimethylindole
Dimethylindole(s)
1-Ethylindole
3-Ethylindole
Indole
1-Methylindole
2-Methylindole
3-Methylindole (skatole)
5-Methylindole
7-Methylindole
Methylindole(s)
3-Phenylindole
3-Propylindole
1,2,3-Trimethylindole
Trimethylindole(s)

Quinolines

2,6-Dimethylquinoline
Dimethylquinolines(s)
8-Hydroxyquinoline
Isoquinoline
4-Methylquinoline
Methylquinoline(s)
Quinoline

Others

Adenine
Benzimidazole (methylbenzimidazole)
Benzothiazole
2-(2-Formyl-5-hydroxymethylpyrrol-1-yl)-propionic acid lactone
2-(2-Formyl-5-hydroxymethylpyrrol-1-yl)-3-phenylpropionic acid lactone
Guanine
Pyrrolo[2,3-b]pyridine
Quinoxaline

Table 5. (*contd.*)

3,6,6-Trimethyl-5,6-dihydro-7*H*-2-pyridin-7-one
1,3,6,6-Tetramethyl-5,6,7,8-Tetrahydroisoquinoline-8-one
Tetrahydroisoquinoline-8-one

Three Fused Rings

Acridans

9,9-Dimethylacridan
2-Isopropyl-9,9-dimethylacridan

Carbazoles

Alkylcarbazole(s)
9-Alkylcarbazole(s)
Carbazole
1,9-Dimethylcarbazole
2,9-Dimethylcarbazole
3,9-Dimethylcarbazole
4,9-Dimethylcarbazole
Dimethylcarbazole
Methylcarbazole
1-Methylcarbazole
2-Methylcarbazole
9-Methylcarbazole
3-Methylcarbazole
4-Methylcarbazole
Methylcarbazole(s)

Others

Azacarbazole
4-Azafluorene
7,8-Benzoquinoline
Harmane
Methylpyrocoll
Norharmane
Pyrido[2,3]indole
Pyrocoll

More than Three Fused Rings

Dibenz[*a,h*]acridine
Dibenz[*a,j*]acridine
7*H*-Dibenzo[*c,g*]carbazole

---

[a] From ref. 51.

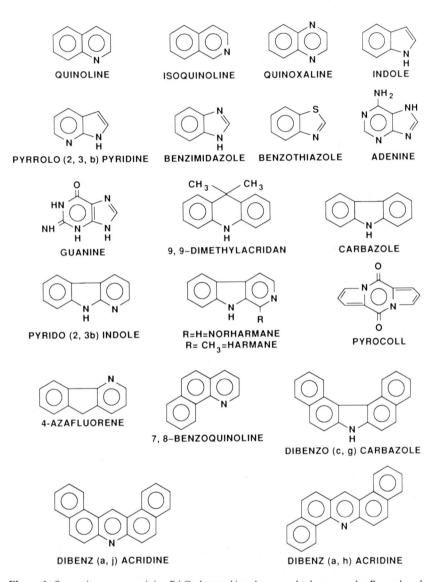

Figure 1. Some nitrogen-containing PACs detected in tobacco and tobacco smoke. Reproduced, with permission, from ref. 51.

of undergoing acid-catalyzed decomposition to yield a covalent adduct at position 8 of guanine. 1-Nitropyrene has been extensively investigated as a representative of nitro-PACs with respect to mutagenicity in *Salmonella* assay systems (58), formation of adducts with DNA (59), and animal carcinogenicity

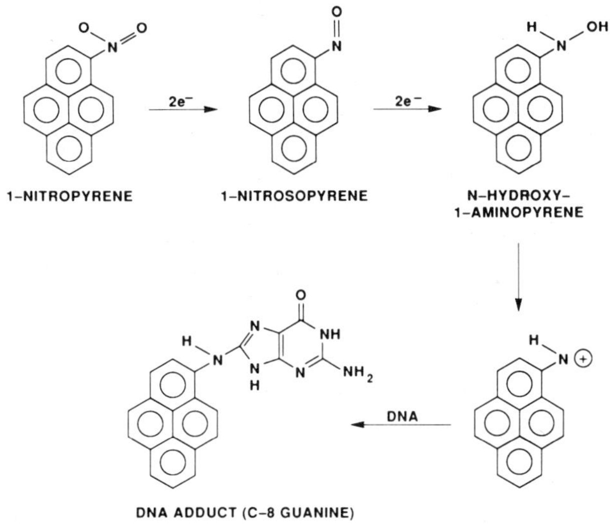

Figure 2. Suggested pathway for metabolic activation of 1-nitropyrene. Reproduced, with permission, from ref. 57.

(60). 3-Nitrofluoranthene has been reported as a potent bacterial mutagen (61) and possibly an animal carcinogen (62).

Several studies have shown that nitrogen-containing PACs in coal- and shale-derived petroleum substitutes are responsible, at least in part, for the overall higher bacterial mutagenicity of these materials relative to petroleum liquids (63, 64). The degree of substitution at the nitrogen atom has been shown to significantly affect the biological activity of nitrogen-containing PACs (65). For example, primary aromatic amines exhibit greater bacterial mutagenic activity than do aza-arenes. It is therefore important to have analytical techniques capable of determining the type of nitro-PACs present for accurate assessment of these materials with respect to mutagenic activity.

In several classes of nitro-PACs, the orientation of the nitro group with respect to the aromatic ring system has been used to predict their direct-acting mutagenicity in *Salmonella* assays (66, 67). Nitro-PACs that have two *peri* substituents adopt a conformation in which the nitro group is perpendicular (or nearly perpendicular) to the aromatic moiety, whereas nitro-PACs that have one or no *peri* substituents adopt conformations in which the nitro group is coplanar (or nearly coplanar) with the aromatic system (66). Furthermore, those compounds in which the nitro substituent is oriented perpendicular to the aromatic moiety exhibit little or no direct acting mutagenicity, while those in which the nitro group is coplanar with the aromatic ring system range from nonmutagens to strong direct-acting mutagens (67).

Quinolines and benzoquinolines are among the major aza-arenes that have been detected in the environment. In addition to being major industrial chemicals, quinoline and isoquinoline have been detected as components of urban air pollution, cigarette smoke, shale oil, and coal liquefaction products (68–77). The various isomeric benzoquinolines have also been shown to be major components in the basic portion of automobile exhaust, urban air particles, and cigarette smoke (69–79). Quinoline is a hepatocarcinogen in both rats and mice (80). Quinoline and both 4- and 8-methylquinoline are active as tumor initiators on mouse skin (81).

5.2. Oxygen-Containing PACs

Of great significance to environmental health is the observation that oxygenated fractions of atmospheric samples can be carcinogenic (82). Indeed, particulate-absorbed PACs are believed to be major contributors to the observed higher death rates caused by lung cancer in urban areas as compared with rural areas (83). Mechanisms by which PAC can be converted to oxidation products in the atmosphere have been extensively investigated. In particular, oxidation processes might "activate" PAC, that is, produce oxidation products similar to those known to be involved in the *in vivo* metabolic activation to carcinogens. Of interest in this connection is the report by Pitts et al. that BaP can be converted to its K-region oxide by ozone (84).

Many carbonyl PACs play an important role in biochemistry. For example, vitamin K_1 (2-methyl-3-phytyl-1,4-naphthoquinone) has been known for its blood-clotting effect (85). Daunorubicin, an anthraquinone derivative, is a known anticancer drug (86).

5.3. Chlorine-Containing PACs

The importance of chlorinated PACs was illustrated in experiments in which pyrene was treated with chlorine in carbon tetrachloride in the presence of $AlCl_3$ as a catalyst (87). The reaction took place at room temperature and was terminated after 1 min. The crude product of synthesis was tested for mutagenicity of *Salmonella typhimurium* strains TA98 and TA100 in the absence and presence of a metabolizing system (liver S9 fractions from Aroclor 1254 pretreated male Sprague–Dawley rats). A mutagenic effect was seen in both the absence and presence of S9 (87). Thus, from a compound with no or very low mutagenic effect, two types of mutagens were formed upon chlorination.

2-Chlorophenothiazine and other derivatives of the neuroleptic drug phenothiazine are effective inhibitors of mutation in the Ames test (88). Phenothiazines are metabolized by cytochrome P-450 mixed-function oxygenases (89), suggesting that these drugs might interfere with BaP metabolism.

5.4. Sulfur-Containing PACs

Sulfur-containing PACs such as thia-arenes have been detected in various environmental samples such as crude oil (90), engine oils, and vehicle exhaust. They occur in considerable concentrations in the emissions of brown-coal and hard-coal combustion (see ref. 90 and references therein). Some thia-arenes have been shown to be potent carcinogens (91). Croisy et al. reported that the S-analogues of chrysene, benzo[c]chrysene and dibenz[a,j]anthracene were even more potent carcinogens than the parent PAHs (92). Evidence for the mutagenic activities of various thia-arenes in the Ames *Salmonella typhimurium* test system has also been reported (93).

6. BIOINDICATORS FOR PACs: THE NEW FRONTIER IN CHEMICAL ANALYSIS

6.1. Bioindicators

As the research developed over the years, scientists applied analytical techniques to the measurement of "biomarkers" of PACs. This trend is a natural progression because the research community is not only interested in detecting PACs in the environment but is now attempting to relate estimates of human exposure to chemicals to some biological responses. Chemicals such as PACs that are active as carcinogens and mutagens have electrophilic properties or are metabolically converted into electrophiles. These reactive derivatives undergo attack by nucleophilic centers in nucleic acids and proteins. PACs enter the body via different pathways (i.e., inhalation, ingestion, dermal adsorption), absorb to phospholipids, and are distributed to various organs in the body. PACs are generally oxidized to epoxides by mixed-function oxidases, are hydrolyzed further to dihydrodiols by aryl epoxide hydrolases to ultimate carcinogens, and react with nucleic acids and proteins, resulting in the formation of covalent adducts (28, 30–32). Special focus has been on DNA adducts, since these species are believed to represent initiating events leading to cell mutation and/or malignant transformation.

Figure 3 shows the formation of ultimate carcinogenic forms of BaP (94). Shown in Figure 3a are the enzymatic steps involved in the formation of BaP-7,8-diol-9,10-oxide. Mono-oxygenases oxidize BaP to form the epoxide

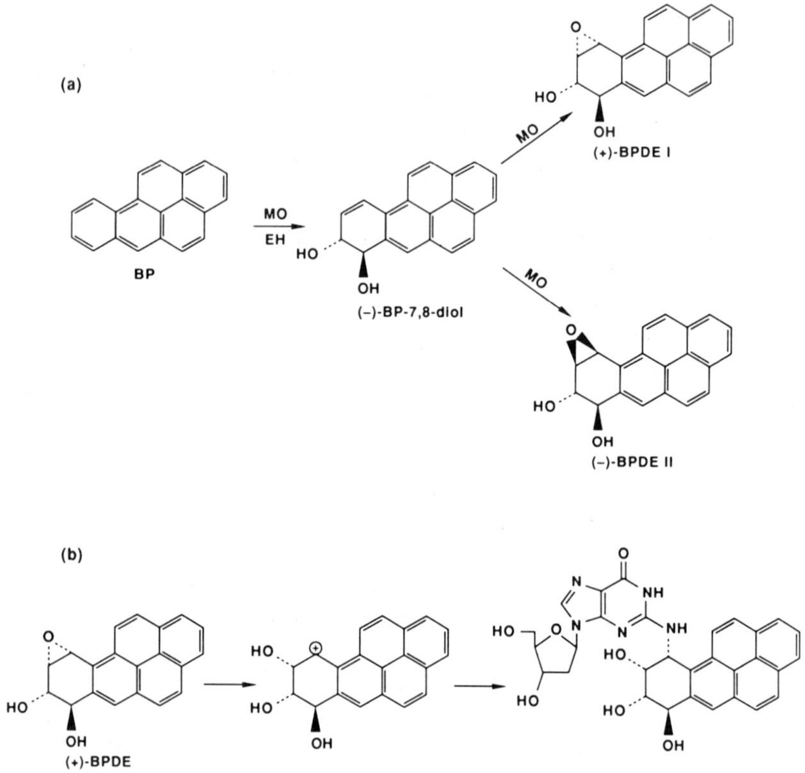

Figure 3. (*a*) Formation of carcinogenic forms of benzo[*a*]pyrene. (*b*) Formation of DNA adduct with BPDE I. Reproduced, with permission, from ref. 94.

intermediate, and epoxide hydrase converts the epoxide to the dihydrodiol. The (−)-BaP-7,8-diol form shown here is the predominant stereoisomeric form produced from this reaction. The second oxidation in the two-step oxidation process is also catalyzed by mono-oxygenases and results in the production of (+)-BPDE I, the major metabolite that binds to the nucleophilic sites on DNA, and (−)-BPDE II, another metabolite that binds to DNA. Diol epoxides formed from the (+)-BaP-7,8-diol enantiomer are (−)-BPDE I and (+)-BPDE II and are found to bind to DNA to a lesser extent. Figure 3*b* illustrates the interaction of (+)-BPDE I and deoxyguanosine to form a DNA adduct. (+)-BPDE I spontaneously decomposes to the carbonium ion triol intermediate. The electrophilic intermediate then interacts nonenzymatically with the nucleophilic sites on the DNA. Illustrated here is (+)-BPDE I binding to the N-2 of guanosine.

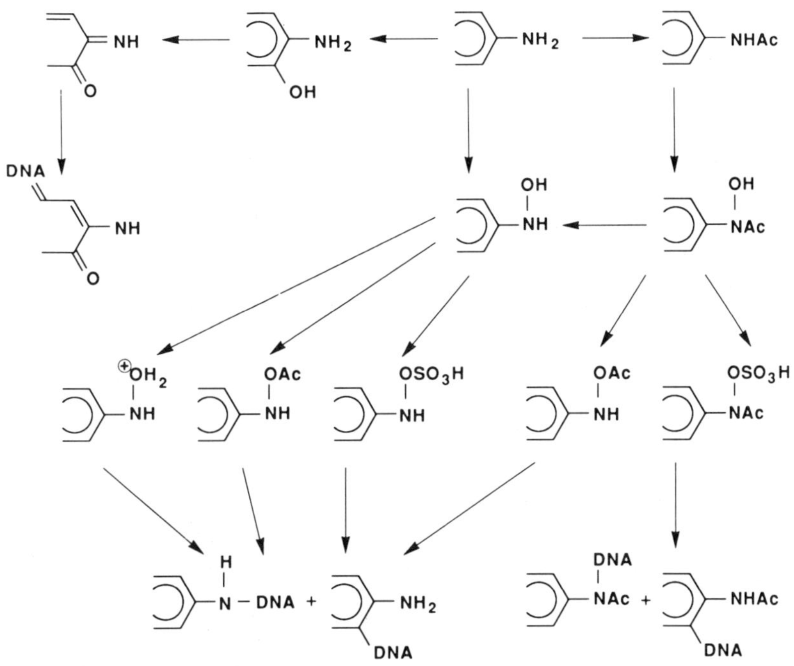

Figure 4. Metabolic pathways leading to DNA adducts from arylamines and arylamides. Reproduced, with permission, from ref. 95.

Figure 4 illustrates another example of metabolic processes for aromatic amines and amides leading to DNA adduct formation (95). Aromatic amines and amides were originally associated with the induction of bladder tumors in humans and have subsequently been found to be carcinogenic at a number of sites in a variety of experimental animals. The initial step in the metabolic activation of these compounds usually involves an N-oxidation (Figure 4), which can be catalyzed by cytochrome P-450, the flavin-containing mono-oxygenase (FMO), or peroxidases such as prostaglandin H synthetase (PHS). Primary arylamines are N-oxidized to N-hydroxy arylamines that can undergo an acid-catalyzed reaction with DNA or that can be further metabolized to a reactive electrophile through acetyl coenzyme A (AcCoA)-dependent O-acetylation, ATP-dependent O-aminoacylation, or 3'-phosphoadenosine 5'-phosphosulfate (PAPS)-dependent O-sulfonylation. The N-oxidation of secondary arylamines results in N-hydroxy metabolites that do not react directly with DNA but that are further converted into electrophilic derivatives by O-sulfonylation (95). Figure 5 depicts various DNA

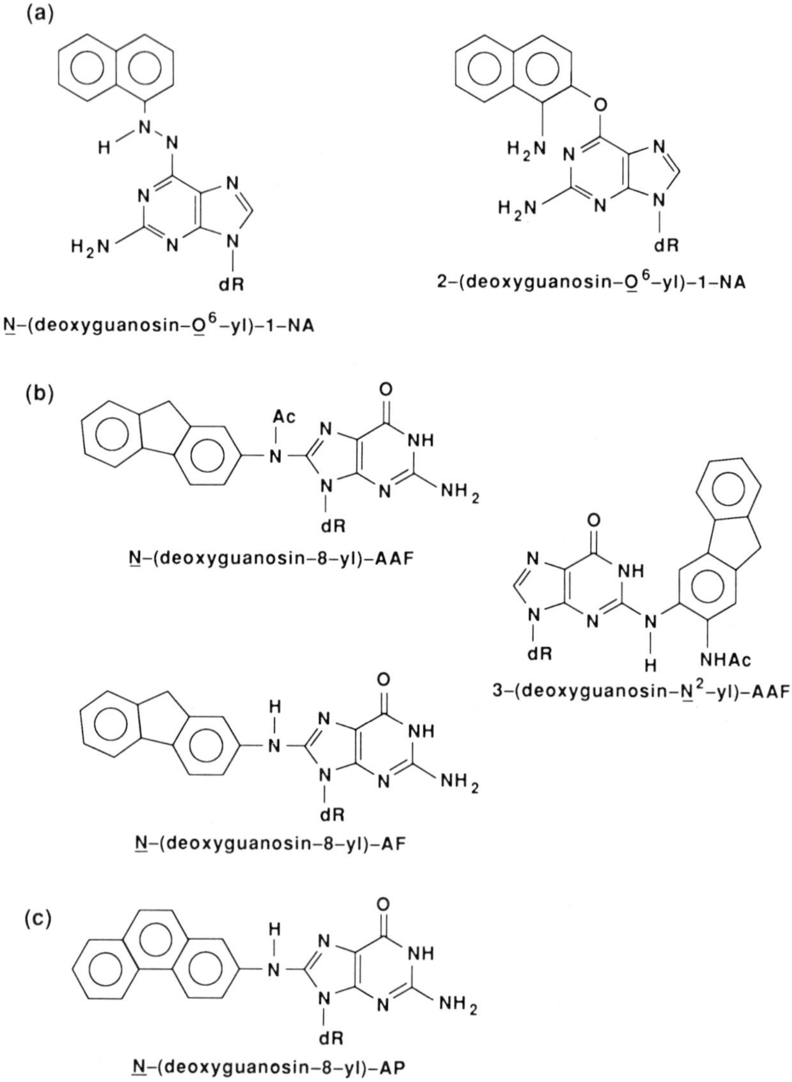

(a)

N-(deoxyguanosin-O^6-yl)-1-NA

2-(deoxyguanosin-O^6-yl)-1-NA

(b)

N-(deoxyguanosin-8-yl)-AAF

3-(deoxyguanosin-N^2-yl)-AAF

N-(deoxyguanosin-8-yl)-AF

(c)

N-(deoxyguanosin-8-yl)-AP

Figure 5. Various types of DNA adducts derived from: (*a*) 1-naphthylamine (NA); (*b*) 2-acetylaminofluorene (AAF); and (*c*) 2-acetylaminophenanthrene (AP). Reproduced, with permission, from ref. 95.

adduct forms derived from several PACs such as 1-naphthylamine, 2-acetylaminofluorene, and 2-acetylaminophenanthrene (95).

The development of improved and/or new analytical techniques for the measurement of DNA adducts in the DNA of cells is critical for providing

Table 6. Carcinogens Activated to Form DNA Adducts by Cultured Human Cells of Bronchus, Colon, Esophagus, Pancreatic Duct, and Bladder[a]

Carcinogen	Bronchus	Colon	Esophagus	Pancreatic Duct	Bladder
Polynuclear aromatic hydrocarbons					
Benzo[a]pyrene	+[b]	+	+	+	+
7,12-Dimethylbenz[a]anthracene	+	+	+	+	0
3-Methylcholanthrene	+	+	+	0	0
Dibenz[a,h]anthracene	+	+	+	0	0
Mycotoxins					
Aflatoxin B_1	+	+	+	0	+
T-2 toxin	0	0	+	0	0
Aromatic amines					
2-Acetylaminofluorene	+	0	+	0	+
Trp-P-1 (3-amino-1,4-dimethyl-5H-pyrido[4,3-b]indole)	0	+	0	0	0

[a] From ref. 97.
[b] Key: +, detection of carcinogen binding to DNA; −, binding not detected; 0, not tested.

21

information about genotoxic exposure to PACs. It is noteworthy that covalent adducts formed in RNA and proteins have no putative mechanistic role in carcinogenesis but may relate quantitatively to total exposure and activation. DNA adducts exhibit different levels of stability. Some are removed spontaneously through chemical depurination, whereas others are removed enzymatically in the process of cellular repairs. A portion remains in DNA over long periods of time and accumulates with multiple exposures (96). Table 6 gives a list of several PACs activated to form DNA adducts by cultured cells of human bronchus, colon, esophagus, pancreatic duct, and bladder (97). It should be emphasized that because cancer is the result of complex interactions among multiple environmental factors and both acquired and inherited host factors, one should consider measurements of carcinogen-DNA adducts as only one part of a battery of other assays involving other biomarkers (e.g., chromosomal abnormalities, markers for point mutations, altered gene expression, reproductive toxicity, and membrane change).

6.2. Analytical Techniques

In most earlier studies of the chemicals of interest, DNA interactions were investigated using radiometric techniques. The interactions of chemicals with DNA were measured both at a gross level (i.e., total binding of compound) or at the molecular level (i.e., reactions formed with specific DNA products). Radiometric techniques generally suffer from several disadvantages: radioisotopes have limited shelf-life (~ 60 days for ^{125}I half-life); instruments for detection and reagents are relatively expensive; and special precautions are required for the shipping, handling, and waste disposal of radioactive materials. Confounding deleterious biological effects resulting from the radioactive labels are also of concern.

Various other techniques have been developed for chemical analysis of PAC–DNA adducts (98). The most widely used methods involve immunochemical techniques (99, 100). A commonly used immunoassay is the enzyme-linked immunoadsorbent assay (ELISA). The ELISA technique gains its sensitivity from the amplification effect of the enzymes conjugated to one of the immunoreactants, usually the antibody (99, 100). Immunoassay techniques are very powerful monitoring tools for DNA adducts because of their excellent specificity and reasonable sensitivity. The immunological principle can be combined with laser and fiberoptic technology to develop a new generation of sensors for PACs and related DNA adducts (101–103). A limitation of immunochemical techniques is the need to produce and characterize the antibody for each specific compound of interest. The specificity of the antibodies is not always perfect and some degree of cross-reactivity occurs, especially for situations involving complex mixtures.

A sensitive radiochemical method for PAC–DNA adduct revolves around the ^{32}P post-labeling technique (104). The method involves hydrolysis of the DNA down to the base/carcinogen state and labeling the nucleotides with a ^{32}P-containing phosphate group. These labeled nucleotides are then separated using thin-layer chromatography (TLC) and are detected by autoradiography. So far the ^{32}P post-labeling technique appears to be the most sensitive method. However, the use of ^{32}P radioactive labels is a major limitation of this method.

Techniques for measuring DNA adducts involve the use of gas chromatography (GC) and gas chromatography/mass spectroscopy (GC/MS) (105). Fourier transform mass spectroscopy (FTMS) is a new development that offers the ability to generate exceptionally high-resolution mass spectra. Electron capture detection (ECD) or negative ion chemical ionization (NICI) mass spectroscopy can be used for detection for GC. In general, only the smaller hydrolyzed adducts are likely to pass through the GC column, thus limiting the technique to the simple alkylated bases. The combination of high-performance liquid chromatography (HPLC) with fluorescence detection has been used to improve the detection of DNA adducts (106).

Fluorimetric methods have been widely used for PAC–DNA adducts because the polyaromatic structures are generally strongly fluorescent. The single-photon counting technique was used to improve the limit of detection (107). Low-temperature matrices have also been used to enhance the sensitivity of detection for PAC–DNA adducts (108). Another improvement was the synchronous luminescence (SL) technique (109, 110). Conventional luminescence spectrometry uses either a fixed excitation or fixed emission wavelength. With SL, both excitation and emission wavelength are scanned synchronously, producing a spectrum with a more resolved structure and more readily identified peaks. The combination of low temperature with synchronous scanning has shown to further improve the sensitivity (111). The SL technique has been used to detect low levels of carcinogen–DNA adducts in human populations (e.g., coke-oven workers) due to environmental exposure to PACs (112, 113).

Another significant development is the fluorescence line-narrowing (FLN) technique, which was used to detect DNA adducts formed from diol epoxides of benzo[a]pyrene, chrysene, 5-methylchrysene, and benzo[a]anthracene (114). In the FLN technique, a narrow-line laser was used to select a small subset of molecules that absorb at the laser frequency. Only the excited molecules fluoresce, exhibiting very sharp emission spectra. One advantage of the FLN technique is the high specificity of the spectral analysis and the ability to study intact DNA samples. The FLN technique requires (a) tunable lasers as excitation sources and (b) cryogenic equipment for measurements at 4.2 K.

Nuclear magnetic resonance (NMR) has also been used to investigate DNA alkylation products (115). The NMR technique is generally useful for structure determination but is not practical for routine analysis because it requires sophisticated instrumentation and skilled workers and has limited sensitivity. Other methods used to study conformation of DNA adducts include electric dichroism (116) and X-ray crystallography (117).

Room-temperature phosphorimetry (RTP) has recently been used to detect BPDE–DNA adducts (118). Unlike conventional low-temperature phosphorimetry, the RTP technique does not require cryogenic equipment and low-temperature refrigerant. The detection limit of BPDE in *in vitro* modified BPDE–DNA was about 15 fmol. The method is characterized by its simplicity and does not require sophisticated and expensive instrumentation and cryogenic refrigerant; thus it is well suited for routine analyses and screening procedures. The RTP approach could provide an additional practical tool for monitoring human exposure to carcinogenic and mutagenic PAC species.

Another novel technique of great promise is surface-enhanced Raman scattering (SERS). The application of Raman spectroscopy for the study of biological systems is rapidly expanding because of the specificity of this analytical technique for chemical identification. Conventional Raman spectroscopy, however, has limited applicability for trace organic detection because of the inherently weak Raman-scattering cross section. The SERS effect, whereby there is an enhancement factor of up to 10^7 in Raman signals from molecules adsorbed on rough metallic surfaces (119), has recently generated increasing interest in the Raman technique. BPDE–DNA adducts have been recently analyzed by SERS (120). For biological samples, conventional Raman spectroscopy has the disadvantage of requiring large samples (usually 10–100 mg of the bulk pure specimens). The increased sensitivity provided by the SERS effect has eliminated this major limitation. The study also indicates that the adduct can be identified in the DNA sample without requiring chromatographic separation of the BaP-tetrol species from the DNA products.

Recently, fiberoptic antibody-based fluoroimmunosensors (FISs) have been developed for BaP and related adducts (101–103). Polyclonal or monoclonal antibodies produced against BaP are immobilized at the terminus of a fiberoptics probe or are contained in a microsensing cavity within the FIS for use both in *in vitro* and *in vivo* fluorescence assays. High sensitivity is provided by laser excitation and optical detection. The FIS device utilizes the back-scattering of light emitted at the remote sensor probe. A single fiber is used to transmit the excitation radiation into the sample and to collect the fluorescence emission from the antigen. The laser radiation reaches the sensor probe and excites the BaP bound to the antibodies immobilized at the fiberoptics probe. The excellent sensitivity of this device illustrates that it has considerable potential to perform trace analyses of chemical and biological samples in complex matrices. Measurements are simple and rapid (~ 12 min), and the technique is applicable to other compounds provided that the appropriate antibodies are used. The FIS instrument can detect 1 fmol of BaP (101) and 40 amol of BaP-tetrol (121, 122). For the past few years, biosensor technology has been at the forefront of analytical instrumentation research. The integration of biological methods, laser systems, and advanced optical sensor technology promises to open new horizons in environmental and biological monitoring of PACs and related bioindicators.

ACKNOWLEDGMENT

This work was sponsored by the Office of Health and Environmental Research, U.S. Department of Energy, under contract DE-AC05-84OR21400 with Martin Marietta Energy Systems, Inc.

REFERENCES

1. E. L. Kennaway and I. Hieger, *Br. Med. J.*, **ii**, 1044 (1930).
2. J. W. Cook, C. L. Hewett, and I. Hieger, *J. Chem. Sci.*, 395 (1933).
3. M. Badger, J. W. Cook, C. L. Hewett, E. L. Kennaway, R. H. Morton, and A. M. Robinson, *Proc. R. Soc. London Ser. B.*, **129**, 439 (1940).
4. L. F. Fieser and M. S. Newman, *J. Am. Chem. Soc.*, **58**, 2376 (1936).
5. M. Badger, J. W. Cook, C. L. Hewett, E. L. Kennaway, N. M. Kennaway, and R. H. Martin, *Proc. R. Soc. B*, **131**, 170 (1952).
6. E. Bayland and E. Sims, *Biochem. J.*, **84**, 571 (1962).
7. P. Sims, *Biochem. J.*, **92**, 621 (1964).
8. E. Boyland and P. Sims, *Biochem. J.*, **91**, 493 (1964).
9. P. Sims, *Biochem. J.*, **105**, 591 (1967).
10. E. Boyland and P. Sims, *Biochem. J.*, **95**, 730 (1965).
11. P. Sims, *Biochem. Pharmacol.*, **16**, 613 (1967).
12. J. F. Waterfall and P. Sims, *Biochem. Pharmacol.*, **22**, 2469 (1973).
13. E. Boyland and P. Sims, *Biochem. J.*, **97**, 7 (1965).
14. P. Sims, *Biochem. Pharmacol.*, **19**, 795 (1970).
15. P. Sims, *Biochem. J.*, **98**, 215 (1966).
16. P. L. Grover and P. Sims, *Biochem. J.*, **110**, 159 (1968).
17. D. M. Jerina, J. W. Daly, B. Wotkop, P. Zaltzman-Nirenberg, and S. Udenfriend, *J. Am. Chem. Soc.*, **90**, 6525 (1968).

18. A. Pullman and B. Pullman, *Adv. Cancer Res.*, **3**, 117 (1955).
19. D. M. Jerina and R. E. Lehr, in *Microsomes and Drug Oxidation*, V. Ulbrick, I. Roots, A. Hildebrandt, and R. N. Estabrook, Eds., Pergamon Press, Elmsford, N.Y., p. 709 (1978).
20. R. G. Harvey, *Acc. Chem. Res.*, **14**, 218 (1981).
21. National Academy of Sciences, *Polycyclic Aromatic Hydrocarbons: Evaluations of Sources and Effects*, Washington, D.C. (1983).
22. D. Schuetzle, T. Riley, T. J. Prater, T. M. Harvey, and D. F. Hunt, *Anal. Chem.*, **54**, 265 (1982).
23. T. B. Atherholt, G. J. McGarrity, J. B. Lewis, J. McGeorge, P. J. Lioy, J. M. Daisey, A. Greenberg, and F. Darack, in *Short-Term Bioassays in the Analysis of Complex Environmental Mixtures*, by M. D. Waters, S. S. Sandhu, J. Lewtas, L. Claxton, G. Strauss, and S. Nesnow, Eds., pp. 211–231, Plenum, New York (1985).
24. J. Siak, T. L. Chan, T. L. Gibson, and G. T. Wolff, *Atmos. Environ.*, **19**, 369 (1985).
25. E. J. Baum, in *Polycyclic Hydrocarbons and Cancer*, H. V. Gelboin and P. O. P. T'so, Eds., Academic Press, New York, Chapter 2, pp. 45–70 (1978).
26. G. J. Badger, R. W. L. Kimber, and J. Novotny, *Aust. J. Chem.*, **17**, 778 (1964).
27. U. S. Environmental Protection Agency, "Preferred Standard Path Report for Polycyclic Organic Matter," *U.S. Environmental Protection Agency, Office Air Quality Planning and Standards*, Strategies Air Standards Division, Durham, N.C. (1974).
28. H. V. Gelboin and P. O. P. T'so, Eds., *Polycyclic Hydrocarbons and Cancer*, Academic Press, New York (1978).
29. A. Bjorseth, Ed., *Handbook of Polycyclic Aromatic Hydrocarbons*, Marcel Dekker, New York (1983).
30. G. Grimmer, Ed., *Environmental Carcinogens: Polycyclic Aromatic Hydrocarbons*, CRC Press, Boca Raton, Fla. (1983).
31. P. L. Grover, Ed., *Chemical Carcinogens and DNA*, Vols. 1 and 2, CRC Press, Boca Raton, Fla. (1979).
32. R. G. Harvey, Ed., *Polycyclic Hydrocarbons and Carcinogenesis*, ACS Symposium Series 283, American Chemical Society, Washington, D.C. (1985).
33. M. L. Lee, M. V. Novotny, and K. D. Bartle, *Analytical Chemistry of Polycyclic Aromatic Compounds*, Academic Press, New York (1981).
34. D. J. Futoma, S. R. Smith, T. E. Smith, and J. Tanaka, *Polycyclic Aromatic Hydrocarbons in Water Systems*, CRC Press, Boca Raton, Fla. (1981).
35. C. M. White, Ed., *Nitrated Polycyclic Aromatic Hydrocarbons*, Huethig Verlag, New York (1985).
36. J. Lesage, G. Perrault, and P. Durand, *Am. Ind. Hyg. Ass. J.*, **48**, 753 (1987).
37. D. L. Dimick, "Prospects for Diesel Passenger Cars and the Need for an Improved Fuel," *API Automobile & Industry Forum*, American Petroleum Institute, Washington, D.C. Jan 23 (1980).
37a. T. Vo-Dinh and P. R. Martinez, *Anal. Chim. Acta.*, **125**, 13 (1981).
37b. D. W. Abbott, R. L. Moody, R. M. Mann, and T. Vo-Dinh, *Am. Ind. Hyg. Assoc. J.*, **47**, 379 (1986).

37c. T. Vo-Dinh, T. J. Bruewer, G. C. Colovos, T. J. Wagner, and R. H. Jungers, *Envir. Sci. Technol.*, **18**, 477 (1984).

37d. T. Vo-Dinh, R. B. Gammage, and P. R. Martinez, *Anal. Chem.*, **53**, 253 (1981).

38. G. M. Breuer, *Anal. Lett.*, **17** (A11), 1293 (1984).

39. M.-L. Yu and R. A. Hites, *Anal. Chem.*, **53**, 951 (1981).

39a. T. Vo-Dinh, *Envir. Sci. Technol.*, **19**, 997 (1985).

39b. T. Vo-Dinh and D. W. Abott, *Envir. Int.*, **10**, 299 (1984).

40. S. Budiansky, *Environ. Sci. Technol.*, **14**, 1026 (1980).

41. D. J. Moschandreas, J. Zabransky, and H. E. Rector, *Environ. Intern.*, **4**, 413 (1980).

41a. T. Vo-Dinh, D. A. White, M. A. O'Malley, P. J. Seligman, and R. C. Beier, *J. Agr. Food Chem.*, **36**, 333 (1988).

42. N. T. Crosby, D. C. Hunt, L. A. Philip, and I. Patel, *Analyst*, **106**, 135 (1981).

43. B. P. Dunn and H. F. Stich, *Proc. Soc. Exp. Biol. Med.*, **150**, 49 (1975).

44. B. P. Dunn, *Environ. Sci. Technol.*, **10**, 1018 (1976).

45. R. C. Beier, G. W. Ivie, and E. H. Oertli, in *Xenobiotics in Foods and Feeds*, J. W. Finley and D. E. Schwass, Eds., ACS Symposium Series No. 234, American Chemical Society, Washington, D.C., p. 295 (1983).

46. D. A. Pearlman, S. R. Holbrook, D. H. Pirkle, and S. H. Kim, *Science*, **227**, 1304 (1985).

47. D. Schuetzle, F. S.-C. Lee, T. J. Prater, and S. B. Tejada, *Int. J. Environ. Anal. Chem.*, **9**, 93 (1981).

48. M. Oehme, S. Mano, and H. Stray, *J. High Resol. Chromatogr.*, **5**, 417 (1982).

49. T. L. Gibson, *Atmos. Environ.*, **16**, 2037 (1982).

50. H. S. Rosenkranz and R. Mermelstein, *Mutat. Res.*, **114**, 217 (1983).

51. I. Schmeltz and D. Hoffman, *Chem., Rev.*, **77**, 295 (1977).

52. J. N. Pitts, Jr., *Environ. Health Perspect.*, **47**, 115 (1983).

53. T. Nielson, B. Seitz, and T. Ramdahl, *Atmos. Environ.*, **18**, 2159 (1984).

54. M. Hirose, M. S. Lee, C. Y. Yang, and G. M. King, *Cancer Res.*, **44**, 1158 (1984).

55. H. D. Rosenkranz, *Mutat. Res.*, **101**, 1 (1982).

56. P. C. Howard and F. A. Beland, *Biochem. Biophys. Res. Commun.* **104**, 727 (1982).

57. R. F. Heflich, P. C. Howard, and F. A. Beland, *Mutat. Res.*, **149**, 25 (1985).

58. K. A. Saito, T. Shinohara, S. Kamataki, and P. Kato, *Arch. Biochem., Biophys.*, **239**, 286 (1985).

59. K. El-Bayoumy and S. S. Hecht, *Cancer Res.*, **43**, 3132 (1983).

60. M. V. Reddy, R. C. Gupta, E. Randerath, and K. Randerath *Carcinogenesis*, **5**, 231 (1984).

61. W. A. Vance and D. E. Levin, *Environ. Mutagen.*, **6**, 797 (1984).

62. H. Ohgaki, N. Matsukara, K. Morino, T. Kawachi, T. Sugimura, K. Morita, H. Tokiwa, and T. Hirota, *Cancer Lett.*, **15**, 1 (1982).

63. M. R. Guerin, C.-H. Ho, T. K. Rao, and J. L. Epler, *Environ. Res.*, **23**, 42 (1980).

64. B. W. Wilson, M. R. Peterson, and R. A. Pelroy, and J. T. Cresto, *Fuel*, **60**, 289 (1981).

65. C.-H. Ho, B. R. Clark, M. R. Guerin, B. D. Barkenbus, T. K. Rao, and J. L. Epler, *Mutat. Res.*, **85**, 335 (1981).

66. P. P. Fu, M. W. Chou, D. W. Miller, G. L. White, R. H. Heflich, and F. A. Beland, *Mutat. Res.*, **143**, 173 (1985).
67. W. A. Vance and D. E. Levin, *Environ. Mutagen.*, **6**, 797 (1984).
68. J. D. Adams, E. J. LaVoie, A. Shigematsu, P. Owens, and D. Hoffman, *J. Anal. Toxicol.*, **7**, 293 (1983).
69. M. W. Dong, D. C. Locke, and D. Hoffmann, *Environ. Sci. Technol.*, **11**, 612 (1977).
70. F. F. Shue and T. F. Yen, *Anal. Chem.*, **53**, 2081 (1981).
71. J. D. Ciupek, D. Zakett, R. G. Cooks, and K. V. Wood, *Anal. Chen.*, **54**, 2215 (1982).
72. R. E. Royer, C. E. Mitchell, H. L. Hanson, J. S. Dutcher, and W. D. Bechtold, *Environ. Res.*, **31**, 460 (1983).
73. R. A. Pelroy and B. W. Wilson, *Mutat. Res.*, **90**, 321 (1981).
74. M. Dong, I. Schmeltz, E. Jacob and D. Hoffman, *J. Anal. Toxicol.*, **2**, 21 (1978).
75. M. E. Snook, P. J. Fortson, and O. T. Chortyk, *Beitr. Tabakforsch.*, **11**, 67 (1981).
76. J. M. Schmitter, I. Ignatiadis and G. Guiochon, *J. Chromatogr.*, **248**, 203 (1982).
77. D. Hoffman and E. L. Wynder, in *Air Pollution*, Vol. 2, 3rd edition, A. C. Stern, Ed., Academic Press, New York, pp. 361–455 (1977).
78. E. Sawicki, J. E. Mecker, and C. Morgan, *Arch. Environ. Health*, **11**, 773 (1965).
79. M. Blumer and T. Dorsey, *Science*, **195**, 283 (1977).
80. K. Hirao, Y. Shinohara, H. Tsuda, S. Fukushima, M. Takahashi, and N. Ito, *Cancer Res.*, **36**, 329 (1976).
81. E. J. LaVoie, A. Shigematsu, E. A. Adams, J. Rigotty, and D. Hoffmann, *Cancer Lett.*, **22**, 269 (1984).
82. S. S. Epstein, *Arch. Evniron. Health*, **10**, 233 (1965).
83. L. B. Lave and E. P. Seskin, *Science*, **169**, 723 (1970).
84. J. N. Pitts, D. M. Lokensgard, P. S. Ripley, K. A. V. Cauwenberghe, V. V. Luk, S. D. Shaffer, A. J. Thill, and W. L. Belser, Jr., *Science*, **210**, 1347 (1980).
85. L. Stryer, *Biochemistry*, 2nd edition, W. H. Freeman, New York, p. 249 (1981).
86. R. N. Capps and M. Vala, *Photochem. Photobiol.*, **33**, 673 (1981).
87. A. Colmsjö, A. Rannug, and V. Rannug, *Mutat. Res.*, **135**, 21 (1984).
88. J. D. Kittle, Jr., L. M. Calle, and P. D. Sullivan, *Mutat. Res.*, **80**, 259 (1981).
89. A. H. Beckett and G. E. Navas, *Xenobiotica*, **8**, 721 (1978).
90. J. Jabob, A. Schmoldt, and G. Grimmer, in *Polycyclic Aromatic Hydrocarbons: A Decade of Progress*, M. Cooke and A. J. Dennis, Eds., Batelle Press, Columbus, Ohio, pp. 417–428 (1988).
91. B. D. Tilak, *Tetrahedron*, **9**, 76 (1960).
92. A. Croisy, J. Mispelter, J. M. Lhoste, F. Zajdela, and P. Jacquignon, *J. Heterocycl. Chem.*, **21**, 353 (1984).
93. R. A. Pelroy, D. L. Stewart, Y. Tominaga, M. Iwao, R. N. Castle, and M. L. Loe, *Mutat. Res.*, **117**, 31 (1983).
94. S. J. Stowers and M. W. Anderson, *Environ. Health Perspect.*, **62**, 31 (1985).
95. F. A. Beland and F. F. Kadlubar, *Environ. Health Perspect*, **62**, 19 (1985).
96. G. N. Wogan and N. J. Corelick, *Environ. Health Perspect*, **62**, 5 (1985).
97. C. C. Harris, *Environ. Health Perspect.*, **62**, 185 (1985).

98. U.S. Department of Energy, *Workshop on the Measurement and Characterization of DNA Adducts*, Office of Health and Environmental Research, Rockville, Md., May 15–16, 1986, U.S. Department of Energy, Washington, D.C. (1986).
99. C. C. Harris, R. H. Yolken, and I. C. Hsu, in *Methods in Cancer Research*, H. Bush and I. C. Yeoman, Eds., Academic Press, New York, pp. 213–242 (1982).
100. R. M. Santella, L. L. Hsieh, C. D. Lin, S. Viet, and I. B. Weinstein, *Environ. Health Perspect.*, **62**, 95 (1985).
101. T. Vo-Dinh, B. J. Tromberg, G. D. Griffin, K. R. Ambrose, M. J. Sepaniak, and E. M. Gardenhire, *Appl. Spectrosc.*, **41**, 735 (1987).
102. B. J. Tromberg, M. J. Sepaniak, T. Vo-Dinh, and G. D. Griffin, *Anal. Chem.*, **59**, 1226 (1987).
103. T. Vo-Dinh, G. D. Griffin, K. R. Ambrose, M. J. Sepaniak, and B. J. Tromberg, in *Polynuclear Aromatic Hydrocarbons: A Decade of Progress*, M. Cooke and A. J. Denis, Eds., Battelle Press, Columbus, Ohio, pp. 885–900 (1988).
104. E. Randerath, M. V. Reddy, and R. C. Gupta, *Proc. Natl. Acad. Sci. U.S.A.*, **78**, 6126 (1981).
105. A. L. Burlingam, K. Straub, and T. A. Baillie, *Mass Spectrom. Rev.*, **2**, 331 (1982).
106. R. O. Rahn, S. S. Chang, J. M. Holland, and L. R. Shugart, *Biochem. Biophys. Res. Commun.*, **109**, 262 (1982).
107. P. Vigny and M. Duquesne, *Photochem. Photobiol.*, **20**, 15 (1974).
108. V. Ivanovitch, N. E. Geacintov, and I. B. Weinstein, *Biochem. Biophys. Res. Commun.*, **76**, 1172 (1976).
109. T. Vo-Dinh, *Anal. Chem.*, **50**, 396 (1978).
110. T. Vo-Dinh, *Appl. Spectrosc.*, **36**, 576 (1982).
111. R. O. Rahn, S. S. Chang, J. M. Holland, T. J. Stephans, and L. H. Smith, *Biochem. Biophys. Methods.*, **3**, 285 (1980).
112. K. Vahakangas, G. Trivers, M. Rowe, and C. C. Harris, *Environ. Health Perspect.*, **62**, 101 (1985).
113. A. A. Haugen, G. Becher, C. Benestad, K. Vahakangas, G. E. Trivers, M. J. Newman, and C. C. Harris, in *Polycyclic Aromatic Hydrocarbons, A Decade of Progress*, M. Cooke and A. J. Dennis, Battelle Press, Columbus, Ohio, p. 377 (1988).
114. M. J. Sanders, R. S. Cooper, R. Jankowick, G. J. Small, V. Heisig, and A. Jeffrey, *Anal. Chem.*, **54**, 816 (1986).
115. K. Hemminki, J. Paasivirta, T. Kurkirinne, and L. Virkki, *Chem. Biol. Interact.*, **30**, 259 (1980).
116. F. F. Kadlubar, W. B. Melchior, T. J. Flammang, A. G. Gaghiano, and H. Yoshida, *Cancer Res.*, **41**, 2168 (1981).
117. H. L. Carrell, J. P. Glusker, R. C. Moschel, W. R. Hudgins, and A. Dipple, *Cancer Res.*, **41**, 2230 (1981).
118. T. Vo-Dinh and M. Uziel, *Anal. Chem.*, **59**, 1093 (1987).
119. R. K. Chang and T. E. Furtak, Eds., *Surface-Enhanced Raman Scattering*, Plenum, New York (1982).
120. T. Vo-Dinh, M. Uziel, and A. L. Morrison, *Appl. Spectrosc.*, **41**, 605 (1987).
121. B. J. Tromberg, M. J. Sepaniak, J. P. Alarie, T. Vo-Dinh, and R. M. Santella, *Anal. Chem.*, in press (1988).

122. T. Vo-Dinh, B. J. Tromberg, G. D. Griffin, K. R. Ambrose, and R. M. Santella, "Immunofluroscence Detection for Fiberoptics Chemical and Biological Sensors," *Proceedings of the Symposium on Laser Spectroscopy Technique, Application, Data Bases and Equipment*, SPIE, O-E Lase, Los Angeles, January 10–15 (1988).

CHAPTER

2

THE FATE OF POLYCYCLIC AROMATIC COMPOUNDS IN THE ATMOSPHERE AND DURING SAMPLING

DOUGLAS A. LANE

Environment Canada
Atmospheric Environment Service
Downsview, Ontario, Canada

1. ABBREVIATIONS

Ace	Acenaphthene
Acy	Acenaphthylene
Anth	Anthracene
Anthan	Anthanthrene
BaA	Benz[a]anthracene
BaP	Benzo[a]pyrene
BbF	Benzo[b]fluoranthene
BeAc	Benz[e]acephenanthrylene
BeP	Benzo[e]pyrene
BghiP	Benzo[ghi]perylene
BkF	Benzo[k]fluoranthene
Chrys	Chrysene
Cor	Coronene
Flt	Fluoranthene
Flu	Fluorene
g	Gram
HPLC	High-pressure liquid chromatography
kPa	Kilopascals
m	Meter
min	Minute
Naph	Naphthalene
NAS	National Academy of Sciences
nm	Nanometer
PACs	Polycyclic aromatic compounds
PAHs	Polycyclic aromatic hydrocarbons
Per	Perylene
Phen	Phenanthrene

ppb   Parts per billion (volume/volume)
ppm   Parts per million (volume/volume)
PUF   Polyurethane foam
Pyr   Pyrene
μm   Micrometer
μg   Microgram
$\tau_{1/2}$ Half-life
°C   Degrees Celsius

2. INTRODUCTION

The term *polycyclic aromatic compounds* (PACs) refers to all of the members of
the *polycyclic aromatic hydrocarbons* (PAHs) and to their alkyl-, amino-,
oxygen-, halogen-, and nitro-substituted derivatives, as well as to the nitrogen
and sulfur heteroatom analogues of the PAHs. In a single chapter, it is not
possible to cover all aspects of the fate of all atmospheric PACs. Indeed,
except for the parent PAHs, very little is known, either about the behavior of
the PACs in the atmosphere or about their chemical and physical properties.
In this chapter, the PAHs will be considered as representatives of the PACs
and, where information exists for other PACs, examples will be presented.

In atmospheric particulate matter, a wide variety of parent PAHs and alkyl
PAHs have been identified (1–3), as have many oxygenated PAHs and PAH
ketones (4–8), nitro PAHs (9–16), and aza-arenes (17). A few reports have
indicated the presence of some sulfur analogues of the PAHs (2, 18) and
polychlorinated PAHs (19, 20).

The emission of PACs to the atmosphere is of concern, principally because
many of the compounds are known carcinogens and/or mutagens. During
their migration through the environment, the PACs may be transformed, by
interaction with oxidants and sunlight, into products which, in some cases, are
much more mutagenic than the compounds from which they were derived.

Many factors, both chemical and physical, have significant effects upon the
survival of the PACs during transport from source to receptor and during the
collection process. Amongst these factors are: the exposure to atmospheric
oxidants and sunlight; the atmospheric temperature; the physical properties
of the various PACs such as vapor pressure; and the nature of the substrate
upon which the PACs are adsorbed. Each of these factors will be considered in
an effort to present an overall picture describing the complexity of the
problem of assessing the presence and importance of PAC contaminants in
the atmosphere.

2.1. Formation and Emission of PACs

It is generally accepted that PAHs are produced in combustion sources through the condensation of ethylenic radicals in the gas phase to form the larger polycyclic compounds (21–27), and there is some evidence to show that the alkylated PAHs are formed more readily at lower pyrolysis temperatures than at high temperatures (28). During the incomplete combustion of organic fuel, the formation of oxygenated and nitrated PAHs may also occur (7, 9).

When the stack gas cools, the gaseous PAHs condense to form micro-particles which then agglomerate to form larger particles. The adsorption of PAHs onto co-entrained fly-ash particles has been described by Natusch and colleagues (29, 30). They showed that, in general, the PAHs remain in the gas phase at temperatures above 150°C but rapidly condense onto fly-ash particles below that temperature. At typical ambient temperatures, PAHs exist primarily in the particle phase.

2.2. Vapor Pressure and Temperature

The extent to which a particular PAC molecule will exist in the gas phase or the particle phase is determined by the vapor pressure of the compound and the ambient temperature. Junge (31) showed, to a first approximation for urban air particulate matter, that compounds with vapor pressures in excess of 1×10^{-5} kPa should occur predominantly in the gas phase, whereas those with vapor pressures below 1×10^{-9} kPa should exist exclusively in the particle phase. For clean air aerosols, the corresponding vapor pressures were 1×10^{-6} and 1×10^{-10} kPa. Any compound with a vapor pressure between these upper and lower limits would be expected to occur in both the vapor and particulate phases.

In the literature, there are numerous citations of vapor pressures for a relatively small number of the PAHs. However, for each PAH, the values reported vary greatly—in some cases, by as much as three orders of magnitude (32). In general, the vapor pressure determined is a function of the experimental method. Effusion methods (and HPLC methods) yield the lowest vapor pressures, whereas subcooled liquid measurements result in the highest (32). For example, the vapor pressure of anthracene was determined by Bradley and Cleasby (33), using the effusion method, to be 8.43×10^{-7} kPa, whereas Schlessinger (34), using liquid-phase measurements, determined a value of 3.17×10^{-5} kPa.

Figure 1 shows the range of vapor pressures reported in the literature for 14 PAHs. In addition, the phase (vapor or particle) in which the compound would be expected to appear in an urban aerosol at 25°C (31) is also indicated. From this figure, one would anticipate that (a) compounds such as anthracene

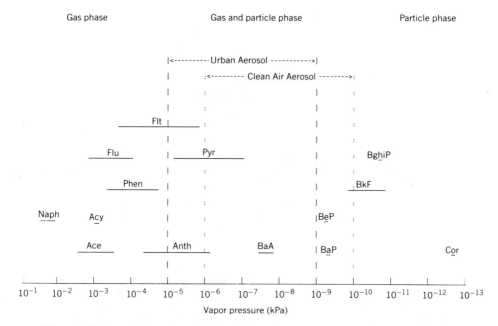

Figure 1. The range of vapor pressures reported in the literature for various PAHs, along with the predicted vapor–particle distribution for urban and clean-air aerosol as reported by Junge (31). Abbreviations: Ace (acenaphthene); Acy (acenaphthylene); Anth (anthracene); BaA (benz[a]anthracene); BaP (benzo[a]pyrene); BeP (benzo[e]pyrene); BghiP (benzo[ghi]perylene); BkF (benzo[k]fluoranthene); Cor (coronene); Flt (fluoranthene); Flu (fluorene); Naph (napthalene); Phen (phenanthrene); Pyr (pyrene).

and phenanthrene should be found primarily in the vapor phase, (b) compounds such as fluoranthene and pyrene should occur in both phases, and (c) compounds such as benzo[a]pyrene and coronene should occur exclusively in the particle phase. This has been observed experimentally by Thrane and Mikalsen (35) and Yamasaki et al. (36). At 25°C, the three-ring PAHs were found primarily in the vapor phase, the four- and five- ring PAHs were found distributed between the particle and the vapor phase, and the six-ring-and-higher PAHs were found almost exclusively in the particle phase.

The ambient temperature has a marked effect on the gas–particle distribution coefficient of the PAHs. It has been shown by Murray et al. (37) that, to a first approximation, the vapor pressure of the PAHs change by an order of magnitude for a temperature change of approximately 15°C. In Figure 1, the range of vapor pressures is shown for 25°C. If the temperature rose to 35°C or 40°C, then PAHs such as BaP and BeP would start to appear in the vapor phase. During the hottest weeks of the year, the five-ring PAHs have been detected (36) in the vapor phase as well as the particle phase. Both BaP and

BeP were observed to have as much as 20% of their total mass appear in the vapor phase. Conversely, if the temperature were to drop to $-20°C$ from $25°C$, then the vapor pressure would drop nearly three orders of magnitude. As a result, many of the vapor-phase PAHs would now occur in both the vapor and particle phases. This phase shift (i.e., higher occurrence of PAHs in the particle phase), resulting from a decrease in temperature, is likely a major contributor to the older observations that PAH concentrations are higher in winter than in summer. Many of the lower PAHs are seen in the particulate matter during periods of extreme cold (36, 37).

2.3. Particle Size Distribution

Many studies have shown that the PAHs are associated primarily with particles less than $3\ \mu m$ in aerodynamic mean diameter (38–45). Miguel (44, 45) has demonstrated a bimodal distribution of PAH in the air of Rio de Janeiro particulate matter in which 60% of BaP in the particulate matter was associated with particles between 0.075 and 0.26 μm, and 30% was associated with particles between 0.26 and 2.0 μm. Similar results were obtained in Pasadena, California (44).

X-ray fluorescence analyses have shown (46) that submicron particles are devoid of any significant transition-metal (iron, copper, and zinc) and crustal–element (aluminum and silicon) content. Since the submicron particles are almost exclusively carbonaceous, and the PAHs are associated primarily with these particles, there seems little possibility that the transition-metal-catalyzed decomposition of PAH could form a major degradation route for PAH on atmospheric particulate matter.

Submicron particles, in the absence of wet deposition, have atmospheric residence lifetimes of between 100 and 1000 h (47), whereas particles in the 1–10-μm size range should have atmospheric lifetimes of 10–100 h. In the absence of wet deposition, therefore, the long atmospheric residence times of the submicron particles suggest that the particle-bound PAC may travel great distances before being removed from the atmosphere. The long-range transport of PACs has been discussed and reviewed by several authors (48–50) and has been shown to be responsible for the deposition of PAC to remote lake systems (50).

3. CHEMICAL AND PHOTOCHEMICAL REACTIONS OF PACs

3.1. Singlet Molecular Oxygen

The existence of the oxidation products of numerous PAHs in atmospheric particulate matter has been taken to indicate that the PAHs react with oxygen or ozone in the atmosphere. The direct interaction of a ground-state PAH molecule (singlet electronic state) and oxygen (triplet electronic state), however, is a spin-forbidden reaction. A mechanism has been proposed by several authors (51–53) in which the PAH molecule acts as a sensitizer to induce the formation of singlet molecular oxygen which, in turn, may react with a ground-state PAH molecule:

$$PAH(S_0) + h\nu \rightarrow PAH(S_1) \tag{1}$$

$$PAH(S_1) \xrightarrow{\text{Intersystem crossing}} PAH(T_1) \tag{2}$$

$$PAH(T_1) + O_2(^3\Sigma_g^-) \rightarrow PAH(S_0) + O_2(^1\Delta_g, {}^1\Sigma_g^+) \tag{3}$$

where S_0 and S_1 denote the singlet ground electronic state and first electronic excited state, respectively, and T_1 denotes the first electronic excited triplet state. The triplet-state PAH formed in reaction (2) can also react with ground-state oxygen (52) to form a sensitizer–oxygen complex:

$$PAH(T_1) + O_2(^3\Sigma_g^-) \rightarrow PAH-O_2 \tag{4}$$

This complex can react with a proximal PAH (PAH_1) to form a peroxide of the proximal molecule:

$$PAH-O_2 + PAH_1 \rightarrow PAH_1-O_2 + PAH \tag{5}$$

Alternatively, the singlet molecular oxygen produced by reaction (3) could react with a proximal PAH molecule to form a peroxide:

$$O_2(^1\Delta_g) + PAH(S_0) \rightarrow PAH-O_2 \tag{6}$$

Although singlet delta molecular oxygen has a mean lifetime of approximately 0.57 s at atmospheric pressure (54) and is therefore likely to engage in atmospheric chemical processes, the singlet sigma state is not produced in the atmosphere by the direct adsorption of sunlight; furthermore, since singlet sigma molecular oxygen has such a short lifetime when produced by other mechanisms, it is not expected to play any significant role in the atmospheric decomposition of the PAC.

Ozone may be photolyzed by solar radiation, yielding both singlet molecular oxygen and singlet oxygen atoms (55):

$$O_3(^1A) + h\nu \rightarrow O_2(^1\Delta_g, {}^1\Sigma_g^+) + O(^1D) \tag{7}$$

The formation of the $^1\Sigma_g^+$ state of molecular oxygen requires excitation by radiation below 260 nm (54, 56) and therefore would not be expected to occur in the atmosphere. The singlet D oxygen atoms formed in reaction (7) may react with ground-state oxygen (57) to form singlet molecular oxygen and triplet P oxygen atoms:

$$O(^1D) + O_2(^3\Sigma_g^-) \rightarrow O_2(^1\Delta_g \text{ or } ^1\Sigma_g^+) + O(^3P) \qquad (8)$$

The quantum yield of $O_2(^3\Sigma_g^-)$ has been shown to increase as the concentration of oxygen molecules increases and is significant at atmospheric pressures.

Singlet molecular oxygen has been implicated in the photodecomposition of many PAHs (56–64) in laboratory photochemical decomposition experiments. Atmospheric particulate matter enriched with anthracene (**I**) and exposed to sunlight (59) was shown to yield the 9,10-endoperoxide (**II**), anthraquinone (**III**), dianthracene dione (**IV**), and 1-hydroxyanthraquinone (**V**) (Scheme 1). Although these products are known to be products of the singlet molecular oxygen reaction with anthracene, it is also possible for them to arise through the reaction of triplet-state anthracene with ground-state oxygen. Frimer (60) suggested that the O–O bond in the endoperoxide (**II**) could cleave, forming a diradical (**VI**) that could rearrange to form a half-oxidized anthraquinone (**VII**), which, in turn, could react with ozone or singlet molecular oxygen to yield anthraquinone (**III**) (Scheme 2).

Numerous PAHs, including phenanthrene, 9,10-diphenylanthracene, fluoranthene, pyrene, benz[*a*]anthracene, chrysene, benzo[*a*]pyrene, and 9-fluorenone, have all been shown to be efficient sensitizers for the production of

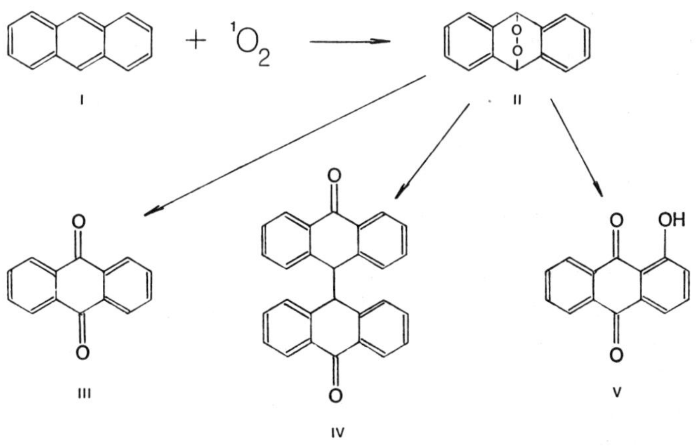

Scheme 1

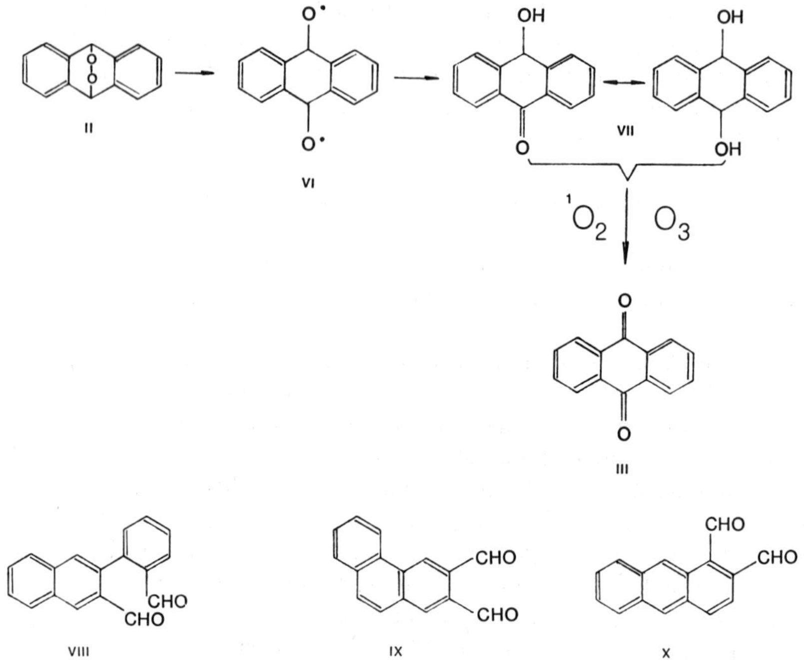

Scheme 2

singlet molecular oxygen (62). The major product of the oxidation of 9,10-diphenylanthracene was identified as the 9,10-diphenylanthracene endoperoxide. Several dialdehydes (**VIII**, **IX**, and **X**) have been isolated from the singlet molecular oxygen decomposition of benz[a]anthracene (63).

The direct conversion of both chrysene and 3-methylcholanthrene to mutagenic derivatives by reaction with singlet molecular oxygen has also been demonstrated (61), although the identity of the products was not established.

Inomata and Nagata (65) demonstrated that the 6-phenoxy-BaP radical (**XII**) was a significant intermediate in the singlet molecular oxidation of BaP (**XI**) to form the BaP quinones (see Scheme 3). The 1,6-, 3,6-, and 6,12-quinones (**XIII**, **XIV**, and **XV**, respectively) of benzo[a]pyrene (**XI**) have been identified in both laboratory experiments and in atmospheric extracts (4, 41, 58, 64). All three of these products exhibit mutagenic activity. Under singlet molecular oxygen oxidation (64), the 6,12-BaP quinone appeared as the major product. However, under one-electron oxidation conditions [using tris(4-bromophenyl)aminium hexachloroantimonate as the one-electron oxidant], the 1,6-BaP quinone appeared as the major product and only a minor amount of the 6,12-BaP quinone was detected. Lee-Ruff et al. (64) also

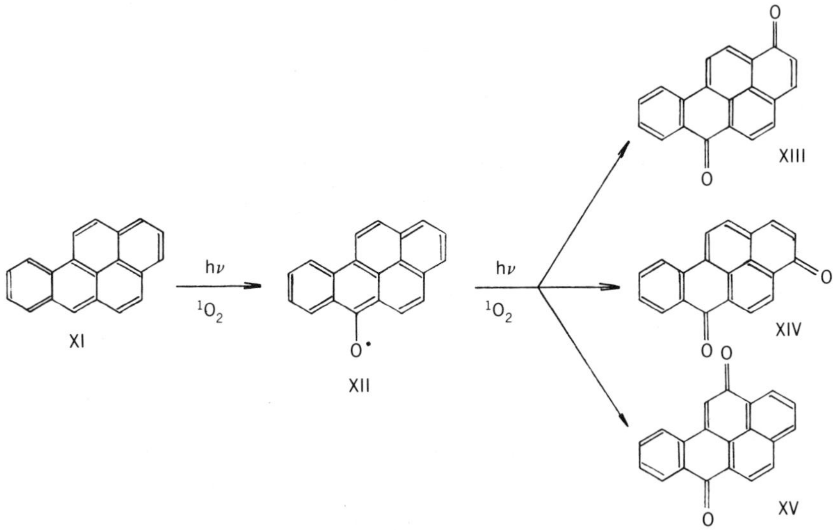

Scheme 3

identified 6-seco-BaP (**XVI**) as one of the decomposition products. This was produced, they suggested, as a result of the formation of either the dioxetan (**XVII**) or the perepoxide (**XVIII**) of BaP (as in Scheme 4).

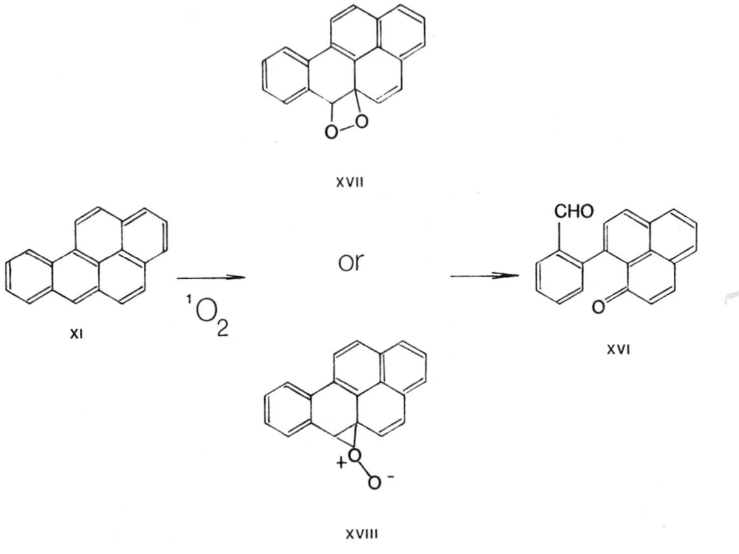

Scheme 4

In preliminary experiments, in which BaP was photooxidized in sunlight, the 6-seco-BaP was detected by mass spectrometric detection as a short-lived intermediate (66). The 6-seco-BaP reached a maximum after 2 h of exposure to sunlight and then decreased until, after 3 h, it was undetectable.

3.2. PAH Oxidation under Simulated Atmospheric Conditions

Many attempts have been made to expose PAH to simulated atmospheric conditions. Glass, silica gel, alumina, cellulose, acetylated cellulose, carbon black, and artificially generated soot have all been used as surrogates for atmospheric particulate matter in experiments in which the PAHs have been deposited, either by a solution coating or vapor deposition procedure, on the surrogate and exposed to gas mixtures including a variety of oxidants. As a result of the different substrates and simulated atmospheres, apparently contradictory results have been obtained. However, some of these differences can be related directly to the experimental conditions. The extent to which these experiments approximate the actual atmospheric condition is not yet established.

The photodecomposition of PAHs under simulated atmospheric conditions was first demonstrated by Falk et al. (67). BaP adsorbed on filter paper underwent a 21% loss in 24 h and a 22% loss in 48 h. When adsorbed on actual atmospheric soot, the BaP showed only a 10% loss over 48 h. Falk was the first to suggest that the soot could stabilize the BaP to photodecomposition.

PAHs were observed to decompose readily on alumina and silica gel G thin-layer chromatographic (TLC) plates (68). The 1,6- and 1,8-pyrene quinones were isolated from a complex mixture that arose from the oxidation of pyrene on the silica gel. However, if the TLC plates were first impregnated with paraffin, a much greater stability of the PAHs to photooxidation was seen (69).

The photooxidation rates for BaP, BbF, and BkF under a variety of simulated atmospheric conditions were reported by Lane (58) and Lane and Katz (70) and these were later augmented with results for several other PAHs by Katz et al. (71). The half-lives, shown in Table 1, indicated that the PAHs were quite reactive when exposed to oxygen in the presence of simulated sunlight. The nature of the photodecompositions suggested (58, 70) that, where PAH existed on particles in a multilayered deposition, the surface layer which was exposed to the atmosphere would react very rapidly, exhibiting the reaction rates in Table 1, while the subsurface PAHs would remain essentially protected from oxidation reactions. Only when the oxidized surface PAHs left the surface could the subsurface PAHs enter into reaction. The rate at which the subsurface layers were accessed was termed the *penetration rate*, and this

Table 1. Half-Lives of PAHs (in hours) under Simulated
Sunlight

PAH	Simulated Sunlight
Anthracene[a]	0.20
Benz[a]anthracene[a]	4.20
Dibenz[a,h]anthracene[a]	9.60
Dibenz[a,c]anthracene[a]	9.20
Pyrene[a]	4.20
Benzo[a]pyrene[b]	5.30
Benzo[e]pyrene[a]	21.10
Benzo[b]fluoranthene[b]	8.70
Benzo[k]fluoranthene[b]	14.10

[a] Data from Katz et al. (71).
[b] Data from Lane and Katz (70).

rate was found to be very similar to the loss rate reported by Falk et al. (67). The role of the surface layer in protecting subsurface PAHs during photo-decomposition has also been discussed by others (103, 104), and, very recently, Valerio et al. (72) have attempted to quantify this theory.

Attempts to simulate more closely the distribution of PAHs in the atmosphere have met with varying degrees of success. Early attempts to generate a synthetic soot through the incomplete combustion of propane, and the subsequent exposure of the soot to irradiation and various oxidants in a flow-through reactor, demonstrated that the PAHs could be decomposed on soot particles (73–75) and that the rapid decomposition of the particle-bound PAHs was caused by the exposure to light.

Rather than generating soot and PAHs *in situ* through the incomplete combustion of some organic material, techniques were developed to coat sieved fly ash with PAHs by a vapor-phase deposition procedure (76). Experiments to investigate the decomposition of PAH on the fly ash (77–80) indicated that the PAHs were stabilized to oxidation and, in particular, BaP, Pyr, and Anth were shown to be very strongly resistant to photooxidation. An unexpected result was the observation (77, 79, 80) of the spontaneous oxidation of Flu, BaFlu, BbFlu, 9,10-DMA, 9,10-DHA, and 4-aza-fluorene (all compounds with a benzylic carbon atom) in the dark. Three possible explanations were offered to explain this observation. First, it was suggested that the dark oxidation could be attributed to a molecular rearrangement of the individual PAH molecules. Alternatively, the availability of some very reactive adsorbed, chemisorbed, or chemically bonded oxygen groups on the surface of the fly ash could have caused the oxidation. The catalytic effect of transition metals on the surface of the fly ash appeared to be an attractive possibility in view of reports (81–83) that transition metals were concentrated on the surface of fly-ash particles. It was suggested that the main pathway for the decomposition of Anth on fly ash was nonphotochemical in nature and that the reactions of PAHs in the atmosphere may be determined primarily by thermal reactions, with photochemical reactions playing only a minor role.

There is a possibility, however, that the dark oxidation reactions are simply artifacts of the experiments and will not be observed on the real atmospheric particulate matter. The fly ash used was sieved to 50 μm in size prior to coating with the PAHs. In the real atmosphere, PAHs are associated predominantly with particles less than 1 μm in size, and these particles have been shown (46) to be carbonaceous and to be devoid of metal content. The vapor coating of fly-ash particles, although yielding interesting results, probably does not give a good representation of the real-world situation.

The acidic or basic nature of coal fly ash has been implicated as being a major factor affecting the photodegradation of particle-bound PAHs (85). On alkaline fly ash the PAHs were stabilized to photodecomposition, whereas on acidic fly ash the photodecomposition of the PAH was enhanced. However, others found no correlation between acidic and basic fly ash and the stabilization of PAH to photodegradation (86, 87), but they did find a correlation between PAH reactivity on fly ash and the color of the fly ash. Dark-colored fly ashes were seen to stabilize PAHs to photodegradation, whereas light-colored fly ashes did not. In addition, BaP and Pyr did not suffer photodegradation if they were adsorbed on fly ash that had more than 10% iron or 0.5% carbon. These observations were explained in terms of an "inner filter" effect (78), which proposes that the PAHs are adsorbed deeply in the pores of a particle and, as a result, are shielded from incident light. This concept has been proposed as a result of earlier investigations (58, 70).

An alternative explanation (88) suggests that the difference in reactivities of the PAHs on fly ash and supports such as alumina and silica gel is a result of the configuration of the PAHs on the coal fly ash. Although coal fly ash showed the least adsorption of the PAH per gram of substrate, it exhibited the greatest adsorption in moles of PAH per square meter. From the adsorption isotherms for Pyr on the various substrates, it was determined that the Pyr occupied the least area per molecule on fly ash, and, furthermore, the plane of the Pyr molecule appeared to be perpendicular to that of the substrate. The greater concentration of the Pyr on the fly-ash surface may enhance the toxicity and/or mutagenicity of the fly-ash particles to cells (88).

In other experiments, photodecomposition of PAHs on particulate matter was simulated in a fluidized bed reactor (89). Various PAH-coated substrates (including Carbosieve S) and coal fly ash were exposed in the reactor to both light and dark conditions. The degradation rates for Pyr on Carbosieve S and on coal fly ash were found to follow apparent first-order kinetics and to exhibit half-lives similar to those reported by Katz et al. (71) for the decomposition of PAHs on cellulose acetate TLC plates. The half-life of Pyr on coal fly ash was estimated to be 6.0–7.5 h.

The half-life for the decomposition of PAHs on coal fly ash was reported in other experiments (90) to range from 29 to 49 h. Half-lives of that magnitude would facilitate the long-range transport of the particle-bound PAHs.

3.3. Ozonolysis Reactions

The formation of ozone in the upper atmosphere and through the interaction of oxidants with air pollutants has been covered in great detail elsewhere (91, 92) and will not be discussed here. Ozone reacts readily with many organic compounds, and its reactions with PAHs in solution have been well documented (93–100).

There are many mechanisms by which ozone may attack the PAH. Considering benz[a]anthracene (**XIX**) as an example, ozone may add across the 5,6-bond to form a molozonide (**XX**), which can proceed to a diacid (**XXI**) as shown in Scheme 5.

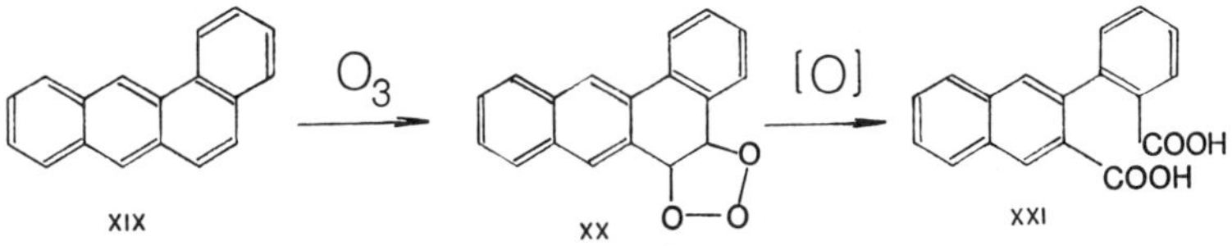

Scheme 5

Another reaction involves attack across the L-region to yield BaA quinone as in Scheme 6.

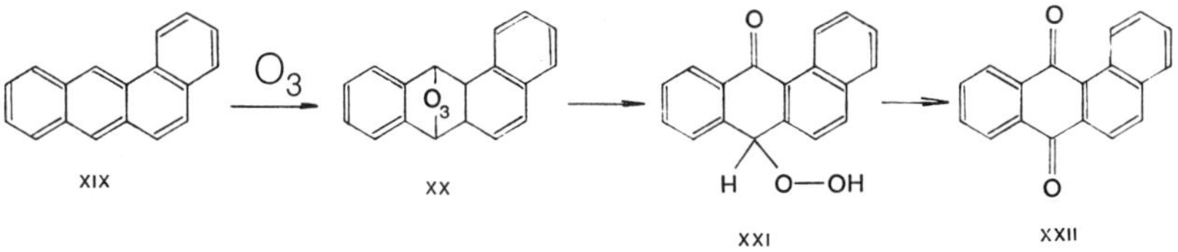

Scheme 6

A third, more complex mechanism involves electrophilic attack and the formation of a hydroxy BaA intermediate that can react further with ozone to form a dihydroxy species and a third ozone molecule, thereby forming a

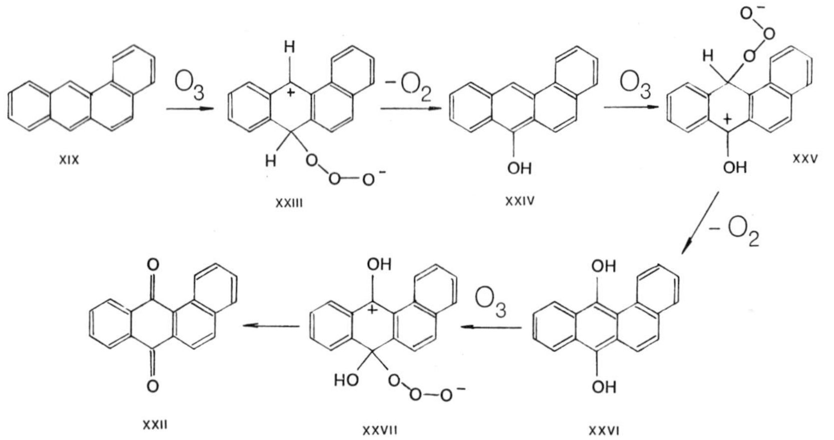

Scheme 7

quinone as shown in Scheme 7. If the oxygen molecules released from the ozonolysis are in the singlet electronic state, then oxidation could also occur through the reaction of singlet molecular oxygen (101).

In the polar fraction of particulate matter extracts, a highly mutagenic product has been identified (102) as the 4,5-oxide of BaP (**XXVIII**):

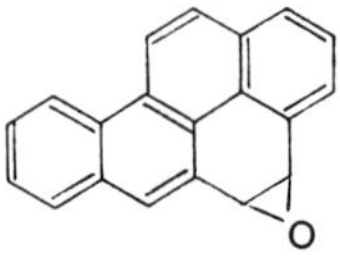

XXVIII

The oxide was then synthesized through the ozonolysis of BaP under laboratory conditions; however, a possible mechanism for the production of this potent mutagen was not presented.

3.4. PAH Ozonolysis under Simulated Atmospheric Conditions

The heterogeneous gas–solid reactions of ozone with PAHs have been carried out for PAHs deposited on many different substrates. In the early experiments of Falk et al. (67), PAHs deposited on filter papers and soot particles were exposed to a synthetic smog containing 30 ppm ozone. The PAHs adsorbed on soot particles were found to be less reactive to ozone than the PAHs on the filter papers. Tebbens et al. (73, 74) also attempted to expose PAHs on soot (generated through the incomplete combustion of propane) to ozone, but the experiments failed when the "gummy" product plugged the sampling filter.

The first kinetic studies of the heterogeneous decomposition of PAHs in the unadsorbed state (on glass and cellulose acetate TLC plates) demonstrated the high rate of reaction of ozone with exposed PAHs (58, 70, 71). The half-lives are shown in Table 2. The most rapid reactions were found for Anth, BaP, and BaA, with half-lives of 0.15, 0.58, and 1.35 h, respectively. In these experiments, a rapid initial reaction (exposed surface PAHs reacting first) was followed by a much slower reaction (the penetration and disruption of the surface layer enabling the oxidants to access the subsurface PAHs). Interestingly, great differences in reaction rate were observed for isomeric structures. For example, BaP had a half-life of 0.58 h, whereas BeP had a half-life of 5.38 h under the same reaction conditions. Except for Pyr, BbF, and BkF, light had only a marginal effect upon the ozonolysis rate. These results have since been duplicated by others (102, 105, 106).

Table 2. Half-Lives of PAHs (in hours) Exposed to Ozone

PAH	Simulated Sunlight plus 0.2 ppm Ozone	Dark Reaction plus 0.2 ppm Ozone
Anthracene[a]	0.15	1.23
Benz[a]anthracene[a]	1.35	2.88
Dibenz[a,h]anthracene[a]	4.80	2.71
Dibenz[a,c]anthracene[a]	4.60	3.82
Pyrene[a]	2.75	15.72
Benzo[a]pyrene[b]	0.58	0.62
Benzo[e]pyrene[a]	5.38	7.60
Benzo[b]fluoranthene[b]	4.20	52.70
Benzo[k]fluoranthene[b]	3.90	34.90

[a] Data from Katz et al. (71).
[b] Data from Lane and Katz (70).

Significant reaction rates were observed for the ozonolysis of PAHs in the dark (see Table 2). As a consequence, one would expect particle-bound PAHs, collected on Hi-Vol filters, to continue to be subjected to ozonolysis reactions from the time the particle is collected on the filter until the sampling procedure has been completed (58, 70, 102). The PAH loadings determined from an analysis of these filters will, inevitably, be lower than the PAH loadings on the particulate matter at the time the particle was trapped on the filter. For some PAHs such as BaP, the measured loadings may be significantly low (102).

The relative humidity of the air appears to have little influence on the rate of ozonolysis of PAHs adsorbed on particulate matter (105, 107). Relative humidity was, however, a factor in the ozonolysis of PAHs on glass-fiber filters and Teflon-impregnated filters. Except for BaP, the degradation of PAH on the filters was much lower at 50% relative humidity than at 1% relative humidity (107).

Conflicting results have been obtained for the ozonolysis of PAHs on diesel particulate. When diesel particulate on a filter was exposed to 0.5 or 1.0 ppm ozone in air, no significant changes in the PAHs or mutagenic activity were detected (108). In other experiments (109), BaP, Per, and 1-nitropyrene were deposited on glass and Teflon filters and on Teflon filters containing fly ash, diesel particles, or ambient air particulate matter. After exposure to humid air containing 100 ppb ozone, SO_2 and nitric acid free NO_2 in the dark for 3 h, no evidence was found for chemical transformations of the PAHs. However, more recent experiments have found a very definite reaction of 1.5 ppm ozone with PAHs on diesel particulate matter (110). The rate of ozonolysis of BaP and Per relative to that of BkF was consistent with the ozonolysis of pure compounds coated on glass-fiber filters. It was further stated (110) that the results of ozonolysis of PAHs on glass plates (70) were directly transferable to the diesel particulate results.

Wood smoke, which has become a major pollutant problem in many areas, has been found to be a rich source of PAHs. As a result, many experiments (111–113) have been carried out to investigate the stability of PAHs, as well as their derivatives and analogues, on wood-smoke particles when subjected to various oxidants. Ozone was found to react readily with the PAHs on wood smoke and did so at a rate twice that of the reaction with NO_2 (111); however, the effect of light on the decomposition of the particle-associated PAHs was considered to be greater than that of either ozone or NO_2 (111, 114). It is not surprising that urban PAH concentrations are on the order of tens of nanograms per milligram of soot, whereas sources such as wood smoke have concentrations that are higher by a factor of 5–10 (111).

The ambient temperature was found to have a significant effect upon the reaction of PAHs adsorbed on wood smoke (111) with NO_x. At $-7\,^\circ$C, the reaction rates were reduced by a factor of 4–10 (relative to their reaction rates at $20\,^\circ$C) for several PAHs. It is reasonable to expect that there will be a similar temperature effect for the ozone reactions, although other experiments suggest otherwise (89).

Another factor that has the potential to affect the availability of the PAHs to light or oxidants is the degree of agglomeration of the particles. It has been suggested that PAHs trapped in the interstitial cavities or pores of carbonaceous particles by the agglomeration of smaller particles to form larger particles may isolate the individual PAH molecules from contact with light or oxidants, thereby increasing their chance of survival during atmospheric transport (58, 70, 115). In view of the apparent reactivity of PAC to atmospheric oxidants, physical isolation of molecules may be a significant factor in the survival of PAC, especially during long-range transport of the particles.

3.5.   OH Radical Reactions

The reaction of hydroxyl radicals with organic molecules in the atmosphere during daylight conditions is considered to be the major reaction of these molecules leading to their removal from the atmosphere (116–119). Hydroxyl radicals abstract hydrogen atoms from C–H bonds in alkanes and carbonyls and from O–H bonds in alcohols and glycols (116). They add to unsaturated C=C double bonds and to aromatic rings, and they interact with primary, secondary, and tertiary amines (116).

It has been suggested (10, 120) that hydroxyl radical attack may be the first step in the nitration of PAHs such as Pyr in both the gas phase and the particle phase. The reaction, leading to 2-nitropyrene (**XXXII**), is shown in Scheme 8. The mechanism may also facilitate the formation of the 2-nitrofluoranthene in the gas phase. Neither the 2-nitropyrene nor the 2-nitrofluoranthene have been identified as combustion products, yet both have been identified in atmospheric extracts (121).

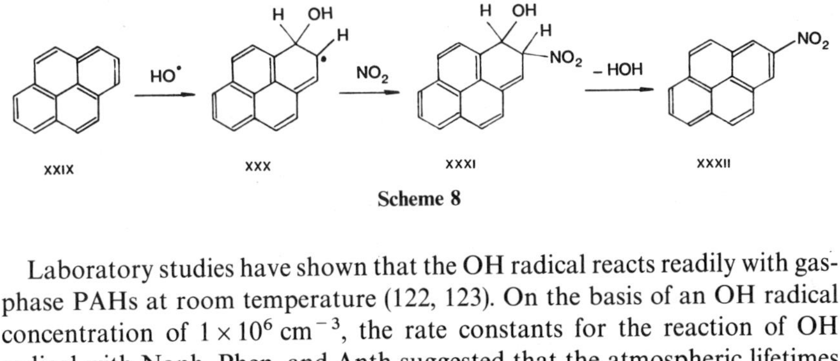

Scheme 8

Laboratory studies have shown that the OH radical reacts readily with gas-phase PAHs at room temperature (122, 123). On the basis of an OH radical concentration of 1×10^6 cm^{-3}, the rate constants for the reaction of OH radical with Naph, Phen, and Anth suggested that the atmospheric lifetimes for these PAHs are approximately 12, 9, and 2 h, respectively. The reaction of 2-methylnaphthylene with the OH radical (124) resulted in a calculated lifetime of only 4 h for this alkyl-PAH.

3.6.   Reactions with Oxides of Nitrogen and Nitric Acid

There now exists in the literature a considerable body of information to show that PAHs are transformed in the atmosphere as a consequence of their exposure to oxidants (119, 125). The reactions of PAHs with the nitrogen oxides and nitric acid to form the nitro-PAH are of considerable interest because many of the resulting products have been shown to be direct-acting mutagens.

Exposure of BaP and Per to 1 ppm NO_2 and ppb concentrations of nitric acid was observed to give rise to 1-nitro-BaP, 3-nitro-BaP, and 6-nitro-BaP (126). In other experiments, smog from Riverside, California was passed through a filter to remove the particulate matter, and the gaseous components passing through the filter were then passed through a filter on which pure BaP had been deposited. Although oxidation products were detected, no nitration products were detected. This was anticipated (126), to some extent, since the NO_x concentration in the smog was low and no nitric acid passed through the filter. It was apparent that the PAHs could not be nitrated by NO_2 alone but required a trace of nitric acid as a catalyst. Substantial supporting evidence for this observation has been obtained by others (109, 127–131).

In much the same way that PAHs adsorbed on fly ash were found to be stabilized to oxidation, PAHs adsorbed on fly ash or diesel soot were found to be stabilized toward nitration reactions (108, 127).

Contradictory results have been obtained regarding the effect of sunlight on the nitration reactions. While Hughes et al. (127) found that light had no effect on the gas-phase nitration of PAH by NO_2, Kamens et al. (111) found that PAHs adsorbed on wood smoke reacted with low concentrations of NO_2 in the presence of sunlight but were unreactive in the dark. The nitration rate for PAHs on wood smoke was, however, a factor of 2 slower than the reaction of the same PAHs with ozone.

Further investigations have demonstrated that PAHs such as Naph, Pyr, and Per do not react with the nitrate radical (132, 133), nor do they react with NO_2 without nitric acid present as a catalyst. They do, however, react with N_2O_5 in the gas phase.

N_2O_5 is formed at night by the reaction of NO_2 and NO_3 (134):

$$NO_2 + O_3 \rightarrow NO_3 + O_2 \tag{9}$$

$$NO_3 + NO_2 \overset{M}{\rightleftharpoons} N_2O_5 \tag{10}$$

Although the atmospheric concentration of N_2O_5 has not been measured directly, it has been calculated from rate constants determined (135) for reaction (10) and from the measured concentrations of NO_2 and NO_3. N_2O_5 concentrations were found to range from approximately 0.001 to 10 ppb (136). During daylight hours the photolysis of NO_3 precludes the formation of N_2O_5.

In laboratory experiments, Pyr adsorbed on a glass-fiber filter was shown to react with N_2O_5, undergoing a 60% decomposition in 50 min and yielding

1-nitro-Pyr as the major product (133). In similar experiments, Per (133) reacted only marginally with N_2O_5. In detailed kinetic studies, Pitts et al. (133) verified that the nitration was, in fact, through reaction with N_2O_5 and not by nitration with nitric acid resulting from the hydrolysis of the N_2O_5.

The indirect mutagenic activity of PAHs extracted from wood-smoke particles was observed to decrease when the smoke particles were exposed to NO_2 and ozone in an outdoor photoreaction chamber (111). Pitts et al. (133) suggested that this observation was due to nitration of the PAHs on the wood-smoke by N_2O_5 that had formed in the photoreactor by reactions (9) and (10) above.

In other experiments, Flt, BaP, and BaA were shown to react readily with N_2O_5 but did not react to any appreciable extent in the presence of both NO_2 and nitric acid. When exposed to N_2O_5, the major products isolated from the reaction with Pyr, Per, BaP, and BaA were the 1-nitro-Pyr, 3-nitro-Per, 6-nitro-BaP, and 7-nitro-BaA, respectively. Flt, on the other hand, gave rise to the 1-, 3-, 7-, and 8-nitro-Flt isomers in approximately equal yields (137, 138). None of these PAHs gave rise to any products in greater than 1% yield when exposed to nitric acid alone, and Chrys was found to be unreactive under all nitration conditions.

Both the 2-nitro-Pyr and 2-nitro-Flt have been detected in the ambient air, but neither compound has been seen in emissions from automobile or diesel engines, power plants, or wood stoves. It has therefore been suggested (133, 137, 139–142) that the occurrence of these two nitro species constitutes evidence for the atmospheric nitration of PAHs. In fact, the 2-nitro-Flt isomer has been reported as the only nitration product arising from the gas-phase nitration of Flt and N_2O_5 (140). At present, there is no explanation for the occurrence of the 2-nitro-Pyr isomer in the atmosphere, since Pyr yields only the 1-nitro-Pyr isomer under such nitration conditions.

3.7. Reaction with Sulphur Oxides

Since PAHs are known to react with SO_2, SO_3, and H_2SO_4 in solution to form sulfinic and sulfonic acids, it has been suggested (101) that the concentrations of H_2SO_4 commonly found in the atmosphere are sufficient to result in chemical reactions with the PAHs. If such reactions do, in fact, occur, the products would not be detected in the normal extraction procedures used for the PAHs and their derivatives.

BaP on a synthetic soot (73, 74) and exposed to 50–80 ppm SO_2 suffered a 63% loss under simulated sunlight but only a 49% loss in the dark. When lower concentrations were used (8–10 ppm), a 48% loss was observed under light and no loss was seen in the dark. In similar experiments (130), using the incomplete combustion of propane to simulate a stack gas and keeping the gas at 150 °C, the rate of decomposition of BaP was accelerated when SO_2 was added to NO_2 and NO gases in the reactor. The addition of SO_3 to the NO and NO_2 in the reactor caused an even greater decomposition rate.

Both Pyr and BaP were deposited on various substrates (including fly ash) and were exposed to 100 ppm SO_2 and SO_3 as well as to light (127). Although SO_2 did not cause any decomposition of the PAH, SO_3 had a minimal effect, but no decomposition products were identified. In addition, the presence or absence of light had little effect on the decomposition rates.

SO_2 appears to be totally unreactive with PAHs at concentrations typical of those one would expect to find in the atmosphere of a polluted city. When Phen, Flt, Chrys, BaA, BaP, BeP, Anthan, BghiP, and Cor were adsorbed on soot particles and exposed to 5 ppm SO_2 (84), no reaction was detected even after 99 days. At SO_2 concentrations of 100 ppb (109), SO_2 did not react with BaP, Per, or 1-nitro-Pyr which had been deposited on glass and Teflon filters containing either fly ash, diesel particulate matter, or ambient particulate matter.

3.8. Reactions with Other Species

Very little information exists concerning the reactions of PACs with other reactive atmospheric gases. Peroxyacetyl nitrate (PAN) has been observed to oxidize BaP to the hydroxy and dihydroxy derivatives at low concentrations and to the BaP quinones at higher concentrations (126). Other peroxides have also been shown to oxidize PAHs to a variety of oxidation products (101).

4. IMPLICATIONS FOR ANALYSIS OF PACs

PACs are extracted from atmospheric particulate matter collected by high-volume filtration. Ambient air is sampled at a rate of 1.13 m^3/min for a period of 24 h, and the particulate matter is trapped on 20- × 25-cm filters usually made from glass fibers or from Teflon-coated glass fibers. In the traditional enclosure, the Hi-Vol sampler collects particles with aerodynamic mean diameters of less than 100 μm. Particles collected on the filter continue to be exposed to a large flux of atmospheric oxidants as long as the air is being drawn through the filter. To collect a particulate fraction more representative of that which can be inhaled and respired, size-selective inlets have been fitted to the standard Hi-Vol sampler. These inlets have been designed to collect only those particles whose aerodynamic mean diameters are less than 10 μm. Various cascade impactor inlets have been developed to gain information on the distribution of the PACs with respect to particle size. To collect the gas-phase component of the air sample and to trap PACs that have volatilized from the trapped particulate matter, traps of Tenax, XAD resin, or poly-urethane foam (PUF) have been placed downstream of the filter (35, 36, 111, 143, 144).

The nature of the filter material appears to have some influence upon the survival of the PACs during the collection process. Particle-bound PACs collected on filters constructed of glass fibers or of quartz glass fibers suffer a greater degree of decomposition than do particle-bound PACs collected on filters of Teflon or Teflon-coated glass fibers (109, 145, 152).

It has been suggested that PACs associated with the particulate matter collected on the filter may sublimate from the particle, may be entrained in the flow of air through the filter, and may pass through the filter (146–148). Experimental verification has come from many investigations (35, 110, 143, 144, 148–153). Several investigators have been careful to point out that the distribution of the PACs between the particulate filter and the adsorbent does not represent the particle–gas distribution of PACs in the atmosphere (35, 149, 153). Not only does the sampler not give a true gas–particle distribution coefficient, it does not provide an accurate method for measuring organic particulate loadings (153).

At ambient temperatures (25 °C), Anth and Phen have been observed to pass through the filter and two PUF cartridges in series, indicating a substantial breakthrough and loss of the lighter PAHs (36). At higher temperatures, one can expect losses of higher-molecular-weight PAC.

Depending on their molecular weight, PAH molecules may exist primarily in the gas phase (the three-ring PAHs), primarily in the particle phase (the six-ring PAHs) or in both phases (four- and five-ring PAHs) in ambient air. At colder temperatures, the distribution will be shifted in favor of the particle phase; in the summer, however, the reverse will be true. Although the Hi-Vol sampler with PUF adsorbent back-up does not give an accurate picture of the gas- and particle-phase distribution of the PAHs, the distribution of the PACs between the filter and the adsorbent does reflect major changes in tempera-ture. Using this sampler, a great deal of experimental evidence has been accumulated to show the gas-phase to particle-phase shift of the PACs during cold weather (35, 36, 110, 143, 149).

The extraction of PACs from particulate matter poses a considerable challenge. The amount of PAHs that can be released from particulate matter depends upon the nature of the particulate matter being extracted. Less than 30% of BaP was extracted from coal fly ash, yet 80% was extracted from atmospheric particulate matter (154, 155). Critical losses may arise (a) from incomplete extraction, (b) from irreversible adsorption of the PAHs on column packings during clean-up procedures, and (c) during solvent concen-tration steps where both chemical and photochemical decomposition may occur (148, 154, 155).

Although many sensitive and selective analytical methods have been developed and more are being developed, the weakest link in the chain of sampling and analysis is the sampler used to collect the atmospheric sample. An accurate means to collect the gas- and particle-phase PACs without changing the gas–particle distribution which exists in the atmosphere and without causing decomposition of the PAHs after collection is urgently needed.

REFERENCES

1. R. C. Lao, R. S. Thomas, H. Oja, and L. Dubois, *Anal. Chem.*, **45**, 909 (1973).
2. M. L. Lee, M. Novotny, and K. D. Bartle, *Anal. Chem.*, **48**, 1566 (1966).
3. A. L. Colmsjö, Y. U. Zebühr, and C. E. Östman, *Atmos. Environ.*, **20**, 2279 (1986).
4. R. C. Pierce and M. Katz, *Environ. Sci. Technol.*, **10**, 45 (1976).
5. J. König, E. Balfanz, W. Funcke, and T. Romanowski, *Anal. Chem.*, **55**, 599 (1983).
6. U. R. Stenberg and T. E. Alsberg, *Anal. Chem.*, **53**, 2067 (1981).
7. T. Ramdahl, *Environ. Sci. Technol.*, **17**, 666 (1983).
8. W. Cautreels and K. Van Cauwenberghe, *J. Chromatogr.*, **131**, 253 (1977).
9. T. Nielsen, "Nitro Derivatives of Polynuclear Aromatics: Formation, Presence and Transformation in Stack and Exhaust Gases and in the Atmosphere," Risø report RISØ-R-455, Risø National Laboratory, Roskilde, Denmark, 43 pages (1982).
10. T. Nielsen, B. Seitz, and T. Ramdahl, *Atoms. Environ.*, **18**, 2159 (1984).
11. T. Ramdahl, G. Becher, and A. Bjørseth, *Environ. Sci. Technol.*, **16**, 861 (1982).
12. M. G. Nishioka, C. C. Howard, and J. Lewtas, Detection of Hydroxy-Nitro-PAHs and Nitro-PAHs in Ambient Air Particulate Extract using Bioassay Directed Fractionation, in *Polynuclear Aromatic Hydrocarbons: Chemistry, Characterization and Carcinogenesis*, M. Cooke and A. J. Dennis, Eds., Battelle Press, Columbus, Ohio, p. 701 (1986).
13. J. Jager, *J. Chromatogr.*, **152**, 575 (1978).
14. A. Liberti, P. Ciccioli, A. Cecinato, E. Brancaleoni, and C. Di Paulo, *J. High Res. Chromatogr.*, **7**, 389 (1984).
15. M. C. Paputa-Peck, R. S. Marano, D. Schuetzle, T. L. Riley, C. V. Hampton, T. J. Prater, L. M. Skewes, and T. E. Jensen, *Anal. Chem.*, **55**, 1946 (1983).
16. J. Arey, B. Zielinska, R. Atkinson, and A. M. Winer, *Atmos. Environ.*, **21**, 1437 (1987).
17. T. Nielsen, P. Clausen, and F. P. Jensen, *Anal. Chim. Acta*, **187**, 223 (1986).
18. M. W. Dong, D. C. Locke, and D. Hoffmann, *Environ. Sci. Technol.*, **11**, 613 (1977).
19. G. Eklund, J. R. Pedersen, and B. Strömberg, *Chemosphere*, **16**, 161 (1987).
20. K. Li, R. C. Lao, P. S. Howes, C. Chiu, and R. Thomas, Analysis of PAH and Nitro-PAH in Ambient Air and Incinerator Emission Samples, presented at the Eleventh International Symposium on Polynuclear Aromatic Hydrocarbons, Gaithersburg, Md., September 23–25, 1987.
21. G. M. Badger and R. W. L. Kimber, *J. Chem. Soc.*, 266 (1960).

22. G. M. Badger and R. W. L. Kimber, *J. Chem. Soc.*, 2746 (1960).
23. G. M. Badger, G. E. Lewis, and I. M. Napier, *J. Chem. Soc.*, 2825 (1960).
24. G. M. Badger and T. M. Spotswood, *J. Chem. Soc.*, 4431 (1960).
25. G. M. Badger, J. K. Donnelly, and T. M. Spotswood, *Aust. J. Chem.*, **15**, 605 (1962).
26. G. M. Badger, J. K. Donnelly, and T. M. Spotswood, *Aust. J. Chem.*, **17**, 1147 (1964).
27. G. M. Badger and J. Novotny, *Nature*, **198**, 1086 (1963).
28. J. D. Adams, E. J. LaVoie, and D. Hoffmann, *J. Chromatogr. Sci.*, **20**, 274 (1982).
29. D. F. S. Natusch and B. A. Tomkins, Theoretical Consideration of the Adsorption of Polynuclear Aromatic Hydrocarbon Vapor onto Fly Ash in a Coal-Fired Power Plant, in *Carcinogenesis, Vol. 3, Polynuclear Aromatic Hydrocarbons*, P. W. Jones and R. I. Freudenthal, Eds., Raven Press, New York, 1978, p. 145.
30. M. R. Schure and D. F. S. Natusch, The Effect of Temperature on the Association of POM with Airborne Particles, in *Polynuclear Aromatic Hydrocarbons: Physical and Biological Chemistry*, M. Cooke, A. J. Dennis, and G. L. Fisher, Eds., Battelle Press, Columbus, Ohio (1982). p. 713.
31. C. E. Junge, *Adv. Environ. Sci. Technol.*, **8**(1), 7 (1977).
32. D. A. Lane and D. M. A. McCurvin, A Personal Computer Database for the Chemical, Physical and Thermodynamic Properties of Polycyclic Aromatic Hydrocarbons, in *Polynuclear Aromatic Hydrocarbons: A Decade of Progress*, M. Cooke and A. J. Dennis, Eds., Battelle Press, Columbus, Ohio p. 477 (1988).
33. R. D. Bradley and T. G. Cleasby, *J. Chem. Soc.*, 1690 (1953).
34. G. G. Schlessinger, Vapor Pressures, Critical Temperatures and Critical Pressures of Organic Compounds, in *Handbook of Chemistry and Physics*, 1972–1973 edition, R. C. Weast, Ed., CRC Press, Cleveland, Ohio, p. D-151.
35. K. E. Thrane and A. Mikalsen, *Atmos. Environ.*, **15**, 909 (1981).
36. H. Yamasaki, K. Kuwata, and H. Miyamoto, *Environ. Sci. Technol.*, **16**, 189 (1982).
37. J. J. Murray, R. F. Pottie, and C. Pupp, *Can. J. Chem.*, **52**, 557 (1974).
38. M. Kertész-Sáringer, E. Mészáros, and T. Várkonyi, *Atmos. Environ.*, **5**, 429 (1971).
39. A. Albagli, L, Eagan, H. Oja, and L. Dubois, *Atmos. Environ.*, **8**, 201 (1974).
40. R. C. Pierce and M. Katz, *Environ. Sci. Technol.*, **9**, 337 (1975).
41. M. Katz and C. Chan, *Environ. Sci. Technol.*, **14**, 838 (1980).
42. L. Van Vaeck, G. Broddin, and K. Van Cauwenberghe, *Environ. Sci. Technol.*, **13**, 1494 (1979).
43. A. H. Miguel and S. K. Friedlander, *Atmos. Environ.*, **12**, 2407 (1978).
44. A. H. Miguel, 'Distribution of Polycyclic Aromatic Hydrocarbons with Respect to Particle Size in Pasadena Aerosols in the Submicrometer Range, in *Polynuclear Aromatic Hydrocarbons: Third International Symposium on Chemistry and Biology — Carcinogenesis and Mutagenesis*, Ann Arbor Science Publishers, Ann Arbor, Mich. p. 383 (1979).
45. A. H. Miguel, *Int. J. Anal. Chem.*, **12**, 17 (1982).
46. J. D. Butler, P. Crossley, and D. M. Colwill, *Sci. Total Environ.*, **19**, 179 (1981).
47. N. A. Esmen and M. Corn, *Atmos. Environ.*, **5**, 571 (1971).

48. A. Bjørseth, L. Gulbrand, and A. Lindskog, *Atmos. Environ.*, **13**, 45 (1979).
49. A. Bjørseth and B. S. Olufsen, Long-Range Transport of Polycyclic Aromatic Hydrocarbons, in *Handbook of Polycyclic Aromatic Hydrocarbons*, A. Bjørseth, Ed., Marcel Dekker, New York, p. 507 (1983).
50. R. A. Hites and P. M. Gschwend, The Ultimate Fates of Polycyclic Aromatic Hydrocarbons in Marine and Lacustrine Sediments, in *Polynuclear Aromatic Hydrocarbons: Physical and Biological Chemistry*, M. Cooke, A. J. Dennis, and G. L. Fisher, Eds., Battelle Press, Columbus, Ohio p. 357 (1982).
51. A. U. Khan, J. N. Pitts, Jr., and E. B. Smith, *Environ. Sci. Technol.*, **1**, 656 (1967).
52. C. S. Foote, *Acc. Chem. Res.*, **1**, 104 (1968).
53. J. N. Pitts, Jr., A. U. Khan, E. B. Smith, and R. P. Wayne, *Environ. Sci. Technol.*, **3**, 241 (1969).
54. R. H. Kummler, M. H. Bortner, and T. Baurer, *Environ. Sci. Technol.*, **3**, 248 (1969).
55. W. B. DeMore and O. F. Raper, *J. Phys. Chem.*, **44**, 1780 (1986).
56. R. H. Kummler and M. H. Bortner, *Ann. N. Y. Acad. Sci.*, **171**, Art. 1, 273 (1970).
57. T. P. J. Izod and R. P. Wayne, *Nature*, **217**, 947 (1968).
58. D. A. Lane, Gas Chromatographic Analysis and Photomodification Studies of Atmospheric Polycyclic Aromatic Hydrocarbons, Ph.D. Thesis, York University, Toronto, Ontario, Canada (1975).
59. M. A. Fox and S. Olive, *Science*, **205**, 582 (1979).
60. A. A. Frimer, Singlet Oxygen in Peroxide Chemistry, in *The Chemistry of Functional Groups*, S. Patai, Ed., Wiley, New York (1983).
61. E. C. McCoy and H. S. Rosenkranz, *Cancer Lett.*, **9**, 35 (1980).
62. W. C. Eisenberg, K. Taylor, D. L. B. Cunningham, and R. W. Murray, Atmospheric Fate of Polycyclic Organic Material, in *Polynuclear Aromatic Hydrocarbons: Mechanisms, Methods and Metabolism*, M. Cooke and A. J. Dennis, Eds., Battelle Press, Columbus, Ohio, p. 395 (1985).
63. J. König, E. Balfanz, W. Funcke, and T. Romanowski, Structure–Reactivity Relationships for the Photooxidation of Anthracene and Its Anellated Homologues, in *Polynuclear Aromatic Hydrocarbons: Mechanisms. Methods and Metabolism*, M. Cooke and A. J. Dennis, Eds., Battelle Press, Columbus, Ohio, p. 739 (1985).
64. E. Lee-Ruff, H. Kazarians-Moghaddam, and M. Katz, *Can. J. Chem.*, **64**, 1297 (1986).
65. M. Inomata and C. Nagata, *Gann*, **63**, 119 (1972).
66. D. A. Lane and D. M. A. McCurvin, Unpublished results.
67. H. L. Falk, I. Markul, and P. Kotin, *AMA Arch. Ind. Health*, **13**, 13 (1956).
68. M. Inscoe, *Anal. Chem.*, **36**, 2505 (1964).
69. B. Seifert, *J. Chromatogr.*, **131**, 417 (1977).
70. D. A. Lane and M. Katz, *Adv. Environ. Sci. Technol.* **8**(2), 137 (1977).
71. M. Katz, C. Chan, H. Tosine, and T. Sakuma, Relative Rates of Photochemical and Biological Oxidation (*in vivo*) of Polynuclear Aromatic Hydrocarbons, in *Polynuclear Aromatic Hydrocarbons: Third International Symposium on Chemistry and Biology—Carcinogenesis and Mutagenesis*, P. W. Jones and P. Leber, Eds., Ann Arbor Science Publishers, Ann Arbor, Mich., p. 171 (1979).

72. F. Valerio, E. Antolini, and A. Lazzarotto, *Int. J. Environ. Anal. Chem.*, **28**, 185 (1987).

73. B. D. Tebbens, J. F. Thomas, and M. Mukai, *Am. Ind. Hyg. Assoc. J.*, **27**, 415 (1966).

74. B. D. Tebbens, M. Mukai, and J. F. Thomas, *Am. Ind. Hyg. Assoc. J.*, **32**, 365 (1971).

75. J. F. Thomas, M. Mukai, and B. D. Tebbens, *Environ. Sci. Technol.*, **2**(1), 33 (1968).

76. A. H. Miguel and D. F. S. Natusch, *Anal. Chem.*, **47**, 1705 (1975).

77. W. A. Korfmacher, D. F. S. Natusch, D. R. Taylor, E. L. Wehry, and G. Mamantov, Thermal and Photochemical Decomposition of Particulate PAH, in *Polynuclear Aromatic Hydrocarbons: Third International Symposium on Chemistry and Biology—Carcinogenesis and Mutagenesis*, Ann Arbor Science Publishers, Ann Arbor, Mich., p. 165 (1979).

78. W. A. Korfmacher, E. L. Wehry, G. Mamantov, and D. F. S. Natusch, *Environ. Sci. Technol.*, **14**, 1094 (1980).

79. W. A. Korfmacher, D. F. S. Natusch, D. R. Taylor, G. Mamantov, and E. L. Wehry, *Science*, **207**, 763 (1980).

80. W. A. Korfmacher, G. Mamantov, E. L. Wehry, D. F. S. Natusch, and T. Mauney, *Environ. Sci. Technol.*, **15**, 1370 (1981).

81. R. W. Linton, A. Loh, D. F. S. Natusch, C. A. Evans, Jr., and P. Williams, *Science*, **191**, 852 (1976).

82. R. W. Linton, P. Williams, C. A. Evans, Jr., and D. F. S. Natusch, *Anal. Chem.*, **49**, 1514 (1977).

83. T. R. Keyser, D. F. S. Natusch, C. A. Evans, Jr., and R. W. Linton, *Environ. Sci. Technol.*, **12**, 768 (1978).

84. J. D. Butler and P. Crossley, *Atmos. Environ.*, **15**, 91 (1981).

85. R. Dlugi and H. Güsten, *Atmos. Environ.*, **17**, 1765 (1983).

86. R. A. Yokley, A. A. Garrison, E. L. Wehry, and G. Mamantov, *Environ. Sci. Technol.*, **20**, 86 (1986).

87. E. H. Wehry, G. Mamantov, A. A. Garrison, R. A. Yokley, and R. J. Englebach, Chemical Transformation of Polycyclic Aromatic Hydrocarbons Vapor-Adsorbed on Coal Stack Ash, in *Polynuclear Aromatic Hydrocarbons: Chemistry, Characterization and Carcinogenesis*, M. Cooke and A. J. Dennis, Eds., Battelle Press, Columbus, Ohio, p. 985 (1986).

88. J. M. Daisey, M. J. D. Low, and J. M. D. Tascon, The Nature of the Surface Interactions of Adsorbed Pyrene on Several Types of Particles, in M. Cooke and A. J. Dennis, Eds., *Polynuclear Aromatic Hydrocarbons: Mechanisms, Methods and Metabolism*, Battelle Press, Columbus, Ohio, p. 307. (1986).

89. J. M. Daisey, C. G. Lewandowski, and M. Zorz, *Environ. Sci. Technol.*, **16**, 857 (1982).

90. T. D. Behymer and R. A. Hites, *Environ. Sci. Technol.*, **19**, 1004 (1985).

91. J. Heicklen, *Atmospheric Chemistry*, Academic Press, New York, 406 pages (1976).

92. National Academy of Sciences, *Ozone and Other Photochemical Oxidants*, National Academy of Sciences, Washington, D.C., 719 pages (1977).

93. P. S. Bailey, *Chem. Rev.*, **58**, 925 (1958).

94. P. S. Bailey, J. E. Batterbee, and A. G. Lane, *J. Am. Chem. Soc.*, **90**, 1027 (1968).
95. P. G. Erickson, P. S. Bailey, and J. C. Davis, Jr., *Tetrahedron*, **18**, 388 (1962).
96. E. J. Moriconi, B. Rakoczy, and W. F. O'Connor, *J. Am. Chem. Soc.* **83**, 4618 (1961).
97. E. J. Moriconi and L. B. Taranko, *J. Org. Chem.*, **28**, 1831 (1963).
98. E. J. Moriconi and L. B. Taranko, *J. Org. Chem.*, **28**, 2526 (1963).
99. E. J. Moriconi and L. Salce, *Adv. Chem. Ser.*, **77**, 65 (1968).
100. A. J. Greenberg and F. B. Darack, Atmospheric Reactions and Reactivity Indices of Polycyclic Aromatic Hydrocarbons, in *Molecular Structure and Energetics, Vol. 4, Biophysical Aspects*, A. Greenberg, Ed., VCH Publishers, Deerfield Beach, Fla., p. 1 (1987).
101. National Academy of Sciences, *Particulate Polycyclic Organic Matter*, National Academy of Sciences, Washington, D.C., 361 pages (1972).
102. J. N. Pitts, Jr., D. M. Lokensgard, P. S. Ripley, K. A. Van Cauwenberghe, L. Van Vaeck, S. D. Shaffer, A. J. Thill, and W. L. Belser, Jr., *Science*, **210** (4476), 1347 (1980).
103. R. Rajagopalan, K. G. Vohra, and A. M. M. Rao, *Sci. Total Environ.*, **27**, 33 (1983).
104. J. M. Benson, A. L. Brooks, Y. S. Cheng, T. R. Henderson, and J. E. White, *Atmos. Environ.*, **19**, 1169 (1985).
105. S. C. Barton, N. D. Johnson, B. S. Das, and R. B. Caton, PAH Losses during High-Volume Sampling, in the Proceedings of the International Technical Conference on Toxic Air Contaminants, Niagara Falls, New York, p. 120 (1980).
106. J. Peters and B. Seifert, *Atmos. Environ.*, **14**, 117 (1980).
107. J. N. Pitts, Jr., H.-R. Paur, B. Zielinska, J. Arey, A. M. Winer, T. Ramdahl, and V. Mejia, *Chemosphere*, **15**, 675 (1986).
108. T. L. Gibson, A. I. Ricci, and R. L. Williams, Measurement of Polynuclear Aromatic Hydrocarbons, Their Derivatives and Their Reactivity in Diesel Automobile Exhaust, in *Chemical Analysis and Biological Fate: Polynuclear Aromatic Hydrocarbons*, M. Cooke and A. J. Dennis, Eds., Battelle Press, Columbus, Ohio, p. 707 (1981).
109. D. Grosjean, K. Fung, and J. Harrison, *Environ. Sci. Technol.*, **17**, 673 (1983).
110. L. Van Vaeck and K. Van Cauwenberghe, *Atmos. Environ.*, **18**, 323 (1984).
111. R. M. Kamens, J. M. Perry, D. A. Saucy, D. A. Bell, D. L. Newton, and B. Brand, Factors which influence PAH Decomposition on Wood Smoke Particles, Paper 84-80.6, Presented at the 77th Annual Meeting of the Air Pollution Control Association, June 24–29, 1984, San Francisco.
112. R. M. Kamens, J. N. Fulcher, and Z. Guo, *Atmos. Environ.*, **20**, 1579 (1986).
113. R. M. Kamens, G. D. Rives, J. M. Perry, D. A. Bell, R. F. Paylor, Jr., R. G. Goodman, and L. D. Claxton, *Environ. Sci. Technol.*, **18**, 523 (1984).
114. N. Takeda and K. Teranishi, *Bull. Environ. Contam. Toxicol.*, **36**, 685 (1986).
115. C.-H. Wu, I. Salmeen, and H. Nikki, *Environ. Sci. Technol.*, **18**, 603 (1984).
116. R. Atkinson, *Int. J. Chem. Kinet.*, **19**, 799 (1987).
117. B. J. Finlayson-Pitts and J. N. Pitts, Jr., *Atmospheric Chemistry: Fundamentals and Experimental Techniques*, Wiley, New York, 1098 pages (1986).
118. R. Atkinson and W. P. L. Carter, *Chem. Rev.*, **84**, 437 (1984).

119. R. Atkinson, *Chem. Rev.*, **86**, 69 (1986).
120. T. Nielsen, *Environ. Sci. Technol.*, **18**, 157 (1984).
121. J. Arey, B. Zielinska, R. Atkinson, A. M. Winer, T. Ramdahl, and J. N. Pitts, Jr., *Atmos. Environ.*, **20**, 2339 (1986).
122. H. W. Biermann, H. Mac Leod, R. Atkinson, A. M. Winer, and J. N. Pitts, Jr. *Environ. Sci. Technol.*, **19**, 244 (1985).
123. R. Atkinson, S. M. Aschmann, and J. N. Pitts, Jr., *Environ. Sci. Technol.*, **18**, 110 (1984).
124. R. Atkinson and S. M. Aschmann, *Int. J. Chem. Kinet.*, **18**, 569 (1986).
125. T. L. Gibson, P. E. Korsog, and G. T. Wolff, *Atmos. Environ.*, **20**, 1575 (1986).
126. J. N. Pitts, Jr., K. A. Van Cauwenberghe, D. Grosjean, J. P. Schmid, D. R. Fitz, W. L. Belser, Jr., G. B. Knudson, and P. M. Hynds, *Science*, **202**, 515 (1978).
127. M. M. Hughes, D. F. S. Natusch, D. R. Taylor, and M. V. Zeller, Chemical transformations of Particulate Polycyclic Organic Matter, in *Polynuclear Aromatic Hydrocarbons: Chemistry and Biological Effects*, A. Bjørseth and A. J. Dennis, Eds., Battelle Press, Columbus, Ohio, p. 1 (1980).
128. D. Grosjean, *Atmos. Environ.*, **17**, 2565 (1983).
129. R. Niessner, D. Klockow, F. Bruynseels, and R. Van Grieken, *Int. J. Environ. Anal. Chem.*, **22**, 281 (1985).
130. E. Brorström-Lundén and A. Lindskog, *Environ. Sci. Technol.*, **19**, 313 (1985).
131. R. A. Yokley, A. A. Garrison, G. Mamantov, and E. L. Wehry, *Chemosphere*, **14**, 1771 (1985).
132. J. N. Pitts, Jr., R. Atkinson, J. A. Sweetman, and B. Zielinska, *Atmos. Environ.*, **19**, 701 (1985).
133. J. N. Pitts, Jr., B. Zielinska, J. A. Sweetman, R. Atkinson, and A. M. Winer, *Atmos. Environ.*, **19**, 911 (1985).
134. T. J. Wallington, R. Atkinson, A. M. Winer, and J. N. Pitts, Jr., *Int. J. Chem. Kinet.*, **19**, 243 (1987).
135. E. C. Tauzon, E. Sanhueza, R. Atkinson, W. P. L. Carter, A. M. Winer, and J. N. Pitts, Jr., *J. Phys. Chem.*, **88**, 3095 (1984).
136. U. F. Platt, A. M. Winer, H. W. Biermann, R. Atkinson, and J. N. Pitts, Jr., *Environ. Sci. Technol.*, **18**, 365 (1984).
137. J. N. Pitts, Jr., J. A. Sweetman, B. Zielinska, R. Atkinson, A. M. Winer, and W. P. Harger, *Environ. Sci. Technol.*, **19**, 1115 (1985).
138. B. Zielinska, J. Arey, T. Ramdahl, R. Atkinson, and A. M. Winer, *J. Chromatogr.*, **363**, 382 (1986).
139. J. N. Pitts, Jr., J. A. Sweetman, B. Zielinska, A. M. Winer, and R. Atkinson, *Atmos. Environ.*, **19** (10), 1601 (1985).
140. J. A. Sweetman, B. Zielinska, R. Atkinson, T. Ramdahl, A. M. Winer, and J. N. Pitts, Jr., *Atmos. Environ.*, **20**, 235 (1986).
141. T. Ramdahl, B. Zielinska, J. Arey, R. Atkinson, A. M. Winer, and J. N. Pitts, Jr., *Nature*, **321**, 425 (1986).
142. R. Atkinson, J. Arey, B. Zielinska, J. N. Pitts, Jr., and A. M. Winer, *Atmos. Environ.*, **21**, 2261 (1987).
143. J. F. Galasyn, J. F. Hornig, and R. H. Soderberg, *J. Air Pollut. Control Assoc.*, **34** (1), 57–59 (1984).

144. J. F. Hornig, R. H. Soderberg, A. C. Barefoot III, and J. F. Galasyn, Woodsmoke Analysis: Vaporization Losses of PAH from Filters and Levoglucosan as a Distinctive Marker for Woodsmoke, in *Polynuclear Aromatic Hydrocarbons: Mechanisms, Methods and Metabolism*, M. Cooke and A. J. Dennis, Eds., Battelle Press, Columbus, Ohio, p. 561 (1985).

145. F. S.-C. Lee, W. R. Pierson, and J. Ezike, The Problem of PAH Degradation During Filter Collection of Airborne Particulates—An Evaluation of Several Commonly used Filter Media, in *Polynuclear Aromatic Hydrocarbons: Chemistry and Biological Effects*, A. Bjørseth and A. J. Dennis, Eds., Battelle Press, Columbus, Ohio, p. 543 (1980).

146. B. T. Commins, Interim Report on the study of Techniques for Determination of Polycyclic Aromatic Hydrocarbons in Air, *Nat. Cancer Inst. Monogr.* **9,** 225–233 (1962).

147. C. Pupp, R. C. Lao, J. J. Murray, and R. F. Pottie, *Atmos. Environ.*, **8,** 915 (1974).

148. R. C. Lao and R. S. Thomas, The Volatility of PAH and Possible Losses in Ambient Sampling, in *Polynuclear Aromatic Hydrocarbons: Chemistry and Biological Effects*, A. Bjørseth and A. J. Dennis, Eds., Battelle Press, Columbus, Ohio, p. 829 (1980).

149. W. Cautreels and K. Van Cauwenberghe. Experiments on the Distribution of Organic Pollutants between Airborne Particulate Matter and the Corresponding Gas Phase, *Atmos. Environ.*, **12,** 1133–1141 (1978).

150. T. Handa, Y. Kato, T. Yamamura, T. Ishii, and K. Suda, *Environ. Sci. Technol.*, **14,** 416 (1980).

151. J. König, W. Funcke, E. Balfanz, B. Grosch, and F. Pott, *Atmos. Environ.*, **14,** 609 (1980).

152. K. Nikolaou, P. Masclet, and G. Mouvier, *Sci. Total Environ.*, **36,** 383 (1984).

153. R. G. Lewis, Problems Associated with Sampling for Semivolatile Organic Chemicals in Air, in Proceedings of the EPA/APCA Symposium on Measurement of Toxic Air Pollutants, EPA Report No. 600/9-86-013, APCA publication no. VIP-7, p. 134 (1986).

154. W. H. Greist, J. E. Caton, M. R. Guerin, L. B. Yeatts, Jr., and C. E. Higgins, Extraction and Recovery of Polycyclic Aromatic Hydrocarbons from Highly Sorptive Matrices such as Fly Ash, in *Polynuclear Aromatic Hydrocarbons: Chemistry and Biological Effects*, A. Bjørseth and A. J. Dennis, Eds., Battelle Press, Columbus, Ohio, p. 819 (1980).

155. W. H. Greist, L. B. Yeatts, Jr., and J. E. Caton, *Anal. Chem.*, **52,** 199 (1980).

CHAPTER

3

GAS- AND LIQUID-CHROMATOGRAPHIC TECHNIQUES

JOHN C. FETZER

Chevron Research Company
Richmond, California

1. INTRODUCTION

The polycyclic aromatic compounds (PACs) are now routinely analyzed by chromatographic procedures. Two major types of chromatography are commonly used, namely, gas chromatography (GC) and high-performance liquid chromatography (HPLC). Each technique offers some unique information that the other does not; these techniques are best suited for different types of PACs and should be viewed as complementary—not competitive— techniques. GC is most useful for the smaller PACs, since volatility is the major prerequisite. The extremely high resolution of GC and its compatibility with mass spectrometers and element-specific detectors are strong points in choosing it over HPLC for those compounds where both techniques are possible. HPLC is amenable for the high-molecular-weight or nonvolatile PACs, and HPLC detectors are more isomer selective. The ultraviolet/visible absorbance and fluorescence detectors commonly used in HPLC rely on energy transitions that are controlled by the composition, size, and shape of a PAC, so specific isomers have different, characteristic spectra.

The basic requirement for a sample to be analyzable by GC is that it be volatile in the temperature range used. The criterion for HPLC is that the sample be soluble in the mobile phase. These criteria overlap for most types of PACs. Using the polycyclic aromatic hydrocarbons (PAHs) as an example, GC is generally used for those of up to 24 carbons, whereas HPLC can be used for aromatics as small as the alkyl-substituted benzenes and naphthalenes. The upper limit for HPLC is presently the condensed PAHs with 38 carbons (13 rings).

The degree of condensation of the PAHs affects both their volatility and solubility so that a structure that is less condensed than another PAH with the same number of ring carbons will generally be both more volatile and more soluble (1). Thus, naphtho[8,1,2-*abc*]coronene, $C_{30}H_{14}$, cannot be analyzed

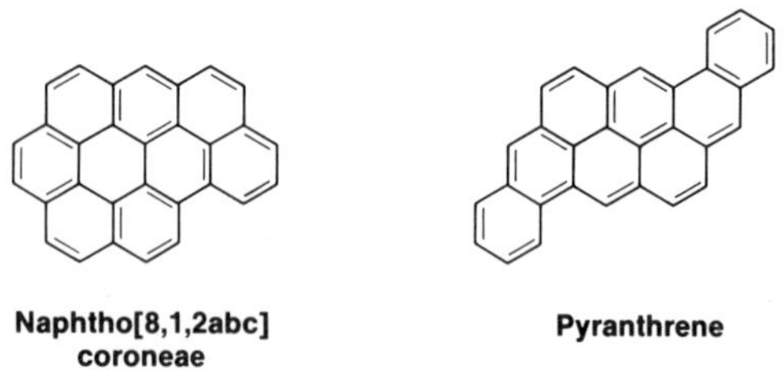

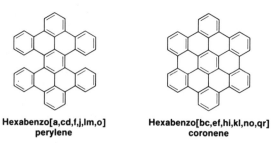

Naphtho[8,1,2abc]
coroneae

Pyranthrene

Figure 1. Structures of two 30-carbon PAHs that differ in degrees of condensation and physical properties.

by GC—except with the use of special high-temperature methods—and is only sparingly soluble in even the stronger HPLC solvents dichloromethane or chlorobenzene. The less condensed pyranthrene, $C_{30}H_{16}$ (structures shown in Figure 1), can be readily analyzed by GC and is very soluble in dichloromethane and chlorobenzene (2). For some PAHs, this effect may be even more accentuated as a result of intramolecular steric strain in less condensed PAHs which will cause the molecules to have a nonplanar conformation. Nonplanar PAHs are much more soluble and volatile than similar planar, more condensed structures. Hexabenzo[a,cd,f,j,lm,o]perylene, $C_{42}H_{22}$, can easily be analyzed by GC and is soluble in the very weak solvents methanol or *n*-hexane; but when this compound is pyrolyzed, the more condensed compound hexabenzo[bc,ef,hi,kl,no,qr]coronene, $C_{42}H_{18}$ (structures shown in Figure 2), is formed. It is so nonvolatile that it sublimes only under high temperature and very high vacuum, is insoluble in dichloromethane or tetrahydrofuran, and is only partially soluble in the extremely strong solvent 1,2,4-trichlorobenzene.

Hexabenzo[a,cd,f,j,lm,o]
perylene

Hexabenzo[bc,ef,hi,kl,no,qr]
coronene

Figure 2. Structures of two similar 42-carbon PAHs, which vary only in the degree of condensation.

2. GAS CHROMATOGRAPHY

2.1 Analysis of PAC by GC

The high resolution and short analysis times of capillary GC have made it the most popular analytical technique for PACs. Capillary columns, with resolutions of several hundred thousand, can separate extremely complex samples such as extracts of carbon blacks, coal tars, and shale oils (3–10). Almost all of the PACs that have been shown to be mutagenically active are volatile. Thus, GC is the easiest and most inexpensive method for routine PAC analysis.

The high resolution is necessary for PAC analyses because of the extreme complexity of most samples. There are many kinds of PACs, and there are various types of isomerization. Fossil-fuel samples contain many PAC isomers, both those with fully alternant structures (containing only six-membered rings) and those with five-membered rings. The alternant PAHs can also be split into two additional classes: the orthocondensed structures such as phenanthrene, anthracene, chrysene, picene, or pentaphene; and the pericondensed structures such as pyrene, perylene, coronene, pyranthrene, or ovalene. The occurrence of each type of PAH is determined by the mechanisms of formation. The most energetic mechanisms produce both types of alternant PAHs, as well as the nonalternant ones. Conditions that are less energetic will tend to favor the pericondensed alternant PAHs because of their higher resonance energies.

In addition, alkylation of the PAC structures is also common. The number of possible isomers increases astronomically with carbon number because of these two factors. For example, there are four PAHs with a formula of $C_{16}H_{10}$, with 28 monomethyl isomers. Adding one ring to the four unsubstituted PAH structures yields 16 $C_{20}H_{12}$ isomers, which have over 150 monomethyl isomers (11). Even more complexity occurs when longer alkyl chains, multiple substitution sites, or saturated-ring substitution are present. This is a general trend for all of the many benzologous PAC series.

2.1.1. Bonded Phases in High-Temperature GC

The stationary phases in capillary GC columns were, until recently, only statically coated onto the column walls. These phases were very thermally unstable, either decomposing or volatilizing at high temperatures. This limited their use to the lower-molecular-weight PACs. This limitation has recently been overcome through the development of cross-linked polymeric or chemically bonded GC phases (12, 13). These new stationary phases extend the usable temperature range of GC upwards by more than 100°. Nine-ring PAHs of molecular weight 374 were found in a dichloromethane extract of a

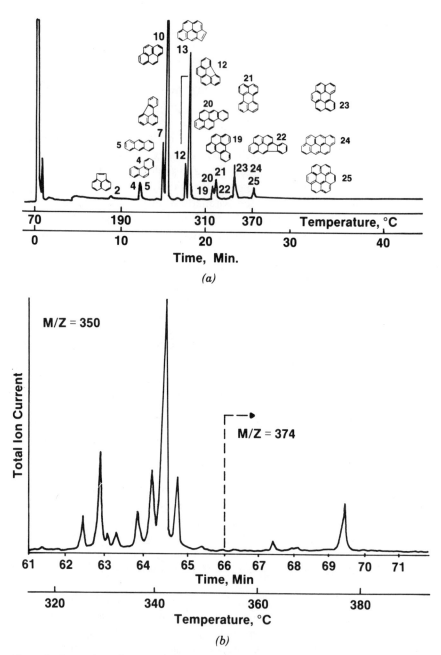

Figure 3. Comparison chromatograms, showing the upper limits in molecular weights of the PAHs eluted for a separation of a carbon-black extract on statically coated (a) and bonded-phase (b) columns. Reprinted, with the permission of the American Chemical Society, from M. L. Lee and R. A. Hites, *Anal. Chem.*, **48**, 1892 (1976) and from W. J. Simonsick and R. A. Hites, *Anal. Chem.*, **58**, 2114 (1986).

carbon black when a column of this type was used (5, 14). Similar statically coated phases could only be used at temperatures that would elute the isomers of molecular weight 302, $C_{24}H_{14}$ (15) (Figure 3).

Bonded phases are also more reproducibly made, more uniformly coated, and more stable than statically coated liquid phases. This makes comparisons between bonded-phase columns more realistic and reliable than those between statically coated columns, which have such large differences in characteristics that meaningful comparison among columns is nearly impossible. There are large, inherent differences in the resolution of different statically coated columns. These are a result of the variations in (a) stationary-phase thickness, (b) uniformity of coverage, (c) bleed rate, (d) oxidation, and (e) thermal stabilities. The reproducibility of bonded phases allows the use of standard retention compound sets, such as a series of orthofused PAHs proposed by Lee et al. (16). A similar series of retention standards have been proposed for the nitrogen-containing PACs (17).

2.1.2. *Liquid-Crystal Phases*

In many cases, the GC separation of PACs has been very difficult because the chemical affinities and boiling points of many isomers are very similar. Other factors must be employed in the separation of these molecules. The shapes of isomeric PACs are slightly different, so a stationary phase that can differentiate PACs by their shapes will be able to resolve isomers. Liquid-crystalline phases have been shown to do this (18–20).

The PACs are retained by their strength of interaction, which, in turn, is mainly a function of the chemical affinity of the molecules to the bonded phase, as well as the effective surface area of the interaction (21, 22). A planar molecule that can permeate between the liquid crystals will have a larger interacting surface area than a similar nonplanar PAC that does not permeate as well (23). The planar PAC will be retained longer than the nonplanar one. The physical properties of the first liquid-crystal stationary phases were very similar to liquid crystals in solution because they were only statically coated on the column walls. These stationary phases suffered from excessive column bleed and solubility in sample solvents, so their lifetimes were short.

Columns that had longer lifetimes became available after the preparation of polysiloxanes with an attached liquid-crystalline moiety. This made the synthesis of bound liquid-crystal phases possible (24). Columns with these phases, however, still suffered from limited useful temperature ranges because the liquid-crystalline phases that resulted were in their orderly state over only a small range of temperatures. These early liquid-crystal phases, however, proved the utility of liquid-crystal phases in achieving a shape-selective PAH separation. These phases were used to separate all of the 12 monomethylbenz[a]anthracene isomers. More recent work has resulted in phases with much wider liquid-crystal ranges, some as wide as from 120°C to 300°C. These newer phases have been used to separate all six monomethylchrysene isomers from each other, the 12 orthocondensed five-ring PAHs, and the three- to five-ring PAHs in a coal tar (25, 26).

Liquid-crystal phases separate the PAHs by their shapes, since the separation is based on permeation of the PAHs between the layered phase. Correlations between various PAH topological parameters (such as connectivity indices and the length-to-breadth ratio) and retention times have been made (27, 28). Prediction of retention times were also made, showing that these columns could be used for analysis of some compounds for which standards cannot be obtained—but the retention order predictions for many structurally similar isomers are not correct (28).

2.2. GC Detectors for PAC Analysis

The flame-ionization detector (FID) is the most used detector in GC because of its universality of response. A newer detector which also responds to most compounds is the photoionization detector (29–31). Both of these detectors rely on conditions in the detection area that are very energetic, resulting in ionization of almost all types of compounds. Because of their universality, these two detectors are not of great use in PAC analyses, except in tandem with other, more selective detectors.

In contrast to the universal detectors, some of the detectors used in the GC analysis of PACs are valued for their selective response to certain elements or compounds. Of the relatively inexpensive detectors, the most commonly used ones are (a) the flame photometric detector (FPD), which responds to phosphorous and sulfur, and (b) the thermionic detector, which responds to nitrogen or phosphorus. The FPD, when used in the sulfur mode, detects the fluorescence of a transient two-sulfur species that is generated in the flame. The thermionic detector is a modified FID whose active sensor is a glass bead doped with mixtures of alkali or alkali earth salts. Although this detector shows high sensitivity and selectivity toward nitrogen compounds, it also is very unstable and the reproducibility of analyses can be a problem (32, 33). This is because the doped cations, particularly rubidium, strontium, and sodium, are active in the detection mechanism and are also relatively volatile under the operating conditions. The responses of the thermionic detector to various classes of compounds also varies greatly. An alternate nitrogen-specific detector is the chemiluminescence detector, which is a combustion detector that produces a chemiluminescent species oxidatively (34). It has a much more uniform response to different types of nitrogen compounds (Figure 4).

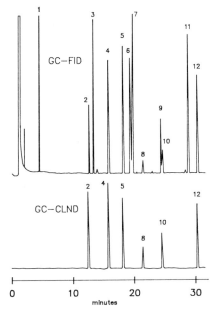

1	Naphthalene
2	2, 6—Dinitrotoluene
3	Fluorene
4	1—Nitronaphthalene
5	2—Nitrobiphenyl
6	Phenanthrene
7	Anthracene
8	Simazine
9	2—Methylanthracene
10	Prometryne
11	Pyrene
12	2—Nitrofluorene

Figure 4. Flame-ionization and chemiluminescence detection of a set of PAC standard compounds (a mixture of PAHs, N-PAHs, and triazine herbicides), showing the high degree of selectivity for nitrogen of the chemiluminescence detector. Reprinted with permission of Antek Instruments, Houston, Texas.

The much more expensive plasma emission spectrometric detectors can be used to monitor many elements because they detect the atomic emission of the individual elements of the sample components. Plasma emission detectors generally have a tunable monochromator, which can be set to the appropriate wavelengths. Microwave plasma detection has been used to selectively detect carbon, deuterium, sulfur, oxygen, and nitrogen in PAC samples (35–37).

The mass spectrometer is probably the second most widely used detector in GC (behind the FID), providing molecular weights and some structural information about the sample components. The fragmentation in electron-impact mass spectrometry (MS), or the use of the various MS/MS techniques, can show which types of heteroatom or alkyl substitution are found on the PAC. The MS fragmentation of a GC peak can also sometimes be used to determine which isomer it might be. The differentiation of various classes of nitrogen-containing PACs has been accomplished by using chemical ionization MS (38). For most isomeric PACs, however, the fragmentation patterns are too similar to be used to distinguish them from each other. The ratios of various ion intensities in chemical ionization methods have been used to

**Table 1. Ion-Intensity Ratios for the
Four-Ring Orthocondensed PAHs**

Isomer	M + 1/M
Naphthacene	0.386
Benzo[a]anthracene ·	0.788
Benzo[c]phenanthrene	1.106
Chrysene	1.193
Triphenylene	1.606

distinguish PAH isomers (Table 1 shows the ratios for the four-ring orthocondensed PAHs), including some PAHs in the 350–374 molecular-weight range (4, 5).

Chemical ionization methods can be used for the selective detection of other classes of PACs. Nitrosubstituted PACs are characterized by using methane as the reagent gas and by using an electron capture mode of detection, similar in mechanism to the electron capture detector (ECD) commonly used to detect halogenated compounds (39, 40). The spectra show high enhancement of the molecular ion and also show the presence of M-16 and M-30 ions for the nitro-PACs. This approach has been used to determine the nitro-PACs in air particulate samples (41).

When MS is the only type of detection used and no selective ionization or derivatization techniques are used, misidentification of the types of PACs found in a sample is common. In an early paper on the GC analysis for large PAHs, the PAHs in a carbon black extract were analyzed by capillary GC/MS. The masses found were given representative structures containing both five- and six-membered rings that were highly phenyl substituted (42). Subsequent HPLC analysis (with off-line MS and spectrofluorometry) showed the compounds to be predominantly alternant PAHs, with no phenyl substitution on any of the structures. The masses detected corresponded to structures either with or without phenyl substitution; in reality, however, only one type of PAHs was found (43).

The Fourier-transform–infrared (FTIR) spectrometer is becoming more widely used as a GC detector because of the high sensitivity available through the better signal-to-noise ratios from multiple scanning. The most powerful applications of this detector system are for determining different compound types. This approach was used in determining that a series of compounds found in a diesel particulate consisted of fluoren-9-ones and not the two nitrogen-containing species suggested from GC/MS—both series start with a compound of molecular weight 180 (44). This class selectivity is most informative for PACs that are functional derivatives, such as the amino-, nitro-, or hydroxyl-PACs (45). The FTIR can be set up to monitor the bands due solely to certain functionalities, such as the –NH or –OH stretches. High-resolution FTIR spectra can be used to differentiate similar PACs if they are obtained by trapping of the effluent by condensation on very cold surfaces. These matrix isolation techniques have been used to determine several small isomeric PAHs, including mixtures of methylchrysene isomers (46).

Other types of GC detectors that are selective for PACs but that have not been extensively used are those that monitor absorbance or fluorescence (47–50). These can be selective and isomer specific. Fluorometric detection is also extremely sensitive. There are several major problems with these types of detection. Few ultraviolet (UV) and fluorescence spectrometers have flow cells that can be heated to the high temperatures necessary to keep the analyte molecules in the gas phase. If this is not done, deposition of the sample components onto the cell windows and walls occurs, changing both the optical characteristics and flow cell volume (which affect both sensitivity and quantitation). In addition, although many PACs have distinct spectra, the commonly available reference spectra are taken in solution. The few published gas-phase spectra are radically different than the gas-phase GC spectra because the reference spectra are normally collected at very low pressures and GC spectra are at higher than ambient pressures. There is much more fine structure in the low-pressure spectra than in the solution spectra because there is no solvation or collisional broadening of the bands (51, 52), but GC spectra are very broad and featureless because of the collisional deactivation (the solvent cage in solution spectra limit the effect of collisions in that mode). Solution spectra cannot be used for guidance in choosing monitoring wavelengths in gas-phase detection because the optimum wavelength from the maximum of a single band in a solution spectrum will not be the maximum in the gas-phase spectrum and can even turn out to be located at the minimum between two bands or to be very shifted. This is a result of spectral differences from desolvation of the molecules.

In general, since GC detectors are not isomer specific, the identification of a particular GC peak as being one of several possible isomers relies solely on the comparison of the retention times to those of standard compounds. This requires that the analyst collect standards for retention time comparison. There are few pure, well-characterized standards for PACs; of the PAHs of 14 or fewer rings, only about 0.03% have been synthesized (53). Table 2 summarizes the trends in available (versus the theoretical) numbers of isomers. Most of these synthesized compounds are not commercially available. Besides being prohibitively expensive (the price for PAC standards is commonly in the $10–$100/mg range), a PAC collection requires an appropriate storage space, since many are potent mutagens or are air- and light-

Table 2. Standards for PACs and PAHs

No. Carbons	No. Hydrogens						
	8	10	12	14	16	18	20
10	1 (1)	-	-	-	-	-	-
12	-	-	-	-	-	-	-
14	-	2 (2)	-	-	-	-	-
16	1 (1)	-	-	-	-	-	-
18	-	-	5 (5)	-	-	-	-
20	-	-	3 (3)	-	-	-	-
22	-	-	2 (2)	12 (12)	-	-	-
24	-	-	1 (1)	12 (13)	-	-	-
26	-	-	-	27 (37)	5 (9)	-	-
28	-	-	-	4 (8)	24 (62)	-	-
30	-	-	-	2 (3)	11 (58)	23 (123)	-
32	-	-	-	1 (1)	7 (46)	16 (289)	-
34				2 (37)	8 (333)	8 (411)	

sensitive. A good organization scheme is also needed so that the large number of standards can be accessed efficiently.

When standards are used for comparison of retention times, a high degree of chromatographic reproducibility is necessary. The retention times of some isomers are very similar on many phases, and a small variation in retention time makes absolute determination impossible. The series of orthocondensed PAHs (naphthalene, phenanthrene, chrysene, and picene) have been suggested as retention-time marker compounds, to be similarly used as the *n*-alkanes are used to calculate Kovats indices (16). These Lee indices can be used to compare different columns or to compare the precision of an analysis through a repetitive series of standard runs. The next member in the series, benzo[*c*]picene (also called fulvilene) has recently become commercially available, and so the usable retention range has been extended to include many of the compounds that can now be analyzed by GC with the use of high-temperature techniques (54). The use of these same standards in HPLC, however, would be limited because of the wider elution range possible. Many PAHs elute later than benzo[*c*]picene, and so they would have no defined retention.

2.3. Applications of GC to PAC Analysis

The method of choice for the smaller PACs has been GC because of the high resolution and short analysis times. It is the overwhelming choice for use in environmental analyses. The U.S. Environmental Protection Agency has mandated that treated effluent waters be tested for several PACs (55), primarily smaller PAHs. The test methods require the use of capillary GC with selective-ion MS detection. Since most of the proven mutagenic and carcinogenic PACs are smaller molecules, GC is also the primary analytical tool for their determination. That the mutagenic and carcinogenic studies, and most of the analytical work, has been limited to the smaller compounds is not accidental. These are the easiest PACs to synthesize, and so pure, well-characterized standards are only available for the smaller PACs (56).

The high resolution of GC makes it the best choice for studying complex mixtures. Many carbonaceous materials have been analyzed by GC solely for this reason. The many alkyl isomers of PACs can usually be separated. As has been previously discussed, the biggest limitation in GC is the nonspecificity of the detectors. Thus, a capillary GC separation can yield over 100 peaks with mass 266, representing the monomethyl isomers of the isomeric set that includes perylene, the benzopyrenes, the benzofluoranthenes, and the other five-ring PAH isomers. Few of these can be unequivocally identified because there are no available standards, and most of these peaks cannot even be classified as a particular aromatic type. A carbon-black extract yielded a series of clusters of peaks that corresponded to PAHs with varying degrees of

Table 3. Comparison between Conventional HPLC and Micro-HPLC

Molecular Weight	Number of Components Resolved in Conventional HPLC	Number of Components Resolved in Micro-HPLC
228	1	1
252	4	5
276	3	14
278	-	1
282	-	3
300	2	6
302	2	5
304	-	
306	-	1
324	-	1
326	7	16
332	-	4
350	6	16
352	-	1
356	2	4

condensation (3). The majority of the sample was found to be alternant PAHs, but the number of isomers found for some molecular weights indicated that nonalternant PAHs also were present. For example, coronene is the only alternant PAH of molecular weight 300, but five peaks with this mass were separated (Table 3). Assignment of structures to the other four peaks could not be made.

Several of the PAH isomers of mass 302 ($C_{24}H_{14}$) were found in a flue deposit (9). These resulted from combustion of anthracitic coal. Of the 34 possible isomers, 11 were unambiguously identified through comparison with standards. In addition, seven other compounds with that mass were found but could not be positively identified due to a lack of standards.

Multiring sulfur heterocyclics were found in a carbon-black sample by GC/MS techniques and were also seen in a solvent-refined coal liquid (57) (Figure 5). The types of condensation of these thiophenic species was hinted at through the fragmentation of the molecular ions. The fragmentation ions did not show the patterns that are characteristic of thiophenes that are substituted on only one side of the thiophene ring. This was ascribed to the easier

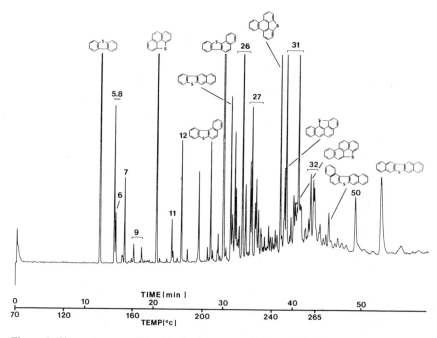

Figure 5. Chromatograms of the thiophenic compounds in a coal liquid sample, as determined by GC/MS and retention time comparison to standards. Reprinted, with the permission of Butterworth, Ltd., from M. Nishioka, M. L. Lee, and R. N. Castle, *Fuel*, **65**, 390 (1986).

hydrogenation that those thiophenes have, making them less likely to survive processing.

The sulfur-containing species in several whole crude oils, as well as their distillation and extraction fractions, were separated and detected with a FPD, and the patterns of peaks were compared to those seen with an FID (58). Additionally, a selective chemical derivatization was used to compare with the unreacted samples. The thiophenic compounds were oxidized to sulfones, were extracted, and were reduced back to thiophenes. The FPD pattern of the treated and unreacted samples indicated that all the sulfur-containing species were thiophenes because, in general, the reaction showed the same patterns. Mass spectral detection was also used in series to determine possible structures for the thiophenic compounds. The samples contained many alkyl-substituted benzothiophenes and dibenzothiophenes.

3. HIGH-PERFORMANCE LIQUID CHROMATOGRAPHY

3.1. The Application of HPLC Versus GC

The larger or more nonvolatile PACs cannot be analyzed by GC because they either will not elute or, if they do, the peaks will be unacceptably broad. In addition, some PACs are thermally unstable and decompose or rearrange pyrolytically to other structures. The separation of these compounds must then be done by a less severe and more universal method. HPLC can accomplish this. The smaller PACs are soluble in the weak normal-phase solvents such as n-hexane and iso-octane or the weak reversed-phase solvents such as methanol or acetonitrile. The PACs that are less soluble in these solvents can generally be dissolved in dichloromethane, ethyl acetate, or tetrahydrofuran—which are strong solvents in either mode of HPLC. These stronger solvents can dissolve PAHs of 12–14 rings. This variation in sample solvents and mobile phases that are possible with HPLC allows it to be used for samples that have PACs with a wide range in the number of aromatic rings or functionality (polarity).

Although GC has much more resolution than HPLC, an optimum PAC separation that uses HPLC is more easily developed than a GC one because many more separation parameters can be changed. With the variation of mobile-phase composition being a major factor in the separation, more variety is available through mixing of the many possible mobile-phase solvents. GC only has the stationary-phase material, the thermal gradient range, and rate of temperature increase as variables. HPLC can utilize all of these options, even though temperature-gradient HPLC is only occasionally used (59–61). HPLC also has several variables that are based on solvent changes. The variation in both mobile and stationary phases makes subtle changes in selectivity easier to achieve.

Besides variations in the solvent gradient shape or rates of change, subtle differences in selectivity can be attained by using similar, but slightly different, solvents (62). For example, methanol and acetonitrile both have similar overall solvent characteristics. There are, however, differences in the affinities for specific interactions. Methanol is an excellent hydrogen-bonding molecule, and acetonitrile is a solvent that has good dipole–dipole interactions (63, 64). Other similar solvent pairs include: n-hexane and 1,1,2-trifluoro-1,2,2-trichloroethane; ethyl acetate and acetone; and dichloromethane and chloroform. One member of each pair has slightly different Hildebrand solubility parameters than the other (63).

The choice of one similar solvent over another can also be necessitated when gradient runs are required that encompass wide ranges in solvent strength. Thus trifluorotrichloroethane is used instead of n-hexane for a normal-phase separation of alkanes, alkenes, and aromatics, since it is miscible with the perfluoroalkane solvents that must be used to separate the alkanes from alkenes. After this separation is completed, a gradient to trifluorotrichloroethane is used, and finally a gradient to n-hexane is used to separate the aromatics by their ring number. An example of this type of separation for a lubricating-oil sample is shown in Figure 6.

The mobile phase in HPLC can affect both the sample molecules and the stationary phase. Selective separations can be made if the change in mobile-phase composition during a gradient separation alters either of the two. For example, dichloromethane and tetrahydrofuran are both strong solvents in the reversed-phase separation of PAHs. At high concentrations, both yield similar retention times for large PAHs. Spectral evidence, however, indicates that dichloromethane interacts strongly with the PAHs and that tetrahydrofuran does not (2, 65, 66). Nuclear magnetic resonance (NMR) spectra of a reversed-phase packing wetted with each of the solvents suggests different interactions between the solvent, and the octadecyl moieties of the reversed-phase packing occurs in the two solvents (67, 68). Separation of two components that are closely eluting in dichloromethane can be achieved with tetrahydrofuran, since the first interacts preferentially with the solutes while the other interacts more with the stationary phase.

For some PACs the deciding criterion for choosing HPLC over GC is not volatility or solubility but is, instead, stability. Certain structures will rearrange or condense at the high temperatures used in GC injection ports or at the end of thermal gradients. Pentacene cannot be analyzed by GC because some of the molecules disproportionate, decomposing to form 6,13-dihydro-pentacene and other products (69). Coal tar pitches, or the synthetic pitches obtained from pyrolysis of naphthalene or other small aromatic molecules,

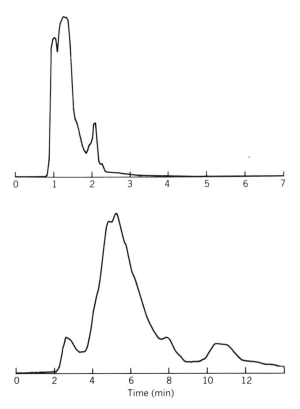

Figure 6. Amino-bonded phase chromatograms of a lubricating oil, monitored at 280 nm, showing the difference in elution strength of *n*-hexane and a weaker mobile phase containing perfluoroheptane. The peaks correspond to various alkylated naphthalenes and phenanthrenes.

contain large amounts of partially condensed structures such as the two binaphthyl isomers (10, 70). When these molecules are exposed to high temperatures, they can be converted to the more condensed structures perylene, benzo[*j*]fluoranthene, or benzo[*k*]fluoranthene (41). These decomposition reactions would obscure the original composition of the samples. Similar losses in functionality of the nitro-PAHs make their analysis by GC impossible (71).

3.2. Separations by Ring Number and Shape

The combination of the complexity of most PAC-containing samples and the lower resolution inherent to HPLC usually makes it necessary to perform several separation steps before the ultimate step of HPLC analysis can be used to determine the concentrations of specific structures. The initial separation steps usually used include such techniques as distillation or vacuum-sublimation, solvent extraction, and adsorption chromatography (72–78). These are usually used to subdivide the original sample into more easily analyzed fractions or to remove interferences.

Once a sample has been fractionated sufficiently to allow HPLC analysis, the analyst has two main options: separating the material either by polarity on a normal-phase column or by carbon content on a reversed-phase column. The first generally results in a separation into classes of PACs by ring number. However, each class of PACs elutes as an independent set of compounds so that the less polar PAHs elute before the more polar nitrogen-containing aza-arenes, which, in turn, elute before the aromatic ketones and quinones. Two-ring aza-arenes thus elute at the same time as the four- or five-ring PAHs, and the three-ring aromatic ketones coelute with the nine-ring PAHs (79, 80).

There are two major types of normal-phase separations of PACs. One type, using underivatized silica, occurs through the strength of hydrogen bonding and acid–base interactions between the sample molecules and the packing (81). The other type, which includes, separations on alumina and on amino-, cyano-, phenyl-, or nitroderivatized silicas (82–86), as well as some exotic packings coated with caffeine (87) or picric acid (88), is based on polarizability and induction effects between the molecules and the packing. This latter group is by far more popular for PAHs because it fractionates a sample roughly by ring number. The separation, however, is not exactly by the number of aromatic rings in the molecules—as is commonly stated—but is more accurately described as being by the number of pi electrons; double-bond equivalents (dbes) have been used in one terminology (89–92). Thus, pyrene (with eight pairs of pi electrons) elutes after phenanthrene (with seven pairs) but before chrysene (with nine pairs).

Superimposed on the separation by the number of pi electrons are differences in retention due to the shape of the individual molecules. This effect can be seen in two forms: For planar PAHs, an isomer that has the longest axis will be retained the most, whereas an isomer that is more compact will elute earlier (93). This type of separation has been said to be based on the length-to-breadth ratio (L/B). The L/B is defined as the ratio of the longest axis of the molecule, L, and its longest perpendicular axis, B. The structures are assumed to be planar (94, 95).

The other major factor affecting retention of the unsubstituted PAHs is planarity. If the compound has a nonplanar conformation (usually due to intramolecular steric effects), it elutes before the planar isomers with the same number of pi electrons (2, 65, 66). Both of the retention effects can be ascribed to the probability of interaction between a molecule's pi electrons and the polar-bonded moiety of the packing. The effective cross-sectional area of the

PAHs is the most important factor. PAHs with a long axis are more likely to interact with several polar-bonded phase moieties, more than a compact molecule. A nonplanar PAH does not effectively interact with several polar moieties simultaneously because it cannot lay flat on the bonded phase surface (85).

Another factor affecting retention in the normal-phase separation of PACs is the alkylation of the molecules. The number and position of substituents, the chain length of substituents, and whether or not saturated rings are present are some of the important factors that can change the retention of a PAC. Alkyl chains typically reduce the retention of PACs, relative to the unsubstituted species. This decreased retention is most likely due to the shielding, by the alkyl chains, of the aromatic rings of the molecules from interaction with the bonded phase. In contrast, saturated rings typically increase retention. The exact reasons for this are not yet known, but perhaps behavior parallel to the small molecules n-hexane and cyclohexane occurs. Cyclohexane is slightly more polar than n-hexane, and an analogous increase in polarity could occur upon substitution of a PAC with a saturated ring. Figure 7 shows a normal-phase separation of three pyrene class standards; the substitution of an alkyl group reduces retention, while a saturated ring increases it.

The reversed-phase separations of PACs can be split into two distinct types: those using either a monomeric phase or a polymeric phase (95). Separations on monomeric and polymeric phases are very distinctly different from each other, and some confusion has occurred because of attempts to compare the two (96–101). Monomeric phases are made by derivatizing silica with a

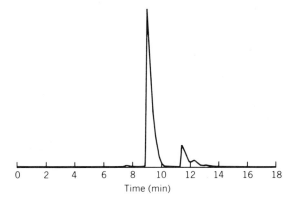

Figure 7. Chromatogram of three pyrene-type PAH standards, showing the earlier elution for n-decyl-substituted pyrene and later elution for the saturated-ring-substituted trihydrobenzo [cd]pyrene.

monochlorotrialkylsilane, resulting in only a single moiety on each de-rivatized silanol. Polymeric phases result from reaction of a dichlorodialkyl- or trichloroalkylsilane with the silanols. When one of these reacts with a silanol, further reaction can occur through the remaining chlorine atom(s). Addition of water to the reaction medium substitutes the residual chlorines

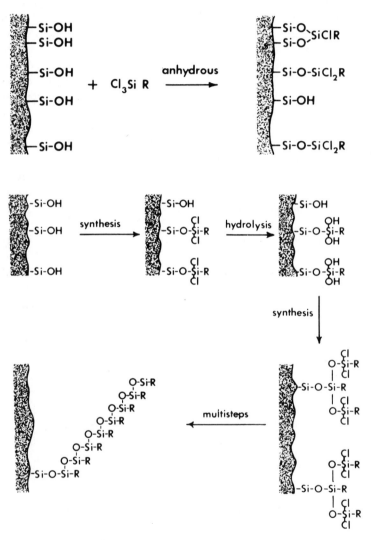

Figure 8. Representations of monomeric and polymeric reversed-phase derivatization of silica. Reprinted, with the permission of the American Chemical Society, from L. Sander and S. Wise, *Anal. Chem.*, **56**, 505 (1984).

with hydroxyl groups, which can then react with another alkylsilane. Repetition of the reaction yields a bonded alkyl-substituted polysiloxane. The most widely used polymeric reversed phases are generated with an octadecylsilane, in which a dichloro- or trichlorosilane is used. These will result in either one or no methyl group on each oligomeric unit. Representations of monomeric and polymeric phases are shown in Figure 8.

A monomeric phase generally separates a sample by the number of carbons in the molecules, whereas a polymeric phase separates the PACs by their three-dimensional shapes, resulting in very effective separations of isomers. The separation on polymeric phases of PACs by their shape occurs because these phases, particularly those made from an octadecyl-methyl dichlorosilane, are made through controlled polymerization. This bonded phase is very structured and orderly because it is formed through a series of octadecyl-methyl siloxane additions. Each of these large moieties hinders the motion of the others. The final bonded phase is a very heavily loaded material made up of octadecyl-methyl polysiloxide chains. Typical commercial packings of this type contain four to six oligomeric units on each bound moiety. These types of packings must be manufactured using wide-pore silicas that allow the oligomeric moiety to move freely. On this type of phase, all 12 of the orthocondensed five-ring PAHs were separated (similarly to the liquid-crystal-phase GC separation—see Figure 9), and several previously unknown 9- and 11-ring isomers that differed in planarity were found in synthesis products that had been thought to be fully characterized (2, 65, 66).

These very orderly polymeric reversed-phase packings can separate very complex mixtures of PACs because two additional variables in PAC structure

Table 4. Comparison between LC Retention on Polymeric and Monomeric Phases

Compound	MW	L/B	LC Retention (Log I)	
			Polymeric	Monomeric
Dibenzo[c,g]phenanthrene	278	1.12	4.07	4.51
Dibenz[a,c]anthracene	278	1.24	4.40	4.73
Benzo[g]chrysene	278	1.32	4.27	4.71
Dibenzo[b,g]phenanthrene	278	1.33	4.33	4.80
Benzo[c]chrysene	278	1.47	4.45	4.85
Dibenz[a,j]anthracene	278	1.47	4.56	4.84
Pentaphene	278	1.73	4.67	4.96
Benzo[a]naphthacene	278	1.77	4.99	4.50
Dibenz[a,h]anthracene	278	1.79	4.73	4.86
Benzo[b]chrysene	278	1.84	5.00	5.00
Picene	278	1.99	5.18	5.02

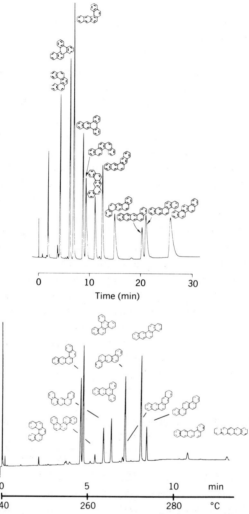

Figure 9. Comparison of the separation of the five-ring orthofused PAH isomers on a liquid-crystal phase by GC and on a polymeric reversed-phase HPLC column. Reprinted, with the permission of the American Chemical Society, from M. Nishioka, H. Chang, and M. Lee, *Environ. Sci. Technol.*, **20**, 1024 (1986) and with the permission of Dr. Alfred Huthig Verlag GmbH, from S. Wise and L. Sander, *J. High Resolut. Chromatogr. Chromatogr. Commun.*, **8**, 248 (1985).

control the separations. In addition to the separation by carbon number that occurs in monomeric packings, both the L/B ratio and the degree of nonplanarity of the molecules are important (Table 4). The degree of nonplanarity in PAHs—or even whether or not a PAH is planar—is generally empirical and can only be inferred from molecular modeling and behavior since the measurement of nonplanarity in PAHs is not yet accurate. Intramolecular steric strain can be estimated by the presence of certain substructures in the molecule (such as a three-sided bay region like that in phenanthrene, the four-sided one in benzo[c]phenanthrene, or the five-sided one in dibenzo[c,g]phenanthrene—see Figure 10) or by the molecule's spectral and chromatographic behavior. Three types of PAHs have been found: those that are always planar, such as coronene or ovalene, which elute in a predictable manner that is based on the L/B ratio and the number of ring carbons; those that are always nonplanar because of intramolecular steric effects, such as hexabenzo[a,cd,f,j,lm,o]perylene or phenanthro[2,3-c]phenanthrene (hexahelicene); and a small number whose degree of nonplanarity seems to change as a function of solvent composition (65, 66).

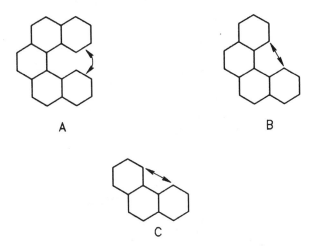

Figure 10. PAH substructures that determine planarity, in decreasing order of effect of the overlapping regions: (A) dibenzo[c, g]phenanthrene; (B) benzo[c]phenanthrene; and (C) phenanthrene.

Mixtures of PAHs, particularly when some are members of this third planarity class, yield very exotic chromatograms when a shape-selective phase is used. As the mobile-phase concentration of strong solvent increases, these PAHs become more nonplanar and their relative retention times decrease proportionally faster than those of planar PAHs. In some instances, retention order changes occur. Figure 11 shows comparative chromatograms of three similar PAHs. At high dichloromethane concentrations the elution order is the reverse of the one at low concentrations and is exactly the opposite of what

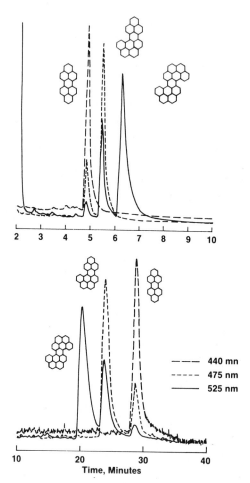

Figure 11. Retention-order reversal with a change from mobile-phase concentration to stronger solvent for a series of similar large PAHs that differ in degrees of planarity. Upper chromatogram represents 35% dichloromethane in methanol; lower chromatogram represents 80% dichloromethane in methanol.

would be expected for a reversed-phase separation. Concurrent changes in both the absorbance, fluorescence, and proton and carbon-13 NMR spectra of these compounds when going from a weak to a strong solvent support the idea that the PAH planarity is changing.

The separation of PAHs on polymeric octadecyl-derivatized silica by their shape, size, and degree of nonplanarity is accomplished by both a dispersive effect (the interaction of the bonded phase with the hydrophobic hydrocarbons) and a permeation effect (91). The very orderly octadecyl-methyl polysiloxane bonded phase can be thought of as forming a slotted structure. NMR studies of various octadecyl-methyl polysiloxane bonded phases have shown that there are drastic differences between the monomeric and poly-

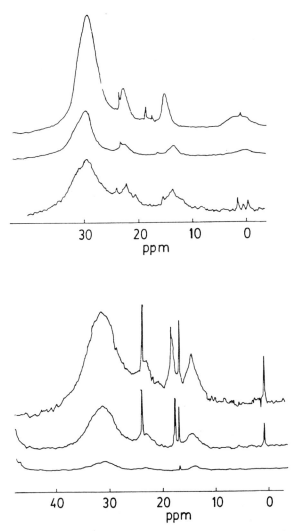

Figure 12. Carbon-13 NMR spectra that show the differences in monomeric and polymeric reversed-phase packings with methanol and chloroform: Upper spectra represent packings in chloroform, middle spectra represent packings in 1:1 chloroform/methanol, and bottom spectra represent packings in methanol. Reprinted, with permission of Elsevier Science Publishing, from P. Shah, L. B. Rogers, and J. C. Fetzer, *J. Chromatogr.*, **388**, 416 (1987).

meric bonded phases (67, 68) (Figure 12). The monomeric phase appears to be very ordered—a brushlike phase (that is spectrally similar to liquid octadecane in methanol solution); but in aqueous mobile phases, the octadecyl moieties are very irregular—a matted phase. The polymeric phases are very different: In methanol/dichloromethane solvent mixtures, each bonded moiety behaves like a sheet of hydrocarbon material. Each octadecyl-methyl polysiloxane moiety is large and hinders the movement and arrangement of the other moieties. Planar PAHs can easily fit into the gaps between these sheets and will then interact with more of the bonded phase. Less planar molecules cannot permeate as well, and so their interaction with the bonded phase is less, and they elute earlier than the planar isomers (Figure 13).

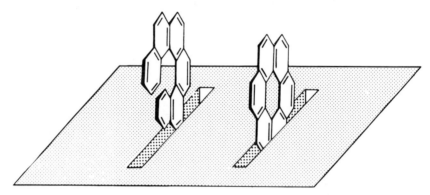

Figure 13. Representation of the slot model of retention showing the differences in permeation of the planar PAH coronene and the nonplanar phenanthro[3,4c]phenanthrene. Reprinted, with permission of Dr. Alfred Huthig Verlag GmbH, from S. Wise and L. Sander, *J. High Resolut. Chromatogr. Chromatogr. Commun.*, **8**, 248 (1985).

Other bonded phases have been synthesized that also show a separation selectivity based on the three-dimensional shape of the PAHs. Monomeric phases with very long alkyl chains (triacontyl, C_{30}) behave similarly to polymeric octadecyl phases (102). Naphthalene- and pyrene-bonded phases yield isomer separations that are almost identical to those from the polymeric octadecyl phases (103). It is postulated that through steric hindrance the bonded planar PAH moieties align and so a permeation mechanism similar to that seen with the polymeric octadecyl phases is the predominant factor in the separations. The solvent strength of the mobile phase controls whether or not a monomeric or polymeric alkyl bonded phase is orderly. If the moiety is small (monomeric octadecyl), aqueous methanol or acetonitrile yield a matted phase. Pure methanol or acetonitrile give a brush phase, but the slots in the bonded phase surface are not deep enough to show very dramatic differences in retention that would be based on PAC shape. Larger moieties (monomeric triacontyl or polymeric octadecyl phases) require stronger solvents to be in a brush conformation and have deep slots which show shape-selective retention.

Although the separation mechanisms of GC with a liquid-crystal stationary phase and HPLC with a shape-selective phase are similar, their utilizations are not. Liquid crystals are usually orderly over only a small range of temperatures. This temperature range must match the GC thermal-gradient range. If not, a mixed separation mechanism will result, since the sample molecules will experience (a) an ordered phase at the tamperatures when the stationary phase is crystalline and (b) a nonordered phase outside the liquid-crystalline temperature range. A shape-selective HPLC phase, however, shows its orderliness over a wide range of solvents since the arrangement of the moieties in the phase is controlled by steric effects rather than phase transitions. Aqueous mixtures and pure concentrations of methanol or acetonitrile, as well as ethyl acetate, acetone, tetrahydrofuran, chloroform, and dichloromethane mobile phases, have all been used with polymeric HPLC phases for shape-selective separations. A series of gradients using these solvents allows shape-selective separations from the alkyl-naphthalenes up through PAHs with 14 rings (3, 43, 56).

The separation of PAC isomers by the shape-selective phases is useful by itself if the sample has a limited range in ring number. If there is a wide range in ring number or there are many isomers of each ring number, the sample will be too complex to analyze easily. The more compact or nonplanar isomers will elute earlier than the more planar isomers or those with a large L/B, so that they will coelute with PACs of smaller ring number. This coelution of PACs with different ring numbers makes identification of components difficult, even when highly selective detectors are used (2, 65, 104).

A separation that is less dependent on planarity could be used as an initial fractionation, isolating PACs by ring number (105). Besides the normal-bonded phase columns previously described, various phenyl-bonded phases have been shown to separate PAHs in this fashion when used in the reversed-phase mode (103). These isolated ring-number fractions could then be separated on a shape-selective reversed-phase column. There is great potential in these multidimensional HPLC separations. Normal-phase HPLC with an amino-bonded column would yield fractions by the number of pi electrons. These, in turn, could be separated on a non-shape-selective reversed-phase column into isomer types by differences in L/B ratio, and finally a shape-selective phase could be used to separate specific alkyl-substituted isomers or PAH isomers that differ in degree of planarity.

Several parameters have been used to correlate the retention behavior of PAHs, with the ultimate goal of predicting retention from molecular structure. Polarizability, molecular moments of inertia, electron densities, and several types of molecular connectivity indices have been used for both normal and reversed phases (106–110). Combinations of these and L/B ratio have given the best results for complex mixtures because of the variety in the retention behavior among the PAHs. Retention on the shape-selective phases, particularly the polymeric reversed phases, can be predicted through a combination of L/B ratio and estimations of nonplanarity based on the number and type of bay regions in the molecules. As would be expected, a five-sided bay region has the most effect on planarity, with four- and three-sided bay regions having less. Similar results were obtained for correlations based on three-dimensional models of PAHs.

3.3. HPLC Detectors for PAC Analysis

The detectors commonly used in HPLC are the refractive index (RI) and single-wavelength UV absorbance detectors. The RI detector and certain other detectors, such as the thermal conductivity and light-scattering detectors, are universal detectors, used like the FID is used in GC. For PAC analyses, however, universal detection capability is not as important as detector selectivity and specificity. The need to (a) differentiate PACs from other classes of compounds, (b) differentiate the various PAC classes from each other, or (c) differentiate specific isomers are the important criteria for detectors.

The HPLC detectors used for PACs analysis have predominantly been single-wavelength UV absorbance and fluorescence detectors. These are used to detect PACs because the pi electron systems of PACs determine the energies of the transitions in the molecules' electronic spectra (111). The electronic (UV absorbance and fluorescence) spectra of the PACs generally have more features than those of other types of organic compounds. This is especially true for the PAHs, where many structures have spectra with a dozen or so maxima. These, and the intervening minima, form a unique characteristic pattern of wavelengths and intensities. The pi electron system also causes the PACs to have fluorescence and phosphorescence behavior that is similar to their UV absorbance behavior. The fluorescence excitation spectrum of a PAH is usually identical to its UV absorbance spectrum, and the fluorescence emission spectrum is usually a mirror image of the excitation spectrum (the mirror plane being the midpoint of the Stokes shift between the two types of fluorescence spectra).

Single-wavelength HPLC detectors, however, can only differentiate isomers if the chromatographic resolution is high and if a series of chromatograms with identical separation conditions are collected using different monitoring wavelengths. Ratioing of the maximum peak heights of the signals can then be used to identify the individual peaks (112). Standards of all the components of interest must also be run to obtain comparison ratios, since these ratios are very unreliable if they are obtained from reference spectra. This approach can be very tedious when the sample is a complex mixture. If there are components for which no standards are available for comparison, assignments can only be tentatively based on the estimated retention times and ratios based on literature spectra—and the effects of differing solvents and instrumental variations can have large effects on ratio values. The absorbance ratios of the unsubstituted PACs are not very useful for identifying the alkylated PACs because the bathochromic shifts and spectral-band-shape changes due to the alkylation will cause the optimum wavelengths for substituted PACs to be very different than those of the unsubstituted PACs.

Some detectors are unique to LC and are not amenable to GC because they require a liquid medium. Dielectric constant (113) and refractive index (114) detectors have been used as universal HPLC detectors. Amperometric and other electrochemical detectors have been successfully used to detect specific types of PACs in complex mixtures. Coal liquids, shale oils, and some environmental water samples have been analyzed for hydroxyl- (phenolic), amino-, and nitro-substituted PACs with electrochemical detectors (115–118). A detector that monitors the optical activity of the HPLC eluent has been used in fingerprinting coal liquids (119, 120).

3.3.1. Full-Spectrum Absorbance and Fluorescence Detectors

If the spectra of PACs and other classes of organic compounds which are obtained from the various types of molecular spectrometric techniques are compared, one singularly important trend can be seen: The absorbance and fluorescence spectra of the PACs contain far more information than those for the other classes of compounds (111). These types of spectra arise from electronic transitions between the ground state and the first excited state of the molecules.

For PACs, and especially for the PAHs, the electronic transitions are determined by the size and shape of the compounds. Electronically, the PAHs are simple molecules whose spectra are only determined by the molecular resonance structures. Isomeric PAHs differ in the locations of electronic density (because the shape of the PAHs are different), and in some cases they can even differ in their numbers of aromatic rings. Clar's annellation theory (111) and Aoki's resonance count (121) theory are based on these arguments and correlate well with observed spectral behavior. As an example, the three alternant condensed PAHs benzo[e]pyrene, benzo[a]pyrene, and perylene (structures shown in Figure 14) differ greatly in the wavelength of their highest absorbance bands. It is about 335 nm for benzo[e]pyrene, 385 nm for

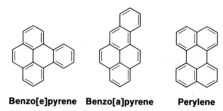

Benzo[e]pyrene   Benzo[a]pyrene   Perylene

Figure 14. Structures of the three five-ring, perifused PAH isomers showing the differences in resonance structure aromaticity.

benzo[*a*]pyrene, and 435 nm for perylene. The resonance structures of the three are also quite different: Benzo[*e*]pyrene has three aromatic rings and one localized double bond, whereas the others have only two aromatic rings and four localized double bonds. Thus, the energy necessary to excite benzo[*e*]pyrene is higher (reflected in a lower wavelength transition). The other two differ from each other because of the differences in molecular shape and the relative locations of the four double bonds.

For specific PAC structures, these types of spectra are also more definitive and informative than other types of spectra. IR, Raman, and NMR spectra are all basically similar for each class of PACs. For example, the IR spectrum of a specific PAH consists mainly of the aromatic C–H stretches and C–C ring deformation bands, but this is true of the spectra of all PAHs. There are few unique characteristics that can be used to differentiate the many types of isomers. The large "information content" of the absorbance and fluorescence spectra is particularly true for isomeric PAHs. This has led to the use by synthetic organic chemists of these spectra as the simplest way to differentiate PAC isomers. The advent of HPLC detectors that are based on photodiode arrays or Vidicon cameras has given this capability to chromatographers as well. These detectors collect and save the spectrum of each eluting HPLC peak and have become a very high-powered tool for PAC analysis, monitoring the total eluent at all wavelengths—not just at one wavelength. This single-wavelength monitoring can easily miss components in a sample (Figure 15).

These "full-spectrum" detectors are radically different than the older single-wavelength detectors, which dispersed "white light" on a diffraction grating and one selected wavelength passed through a monochromator (Figure 16). This single-wavelength beam was shone through the flow cell. The intensity was measured using a photomultiplier, yielding an absorbance or fluorescence profile at only one wavelength. The photodiode (122) or Vidicon (123) absorbance detectors pass the "white light" through the flow cell first and then disperse it. A row of photosensitive elements (the photodiode array) or a television-type camera (the Vidicon) monitor the light intensities. The emitted

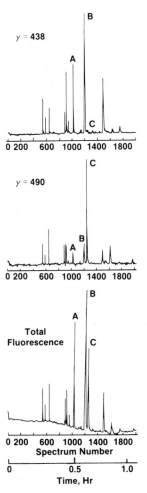

Figure 15. Chromatograms of a carbon-black extract obtained with fluorescence monitoring at two separate single wavelengths and with a photodiode array detector in the total emission mode. Reprinted, with permission of Dr. Alfred Huthig Verlag GmbH, from J. Gluckman and M. Novotny, *J. High Resolut. Chromatogr. Chromatogr. Commun.*, **8**, 675 (1985).

fluorescence in a full-spectrum detector is also collected in this fashion, but perpendicularly to the incident excitation beam (which contrasts with the single-wavelength fluorescence detectors where the emitted light was passed through a monochromator before detection). A more detailed description and comparison of the various types of photodiode array detectors can be found in ref. 123.

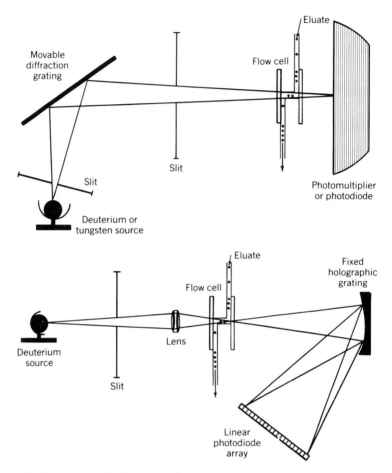

Figure 16. Comparison of the basic optical arrangements in single-wavelength and photodiode array detectors. Reprinted, with permission of Cahners Publishing Co., from R. Ryall and D. Radzik, *Chromatography*, **2**, 29 (1987).

Absorbance and fluorescence emission (if broad-band excitation is used) spectra can be collected in this fashion (124–127). Fluorescence excitation spectra cannot be collected by using diode-array or Vidicon detectors, and these spectra must be collected by stop-flow, fast-scanning techniques. Of the two types that are amenable to full-spectral detection, absorbance spectra are much more informative because there are more bands that arise from the higher-energy (lower-wavelength) transitions. The fluorescence emission spectrum shows only the lowest-energy set of transitions.

The detailed features of electronic transition spectra, such as the number of bands, relative intensities, bandshapes, and wavelengths of the maxima and minima, are dependent on the electronic transitions of the molecules. These, in turn, are determined by the overall size and shape of the molecules (111). Larger PACs generally have higher-wavelength transitions than smaller PACs, but specific sets of isomers can vary greatly. Various isomeric PACs usually have very different spectra.

A photodiode array absorbance detector was used to identify the large PAHs in an extract of a diesel particulate (128). The spectra of two structurally similar seven-ring isomers that were found are shown in Figure 17. These are easily differentiated spectrally. Several PAHs of up to 10 rings were seen in that sample (Figure 18). In similar fashion, various 8–11-ring PAH isomers were identified in synthetic reaction products, allowing identification of

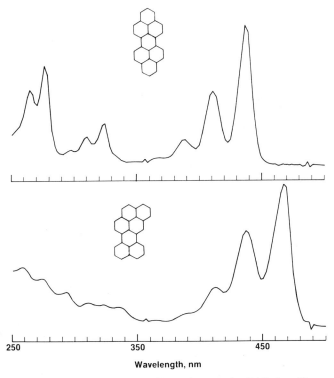

Figure 17. UV absorbance spectra of a pair of isomeric seven-ring PAHs found in an extract of a diesel particulate. Reprinted, with permission of Friedrich Viewig and Sohn Verlaggesellschaft GmbH, from J. C. Fetzer, W. R. Biggs, and K. Jinno, *Chromatographia*, **21**, 440 (1986).

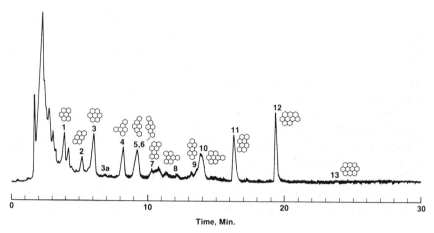

Figure 18. Chromatogram of a diesel particulate extract with peak identifications made from UV absorbance spectra. Reprinted, with permission of Friedrich Vieweg and Sohn Verlaggesellschaft GmbH, from J. C. Fetzer, W. R. Biggs, and K. Jinno, *Chromatographia*, **21**, 441 (1986).

several previously unknown structures (2, 65, 66). The PAHs in an extract of a solid waste sample were also identified by using a photodiode array detector (129).

Because the detection element in photodiode array detectors is a row of photosensitive semiconductor material, there is a trade-off between spectral range and spectral resolution. Only a limited number of photodiodes can fit into the focal plane of the dispersed light (122). If the dispersed light is spread further to allow more photodiodes, the intensity per photodiode drops with a decrease in sensitivity. Most photodiode array detectors are manufactured with either 256 or 512 diodes. Instrument manufacturers have two possible courses to take: They can have higher resolution over a short range of wavelengths, or they can sacrifice resolution in order to monitor a longer range. Both courses have been taken, producing (a) some detectors with a range of 190–375 nm with approximately 0.6-nm resolution and (b) others with a wider range of 190–600 nm but a resolution of only about 2 nm.

Spectral range is generally a more important criterion than spectral resolution for a photodiode array detector that is to be used for PAC analysis. The absorbance spectra of the PACs, particularly the PAHs, contain several bands that are usually 25–50 nm apart (111). There is little fine structure requiring spectra resolutions of less than 1–2 nm. Dibenzo[*def*, *mno*]chrysene is one of the few PAHs that has spectral bands with a separation of less than 15 nm, having a "split band" (at 434 nm and 442 nm in dichloromethane) that is easily seen even with 2-nm resolution (Figure 19). The spectral range of a

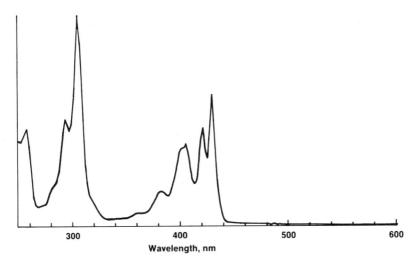

Figure 19. Spectrum of dibenzo[*def*, *mno*]chrysene, collected with a photodiode array detector, showing the high-wavelength spectral features that are discernible with an array that has 2-nm resolution.

photodiode array detector, however, must be large because of the absorbance at high wavelengths of many compounds. An example is the five-ring alternant PAH isomers benzo[*e*]pyrene, benzo[*a*]pyrene, and perylene. The latter two have maximum absorbances at 385 nm and 435 nm, respectively, both outside of the observing range of some commercially available detectors. The highest wavelength of absorbance for the PAHs generally increases with ring number, so the upper limit of spectral range becomes even more important if large PACs are to be studied (for some larger PAHs, even the 600-nm upper limit, which is available with a few detectors, is exceeded). Therefore, a resolution of 1–2 nm is sufficient for PAC analyses (such as determining whether or not a PAC is alkylated), but a limited spectral range is not acceptable since many isomers can only be differentiated at higher wavelengths.

For the alkyl-substituted PACs, full-spectrum detectors also provide information complementary to the retention data. Alkyl substitution usually does not change the pattern of maxima and minima in the absorbance spectrum of a compound, other than a slight bathochromic shift (130, 131). Methylation usually causes a 3–6-nm shift, whereas a saturated ring can cause a change of 10–15 nm (Figure 20). The overall pattern of maxima and minima does not change, so a particular peak can easily be identified as an alkylnaphthalene, an alkylpyrene, or an alkylacridine. When retention data are coupled with spectral data, identification of isomeric alkyl-substituted

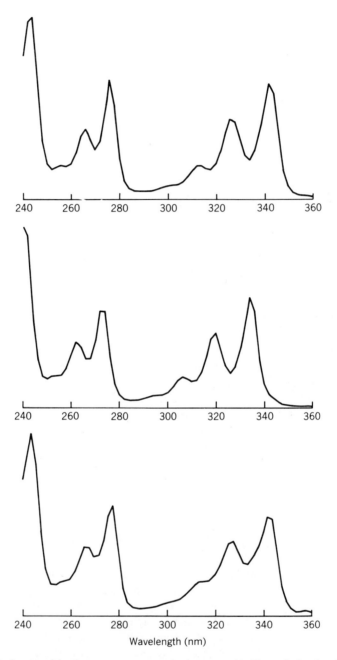

Figure 20. Spectra of the three pyrene-type standards separated in Figure 3, showing the effects of alkylation. Upper spectrum represents 1-*n*-decylpyrene, middle spectrum represents pyrene, and lower spectrum represents trihydrobenzo[*cd*]pyrene. Note that the two alkylated pyrenes have similar spectra but that the retention times (in Figure 3) are very different.

PACs can be made without the stringent need for standards that are necessary when isomeric unsubstitued PACs are being identified. Quantitation, when standard compounds of the alkyl-substituted PACs are not available, can also be more accurate, since the changes in spectral maxima and retention times can be compared to those of the unsubstituted species. These can be used to estimate the degree of alkylation. Adjusted, estimated response factors can then be used to quantitate the peaks.

The spectral output of a photodiode array detector can be used to augment the resolution of a chromatographic separation. Two components that are not completely resolved can be spectrally separated by deconvolution of the eluent spectra. Mixtures of closely eluting PAHs were used to generate spectra that had been separated and quantitated using a deconvolution program (132, 133) (Figure 21).

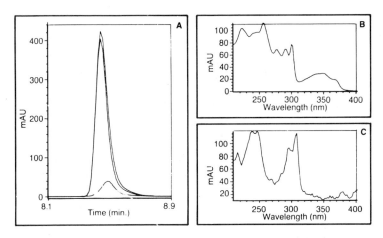

Figure 21. Deconvolution of the UV spectra of two benzofluoranthene isomers from a coeluting peak. Reprinted, with permission of Cahners Publishing Co., from L. Ramos, R. Stewart, and B. Rohrback, *Chromatography*, **2**, 57 (1987).

One problem with the full-spectrum detectors is the large, and often overwhelming, amount of information available to the analyst from a chromatogram (134). Since each peak generates at least one spectrum, the analyst is typically left with the task of matching several dozen spectra to those previously collected with standards. This could be a very time-consuming task. However, the data are already stored in a form that is available to a computer system, which makes automatic spectral matching possible. A reference spectra library can easily be created by storing the spectra of standards in one file, which can be expanded as additional standards are obtained (135, 136). The computer can, in turn, compare these to those spectra found in a sample. The library can combine retention time information with spectral matching, if standard chromatographic conditions are used, yielding very definitive identifications. Automatic identification of the PAHs in a coal tar that was separated by reversed-phase HPLC has been performed using a spectral matching program with a library of 170 standard PAH spectra (136, 137). Full-spectrum detectors have also been used for characterization of the aromatic hydrocarbons in petroleum fractions and coal liquids (138–141).

3.3.2. The Use of Laser-Based Detectors

The sensitivity of fluorometric detection is higher than that of absorbance detection because of the direct measurement of the emitted light intensity with little background or interferences. Even lower detection levels, or greater selectivity for the compounds of interest over interferences, can be achieved through the use of laser-based excitation. With their high intensity, lasers can achieve much greater efficiency of excitation than other light sources. This makes possible detection using spectrometric techniques which are not possible without the use of lasers. These include two-photon excitation fluorescence and a variety of Raman spectrometries.

Two-photon excitation involves the use of one laser to excite the molecules of the component of interest to its first excited state. Simultaneously, a second laser excites the excited molecules to their second excited state by using an exactly selected wavelength that corresponds to that energy transition (142–144). This technique is very selective and sensitive (although current systems have extremely high selectivity, their sensitivity is only slightly better than UV absorbance detection because of the low intensities presently available from dye lasers), but it requires determination of the two excitation energies from a solution of a pure standard. This has two major drawbacks: There are no commercially available pure standards for most PACs; and even with the use of tunable lasers and a precision, computer-controlled apparatus, only a few components can be analyzed during a chromatographic run.

Normal (nonlaser) fluorescence detection has extremely high sensitivities, on the order of parts per billion. Laser excitation increases this by up to 100-fold (145, 146). Sensitivity is increased even further by specially designed detection cells or on-column detection. On-column detection is possible in capillary HPLC through the use of a segment of the fused-silica capillary as the detection cell (147–149). The colored polyimide coating of the capillary is burned away, exposing the transparent silica. Laser light is focused on this spot, yielding an effective cell volume of only a few nanoliters (a similar approach has also been used when a high-intensity arc source lamp is used as the light source with appropriate focusing optics).

A laser-based detection system that measures the fluorescence lifetimes of individual peaks has been used to identify PAH isomers (150–152). The lifetimes of very structurally similar isomers can vary greatly. This type of detection, however, requires standards for comparison to unknown peaks, since fluorescence lifetimes are solvent dependent and therefore must be known for the chromatographic mobile phase used.

Many PACs are highly symmetrical, and so they do not have a large number of allowed IR transitions. Symmetrical deformations, however, can be observed using Raman spectrometry (153, 154). The use of Raman spectrometry for detection requires laser excitation because the transitions are forbidden and they occur through tunneling. High intensities are therefore necessary. Coronene and other large PAHs were identified at parts-per-million levels in a heavy cycle oil sample by the use of coherent anti-Stokes Raman (CARS) spectral detection (155, 156). This system also contained a Vidicon-based, full-spectrum detector for collecting (a) UV absorbance and (b) fluorescence excitation and emission spectra.

A universal detection system using laser-excited fluorescence, photoionization, and photoacoustic spectrometry has been tested using some PACs as model compounds (157). In the photoionization mode, the system detected all species—similarly to the use of FID or PID in GC. The photoacoustic and fluorescence detection modes displayed selective responses for certain classes of compounds.

One potential problem for all absorbance or fluorescence detectors is the possibility of photodecomposition of sample components when exposed to the high-intensity light in the flow cell (158). This would especially be true for laser-based detectors because of their more intense light sources. Certain polar solvents, particularly the commonly used reversed-phase solvents methanol and acetonitrile, appear to accelerate the photolytic breakdown of smaller PAHs.

3.3.3. The Use of GC Detectors in Micro-LC

The combination of HPLC separations with detection by GC detectors has become possible through the recent development of HPLC equipment that can operate at flow rates much lower that conventionally used (159, 160). These "micro-LC" techniques are based mainly on the miniaturization of HPLC columns. Micro-LC columns have internal diameters of 1 mm or less, much smaller than the 5–10 mm found in analytical-scale HPLC. This reduces the flow rates from the 1–5-ml/min range used in analytical-scale HPLC to the 10–100-μl/min range, and it necessitates the reduction in size of injection valves, detector cells, and any other extracolumn volumes in order to minimize losses in resolution due to band spreading. These reductions result in a great increase in separation efficiency. Micro-LC columns typically have more than 10 times as many theoretical plates as conventional HPLC columns.

This miniaturization makes it possible to use GC detectors, since micro-liter-per-minute flow rates will not quench the flame in an FID or FPD detector (161) nor will they overwhelm the vacuum system on a mass spectrometer (3, 162, 163). The area in which a traditional GC detector has been most widely used in micro-LC is mass spectrometry (MS). Analytical-scale HPLC with MS detection has been utilized for some PAC analyses, but it suffers from inherent problems because the vacuum system of the spectrometer cannot be easily coupled to the exit of the HPLC. Micro-LC/MS has been used for the PAHs (3, 162, 163) and the phenol-, nitrogen-, and sulfur-

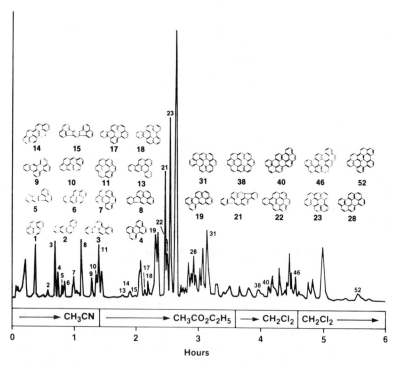

Figure 22. Some of the large PAHs in a carbon-black extract, separated on a shape-selective, reversed-phase HPLC. Reprinted, with permission of the American Chemical Society, from P. Paeden, M. Lee, Y. Hirata, and M. Novotny, *Anal. Chem.*, **52**, 2269 (1980).

containing PACs in a refinery effluent sample (164, 165), coal liquids, and extracts from carbon-black and diesel particulates. Figure 22 is a sulfur-specific chromatogram of a thiophenic PAC containing coal liquid. The high resolution of micro-LC is offset by the very long analysis times. For the separation of the carbon-black extract, the chromatographic runs were over 10 h long. Micro-HPLC has also been used with FTIR (166) and NMR (167) detection for specific compound classes.

3.3.4. Preparative-Scale HPLC with Secondary Detection

The scale-up of HPLC separations from the typical "analytical scale" (usually 0.5–5-ml/min flow rates on a 25-cm-long column with a 0.5-cm internal diameter and a sample loading of less than 1 mg) can be easily made through proportionally enlarging the column. Some currently available commercial columns contain kilograms of packing and have loading capacities that are several thousandfold greater than those of analytical-scale HPLC. Several grams of sample can be injected. Peak identification through the use of secondary (i.e., off-line) techniques thus becomes possible with this "preparative-scale" HPLC. The large quantities of material collected for each peak can be characterized by various mass spectral (168) or NMR (169) methods which require large amounts of sample. Two-dimensional NMR, for example, requires several milligrams of a purified component if the spectra are to be collected in a reasonable length of time. NMR is the only technique that can yield information on the carbon and hydrogen skeletal arrangement of the molecule, and so it is invaluable when specific isomeric structures need to be determined. High-resolution MS can differentiate molecules of the same nominal mass but of different composition. This is especially useful in the analysis of high-molecular-weight PACs, since it is impossible to tell if a mass is due to a structure with two oxygen atoms or one sulfur atom (both are nominally 32 daltons) or whether the structure has a carbon atom or is much less condensed and has 12 hydrogen atoms that contribute to the molecular weight. Field ionization MS yields a spectrum that is usually composed of only the molecular ions of the sample, and fragmentation is minimal (170–176), but direct introduction of a chromatographic effluent is not possible.

A variety of fluorescence and phosphorescence methods are not amenable to direct coupling to a chromatograph. Matrix isolation and Shpol'skii fluorescence (177–180) and room-temperature phosphorescence (181) require the analytes to be in a solid matrix. Additionally, fluorescence excitation spectra can only be collected by stop-flow methods during a separation, but collection of the peaks for off-line spectrometry is not difficult and does not disturb the determination of retention times.

3.4. Applications of HPLC to PAC Analysis

HPLC has been widely used for PAC analysis. Two major areas of application have been the foci, namely, compositional analysis of carbonaceous materials (whole petroleum and its distillation and adsorption chromatographic fractions; coal liquids and tars; shale oils; carbon blacks; etc.) and environmental analysis. The variety of PAC types determine which separation options can be used. Some questions that may be asked are as follows: Are the molecules heteroatom containing or not? Do they contain any five-member or only six-member rings? Are they ortho- or perifused? As has been previously discussed, the lower-molecular-weight PACs are more amenable to GC, and the larger ones are more suitably analyzed by HPLC. The mass spectrometer is a better detector for the smaller PAHs because of the fewer isomers of each. mass. The element-specific detectors are useful for heterocyclics but not for PAHs. Photodiode array detectors are better for the fully alternant PAHs because the UV absorbance spectra of the nonalternant PAH are more featureless, without the patterns of maxima that characterize the fully alternant PAH spectra.

HPLC is the separation method of choice for samples with a wide range in PAC polarity or molecular weight or for the larger, less volatile PACs. The large majority of separations of PACs by HPLC have been of PAHs, but most of these have been of standard compound mixtures. In one of the earliest uses of HPLC for a complex PAH-laden sample, the dichloromethane extract from a carbon black was found to contain PAHs of 16–36-ring carbons by reversed-phase HPLC with a sequential series of solvent gradients (Figure 22). Off-line fluorescence and mass spectrometries were used for characterization (43).

The PACs in other combustion particulates have also been studied by HPLC techniques. Some large PAHs of up to 10 rings were found in a diesel particulate extract (128). A variety of detection systems, including mass spectral, electrochemical, and fluorometric, were used. The smaller PAHs in carbon black, coal tar, and diesel particulate have been separated using reversed-phase HPLC with UV absorbance or fluorometric detection (178–182). The highly mutagenic nitro-PAHs in particulates have been separated with both normal and reversed phases with fluorescence, electrochemical, or mass spectral detection (183–186).

The use of HPLC for environmental analyses relies mainly on isomer-specific separation selectivity and detection. Aqueous samples are also compatible with reversed-phase columns, making sample preparation much simpler. The U.S. Environmental Protection Agency Method 610 is a reversed-phase separation for 16 smaller PAHs using fluorescence and UV absorbance detection (187). Similar methods have been used for analyzing the

PAHs and hydroxy-PAHs in petroleum-refinery effluent waters (192–194). Reversed-phase HPLC was used to separate a variety of hydroxy- and dihydroxy-substituted PAH metabolites (195, 196).

Liquid fuels, lubricating oils, and other petroleum distillates, on the other hand, are much more easily analyzed for PAHs directly using normal-phase HPLC because those types of sample materials are soluble in the saturated hydrocarbon solvents that are commonly used as mobile phases.

The amino-bonded phase separation of PAHs by the number of pi electrons has become widely used for crude petroleums, their distillates, and various petroleum products (76). Combined use of that type of separation and photodiode array detection has yielded very detailed breakdowns of sample aromatic types. The distribution of mono-, di-, tri-, and tetra-aromatics (one- to four-ring PAHs) in crude petroleum residua were separated in this fashion and were characterized by field ionization MS. Coal liquids and shale oils have also been analyzed in this fashion, but extensive clean-up of the hydrocarbon fraction through adsorption chromatography was required (57, 171–173).

Although the amino-bonded phase separation yields PAH fractions by the number of pi electrons, a mixture of PAC types will give fractions that contain PACs of different ring numbers. The low-polarity sulfur heterocyclics will elute slightly later than their PAH analogues. Thiophene (which has six double bonds) elutes in the pyrene (which has eight double bonds) fraction, and the azarenes and aromatic ketones start coeluting in the fractions of even larger PAHs (83, 172). This can complicate the use of photodiode array detectors to determine individual aromatic types, but MS techniques easily differentiate the very dissimilar types of PACs that coelute (197, 198). The following have all been analyzed by preparative-scale normal-phase HPLC: the PAHs in diesel fuel and lubricating (199) and other mid-distillate (200, 201) oils; phenolic PAHs in heavy petroleum fractions, shale oils, and coal liquids (202, 203); and a variety of nitrogen- and sulfur-containing PACs in coal liquids.

Because of its importance as a fuel source, petroleum has been widely studied (204–218). Coal liquids and shale oils share in this analytical interest because of their potential use (219–229). These three types of materials are extremely complex, containing many types of PACs (usually with additional complexity from the large number of homologous series due to variations in alkyl chain length or the number of aromatic rings) and many types of structural isomerism. This complexity has generated many approaches to analysis.

A normal-bonded phase separation was used to isolate a $C_{24}H_{14}$ fraction as the first step in the determination of the isomers in a coal-tar extract (102). Shape-selective reversed-phase HPLC was used to isolate individual components, which were characterized by retention times, room-temperature and low-temperature (Shpol'skii) fluorescence, and absorbance spectra. Similar work on a residential flue gas condensate (from combustion of anthracite coal) for the same isomer set relied on a reversed-phase separation; GC retention, MS, absorbance, and fluorescence spectra were used for identification (9). Sixteen of the 34 possible isomers were found in the coal-tar extract, while 11 isomers were found in the flue gas condensate.

Several large PAHs in a variety of carbonaceous materials have been separated by HPLC where photodiode array detection was used to identify the peaks (137). The variation in the types of PAHs found in the different materials hinted that the mechanisms of production of the PAHs differed. A carbon black was found to contain very condensed PAH structures (which indicated ring-formation reactions involving additions of only two or four carbons to create single rings), while a coal-tar sample predominantly consisted of structures that resulted from condensation of smaller PAHs. Certain isomers were seen in one sample but were found only in low concentrations, or were not found at all, in the other sample. The dichloromethane extract of a diesel particulate was analyzed in a similar manner and showed similarities to both the carbon black and coal tar.

4. SUMMARY

The analysis for PACs has mainly centered on the use of chromatographic techniques because of the extreme complexity of most samples. GC and LC have each found their specific areas of application, and these techniques generally provide complementary information. The extremely high resolution of GC, and its capability to interface to mass spectrometers or element-specific detectors, is most useful for the smaller PAHs, since volatility is a requirement for transport down the column. HPLC has the advantages of more isomer-specific detectors and a wider variation of separation conditions than does GC, but it does not have sufficient resolution to separate complex samples.

The resolution available from capillary GC is usually large enough to separate isomeric PACs from each other, particularly if shape-selective liquid-crystal stationary phases are used. The recently developed cross-linked or bound stationary phases have extended the range of GC to include some 9- and 10-ring PAH isomers, but the larger PAHs cannot be analyzed with GC at this time.

The past few years have seen a great increase in the use of HPLC for analysis of PACs. This interest has been mainly fueled by a desire to characterize the larger or more polar PACs that are not volatile enough to be amenable to analysis by GC. Although HPLC cannot attain the high resolution that GC can, the wider available variety in separation mechanisms (because of the variability in both mobile and stationary phases) has made ring number or isomer-selective separations possible. GC detectors are more informative than HPLC detectors for most types of compounds because they can give molecular weights (MS detectors) or indicate heteroatom content (the thermionic detector for nitrogen or the flame photometric detector for sulfur are two common examples).

However, some HPLC detectors are better than GC detectors for PAC analysis. The UV/visible absorbance or fluorescence detectors offer greater sensitivity and more information on the specific isomeric PACs found in a sample. Spectrofluorometric detection, particularly with laser-based excitation, can detect picogram (or lower) amounts of PACs. The structure of the PACs, and particularly the PAHs, causes their absorbance and fluorescence spectra to contain much more information than the average organic compound. A typical PAH absorbance spectrum, for example, contains a dozen or so sharp maxima. These and the intermediate minima form a unique pattern characteristic of an individual isomer. Thus isomeric structures with very similar structures can be readily distinguished by an HPLC separation where an absorbance or fluorescence detector is used.

REFERENCES

1. E. Clar, *Polycyclic Hydrocarbons*, Vols. 1 and 2, Academic Press, New York (1964).
2. J. C. Fetzer and W. R. Biggs, *J. Chromatogr.*, **295**, 161 (1984).
3. A. Hirose, D. Wiesler, and M. Novotny, *Chromatographia*, **18**, 239 (1984).
4. W. J. Simonsick and R. A. Hites, *Anal. Chem.*, **56**, 2749 (1984).
5. W. J. Simonsick and R. A. Hites, *Anal. Chem.*, **58**, 2114 (1986).
6. R. M. Campbell and M. L. Lee, *Anal. Chem.*, **58**, 2247 (1986).
7. M. Nishioka, D. G. Whiting, R. M. Campbell, and M. L. Lee, *Anal. Chem.*, **58**, 2251 (1986).
8. M. Nishioka, M. L. Lee, and R. N. Castle, *Am. Chem. Soc., Div. Pet. Chem. Prepr.*, **31**, 827 (1986); *Chem. Abstr.* **105**, 155939 (1986).
9. W. Schmidt, G. Grimer, J. Jacob, G. Dettbarn, and K. W. Naujach, *Fres. Z. Anal. Chem.*, **326**, 401 (1987).
10. M. Nishioka, H. C. Chang, and M. L. Lee, *Environ. Sci. Technol.*, **20**, 1023 (1986).
11. M. L. Lee, M. V. Novotny, and K. D. Bartle, *Analytical Chemistry of Polycyclic Aromatic Compounds*, Academic Press, New York, p. 366 (1981).
12. K. Grob and G. Grob, *J. Chromatogr.*, **213**, 211 (1981).
13. B. W. Wright, P. A. Paeden, M. L. Lee, and T. J. Stark, *J. Chromatogr.*, **248**, 17 (1982).

14. K. Levssen, *Fundamental Aspects of Organic Mass Spectrometry*, Weinheim Publishing, New York, p. 19 (1978).
15. C. M. White, in *Handbook of Polycyclic Aromatic Hydrocarbons*, Vol. 2, A. Bjorseth, Ed., Marcel Dekker, New York, p. 196 (1985).
16. M. L. Lee, D. L. Vassilaros, C. M. White, and M. Novotny, *Anal. Chem.*, **51**, 768 (1979).
17. M. Novotny, R. Kump, F. Merli, and L. J. Todd, *Anal. Chem.*, **21**, 401 (1980).
18. H. Lamparczyk, *J. High Resolut. Chromatogr. Chromatogr. Commun.*, **8**, 90 (1985).
19. K. E. Markides, H. C. Chang, C. M. Schregenberger, B. J. Tarbet, J. S. Bradshaw, and M. L. Lee, *J. High Resolut. Chromatogr. Chromatogr. Commun.*, **8**, 516 (1985).
20. K. E. Markides, M. Nishioka, B. J. Tarbet, J. S. Bradshaw, and M. L. Lee, *Anal. Chem.*, **57**, 1296 (1985).
21. H. Lamparczyk and A. Radecki, *Chromatographia*, **18**, 616 (1984).
22. H. Lamparczyk, *Chromatographia*, **20**, 284 (1985).
23. M. Nishioka, B. A. Jones, B. J. Tarbet, J. S. Bradshaw, and M. L. Lee, *J. Chromatogr.*, **357**, 79 (1986).
24. M. L. Lee, R. C. Kong, C. L. Woolley, and J. S. Bradshaw, *J. Chromatogr. Sci.*, **22**, 136 (1984).
25. B. A. Jones, J. S. Bradshaw, M. Nishioka, and M. L. Lee, *J. Org. Chem.*, **49**, 4947 (1984).
26. J. S. Bradshaw, C. Schregenberger, K. H. C. Chang, K. E. Markides, and M. L. Lee, *J. Chromatogr.*, **358**, 95 (1986).
27. R. Kaliszan and H. Lamparczyk, *J. Chromatogr. Sci.*, **16**, 246 (1978).
28. K. D. Bartle, M. L. Lee, and S. A. Wise, *Chromatographia*, **14**, 69 (1981).
29. L. F. Jaramillo and J. N. Driscoll, *J. High Resolut. Chromatogr. Chromatogr. Commun.*, **2**, 536 (1979).
30. J. N. Driscoll, *J. Chromatogr. Sci.*, **20**, 91 (1982).
31. J. N. Driscoll, *J. Chromatogr. Sci.*, **23**, 488 (1985).
32. J. A. Lubkowitz, J. L. Glajch, B. P. Semonian, and L. B. Rogers, *J. Chromatogr.*, **133**, 37 (1977).
33. B. P. Semonian, J. A. Lubkowitz, and L. B. Rogers, *J. Chromatogr.*, **151**, 1 (1978).
34. Antek Instruments, Houston, Texas.
35. P. C. Uden and D. E. Henderson, *Analyst*, **102**, 889 (1977).
36. M. A. Eckhoff, T. H. Ridgway, and J. A. Caruso, *Anal. Chem.*, **55**, 1004 (1983).
37. M. L. Bruce and J. A. Caruso, *Appl. Spectrosc.*, **39**, 942 (1985).
38. M. M. Boduszynski, *Energy Fuel*, in press.
39. D. F. Hunt and F. W. Crow, *Anal. Chem.*, **50**, 1781 (1978).
40. T. Ramdahl and K. Urdal, *Anal. Chem.*, **54**, 2256 (1982).
41. T. Ramdahl, G. Becher, and A. Bjorseth, *Environ. Sci. Technol.*, **16**, 861 (1982).
42. M. L. Lee and B. W. Wright, *J. Chromatogr. Sci.*, **18**, 345 (1980).
43. P. A. Paeden, M. L. Lee, Y. Hirata, and M. Novotny, *Anal. Chem.*, **52**, 2268 (1980).
44. M. D. Erickson, D. L. Newton, E. D. Pellizzari, and K. B. Tomer, *J. Chromatogr. Sci.*, **7**, 449 (1979).
45. R. S. Brown and L. T. Taylor, *Anal. Chem.*, **55**, 723 (1983).

46. P. Tokousbalides, E. R. Hinton, R. B. Dickinson, P. V. Bilotta, E. L. Wehry, and G. Mamantov, *Anal. Chem.*, **50**, 1189 (1978).
47. W. Kaye, *Anal. Chem.*, **34**, 287 (1962).
48. J. Merritt, F. Comendant, S. T. Abrams, and V. N. Smith, *Anal. Chem.*, **35**, 1461 (1963).
49. M. Novotny, F. J. Schwende, M. J. Hartigan, and J. E. Purcell, *Anal. Chem.*, **52**, 736 (1980).
50. C. S. Creaser and A. Stafford, *Analyst*, **112**, 423 (1987).
51. W. Schmidt, *J. Chem. Phys.*, **66**, 828 (1977).
52. P. Froehlich and E. L. Wehry, in *Modern Fluorescence Spectroscopy*, Vol. 3, E. Wehry, Ed., Plenum Press, New York, p. 79 (1981).
53. J. R. Dias, *Handbook of Polycyclic Compounds—Part 1, Benzenoid Hydrocarbons*, Elsevier, Amsterdam, p. 125 (1987).
54. W. Schmidt, Institute for Environmental Carcinogens, Ahrensberg, West Germany, Personal communication, 1986.
55. B. S. Middleditch, S. R. Missler, and H. B. Hines, *Mass Spectrometry of Priority Pollutants*, Plenum Press, New York (1981).
56. M. Novotny, A. Hirose, and D. Wiesler, *Anal. Chem.*, **56**, 1243 (1984).
57. M. L. Lee and R. A. Hites, *Anal. Chem.*, **48**, 1890 (1976).
58. P. J. Arpino, I. Ignatiadis, and G. DeRycke, *J. Chromatogr.*, **390**, 329 (1987).
59. J. Cmielowiec and H. Sawatzky, *J. Chromatogr. Sci.*, **17**, 245 (1979).
60. U. D. Neue, B. P. Murphy, and J. Crooks, in *Polynuclear Aromatic Hydrocarbons: Physiology, Biology, and Chemistry, 6th International Symposium*, M. Cooke, A. J. Dennis, and G. L. Fisher, Eds., Battelle Press, Columbus, Ohio, p. 596 (1982).
61. K. Jinno and M. Kuwajima, *Chromatographia*, **22**, 13 (1986).
62. W. R. Melander and C. Horvath, in C. Horvath, Ed., *High Performance Liquid Chromatography—Advances and Perspectives*, Vol. 2, Academic Press, New York, p. 176 (1980).
63. A. M. F. Barton, *CRC Handbook on Solubility Parameters and Other Cohesion Phenomena*, CRC Press, Boca Raton, Fla. (1983).
64. C. H. Lochmuller, M. A. Hamzavi-Abedi, and C. X. Ou, *J. Chromatogr.*, **387**, 105 (1987).
65. J. C. Fetzer and W. R. Biggs, *J. Chromatogr.*, **322**, 275 (1985).
66. J. C. Fetzer and W. R. Biggs, *J. Chromatogr.*, **386**, 87 (1987).
67. M. E. McNalley and L. B. Rogers, *J. Chromatogr.*, **331**, 23 (1985).
68. P. Shah, L. B. Rogers, and J. C. Fetzer, *J. Chromatogr.*, **388**, 411 (1987).
69. E. J. Gallegos, Chevron Research Company, Unpublished results.
70. I. M. Lewis, *Carbon*, **18**, 191 (1980).
71. J. A. Sweetman, F. W. Karasek, and D. Schuetzle, *J. Chromatogr.*, **247**, 245 (1982).
72. G. Grimmer and H. Bohnke, *J. Assoc. Off. Anal. Chem.*, **58**, 725 (1975).
73. L. R. Snyder, *Anal. Chem.*, **33**, 1527 (1961).
74. D. Hoffmann and E. Wynder, *Anal. Chem.*, **32**, 295 (1960).
75. H. Stray, S. Manoe, A. Mikalsen, and M. Dehme, *J. High Resolut. Chromatogr. Chromatogr. Commun.*, **7**, 74 (1984).
76. M. M. Boduszynski, *Energy Fuels*, **1**, 2 (1987).

77. E. Clar and C. C. Mackay, *Tetrahedron*, **28,** 6047 (1972).
78. M. Zander, J. Haase, and H. Dreeskamp, Erdoel Kohle, Erdgas Petrochem., **35,** 65 (1982).
79. P. Burchill, A. A. Herod, J. P. Mahon, and E. Pritchard, *J. Chromatogr.*, **281,** 109 (1983).
80. S. C. Ruckmick and R. J. Hurtubise, *J. Chromatogr.*, **321,** 343 (1985).
81. H. Engelhardt and H. Elgass, in *High Performance Liquid Chromatography—Advances and Perspectives*, Vol. 2, C. Horvath, Ed., Academic Press, New York, p. 57 (1980).
82. C. N. Ho, D. L. Karlesky, J. R. Kennedy, and I. M. Warner, *J. Liquid Chromatogr.*, **9,** 1 (1986).
83. J. C. Fetzer and W. R. Biggs, *J. Chromatogr.*, **346,** 81 (1986).
84. S. D. Killops, *J. High Resolut. Chromatogr. Chromatogr. Commun.*, **9,** 302 (1986).
85. P. Smidl and K. Pecka, *Ropa Uhlie*, **27,** 182 (1986); *Chem. Abstr.*, **103,** 1, 341, 610 (1985).
86. S. C. Ruckmick and R. J. Hurtubise, *J. Chromatogr.*, **361,** 47 (1986).
87. G. Felix, C. Bertrand, and F. Van Gastel, *Chromatographia*, **20,** 155 (1985).
88. T. J. Wozniak and R. A. Hites, *Anal. Chem.*, **55,** 1791 (1985).
89. S. A. Wise, W. J. Bonnett, and W. E. May, in *Polynuclear Aromatic Hydrocarbons: Chemical and Biological Effects, 4th International Symposium*, A. Bjorseth and A. J. Dennis, Eds., Battelle Press, Columbus, Ohio, p. 791 (1980).
90. D. Karlesky, D. C. Shelley, and I. M. Warner, *J. Liq. Chromatogr.*, **6,** 471 (1983).
91. K. Jinno and K. Kawasaki, *Chromatographia*, **18,** 44 (1984).
92. P. L. Grizzle and J. S. Thomson, *Anal. Chem.*, **54,** 1071 (1982).
93. S. A. Wise, W. J. Bonnett, F. R. Guenther, and W. E. May, *J. Chromatogr. Sci.*, **19,** 457 (1980).
94. S. A. Wise and L. C. Sander, *J. High Resolut. Chromatogr. Chromatogr. Commun.*, **8,** 248 (1985).
95. L. C. Sander and S. A. Wise, in *Advances in Chromatography*, Vol. 25, J. C. Giddings, E. Grushka, J. Cazes, and P. R. Brown, Eds., Marcel Dekker, New York, p. 139 (1986).
96. R. J. Hurtubise, T. W. Allen, and H. F. Silver, *J. Chromatogr.*, **235,** 517 (1982).
97. S. A. Wise and W. E. May, *Anal. Chem.*, **55,** 1479 (1983).
98. S. A. Wise, L. C. Sander, and W. E. May, *J. Liq. Chromatogr.*, **6,** 2709 (1983).
99. K. Jinno and K. Kawasaki, *Chromatographia*, **17,** 445 (1983).
100. N. Tanaka, Y. Tokuda, K. Iwaguchi, and M. Araki, *J. Chromatogr.*, **239,** 761 (1982).
101. S. A. Wise and L. C. Sander, *J. High Resolut. Chromatogr. Chromatogr. Commun.*, **8,** 248 (1985).
102. S. A. Wise and L. C. Sander, *Anal. Chem.*, **59,** 2309 (1987).
103. K. Jinno, T. Nagoshi, N. Tanaka, M. Okamoto, J. C. Fetzer, and W. R. Biggs, *J. Chromatogr.*, **392,** 75 (1987).
104. J. C. Fetzer, in *Chemistry of Polynuclear Aromatic Hydrocarbons*, L. Ebert, Ed., Advances in Chemistry Series, Vol. 217, American Chemical Society, Washington, D.C., p. 309 (1987).

105. S. A. Wise, B. A. Benner, H. Liu, G. D. Byrd, and A. Colmsjo, *Anal. Chem.*, **60**, 630 (1988).
106. M. N. Hasan and P. C. Jurs, *Anal. Chem.*, **55**, 263 (1983).
107. K. Jinno and K. Kawasaki, *J. Chromatogr.*, **316**, 1 (1984).
108. K. Jinno and K. Kawasaki, *Chromatographia*, **17**, 445 (1983).
109. K. Jinno and K. Kawasaki, *Chromatographia*, **18**, 103 (1984).
110. R. H. Rohrbaugh and P. C. Jurs, *Anal. Chem.*, **59**, 1048 (1987).
111. E. Clar, *The Aromatic Sextet*, Academic Press, New York (1972).
112. A. M. Krstulovic, D. M. Rosie, and P. R. Brown, *Anal. Chem.*, **48**, 1383 (1976).
113. P. C. Hayes and S. D. Anderson, *Anal. Chem.*, **57**, 2094 (1985).
114. D. J. Cookson, C. J. Rix, I. M. Shaw, and B. E. Smith, *J. Chromatogr.*, **312**, 237 (1984).
115. M. G. Khaledi and J. G. Dorsey, *Anal. Chim. Acta*, **161**, 201 (1984).
116. F. Senftleber, D. Bowling, and M. S. Stahr, *Anal. Chem.*, **55**, 810 (1983).
117. W. L. Caudill, M. V. Novotny, and R. M. Wightman, *J. Chromatogr.*, **261**, 415 (1983).
118. G. Chiavari, V. Concialini, and P. Vitali, *J. Chromatogr.*, **249**, 385 (1982).
119. E. S. Yeung, A. Rougvie, D. R. Bobbitt, C. G. Venier, T. G. Squires, and B. F. Smith, *Proc. Int. Conf. Coal Sci.*, 635 (1983); *Chem. Abstr.*, **102**, 169372 (1985).
120. D. R. Bobbitt, B. H. Reitsma, A. Rougvie, E. S. Yeung, T. Aida, Y. Y. Chen, B. F. Smith, T. G. Squires, and C. G. Venier, *Fuel*, **64**, 114 (1985).
121. J. Aoki, *Senryo to Yakuhin* (*Jpn. J. Dyes Pharmaceut.*), **29**, 232 (1984); *Chem. Abstr.* **102**, 184465 (1985).
122. T. Alfredson and T. Sheehan, *J. Chromatogr. Sci.*, **24**, 473 (1986).
123. R. R. Ryall and D. M. Radzik, *Chromatography*, **2**, 28 (1987).
124. J. R. Jadamec, W. A. Saner, and Y. Talmi, *Anal. Chem.*, **49**, 1316 (1977).
125. D. C. Shelly, M. P. Fogarty, and I. M. Warner, *J. High Resolut. Chromatogr. Chromatogr. Commun.*, **4**, 616 (1981).
126. C. J. Appellof and E. R. Davidson, *Anal. Chem.*, **53**, 2053 (1981).
127. J. C. Gluckman, D. C. Shelly, and M. V. Novotny, *Anal. Chem.*, **57**, 1546 (1985).
128. J. C. Fetzer, W. R. Biggs, and K. Jinno, *Chromatographia*, **21**, 439 (1986).
129. D. J. Desilets, P. T. Kissinger, F. E. Lytle, M. A. Horne, M. S. Ludwiczak, and R. B. Jacko, *Environ. Sci. Technol.*, **18**, 386 (1984).
130. R. A. Freidel and M. Orchin, *Ultraviolet Spectra of Aromatic Compounds*, Wiley, New York (1951).
131. R. M. Silverstein, G. C. Bassler, and T. C. Morrill, *Spectrometric Identification of Organic Compounds*, 3rd edition, Wiley, New York, p. 248 (1974).
132. B. Verdeginete, R. Essers, T. Bosman, J. Reijnen, and G. Kateman, *Anal. Chem.*, **57,** 971 (1985).
133. L. S. Ramos, R. J. Stewart, and B. G. Rorhback, *Chromatography*, **2,** 95 (1987).
134. A. C. J. H. Drouen, H. A. H. Billiet, and L. de Galan, *Anal. Chem.*, **57**, 962 (1985).
135. D. W. Hill, T. R. Kelley, and K. J. Langner, *Anal. Chem.*, **59**, 350 (1987).
136. D. Demorest, J. C. Fetzer, I. S. Lurie, S. M. Carr, and K. B. Chatson, *LC/GC*, **2,** 128 (1987).
137. J. C. Fetzer, *Polycyclic Aromatic Hydrocarbons, 11th International Symposium*, Gaithersburg, Md., September 24, 1987.

138. D. R. Choudhury, in *Polynuclear Aromatic Hydrocarbons: Chemistry, Analysis and Biological Fate, 5th International Symposium*—1980, M. Cooke and A. Dennis, Eds., Battelle Press, Columbus, Ohio, p. 265 (1981).

139. K. W. Jost, T. Crispin, and I. Halasz, *Erdoel Kohle, Erdgas Petrochem.*, **37**, 178 (1984).

140. T. Alfredson, T. Sheehan, T. Lenert, S. Aamodt, and L. Correia, *J. Chromatogr.*, **385**, 213 (1987).

141. L. Huber, J. Emmert, A. Gratzfeld-Huesgen, and W. Dulson, *Staub-Reinhalt Luft*, **47**, 22 (1987); *Chem. Abstr.*, **106**, 188224 (1987).

142. M. J. Sepaniak and E. S. Yeung, *Anal. Chem.*, **49**, 1554 (1977).

143. M. J. Sepaniak and E. S. Yeung, *J. Chromatogr.*, **190**, 377 (1980).

144. M. J. Sepaniak and E. S. Yeung, *J. Chromatogr.*, **211**, 95 (1981).

145. J. H. Richardson, in *Modern Fluorescence Spectroscopy*, Vol. 3, E. L. Wehry, Ed., Plenum Press, New York, p. 12 (1981).

146. E. J. Guthrie, J. W. Jorgenson, and P. R. Dluzneski, *J. Chromatogr. Sci.*, **22**, 171 (1984).

147. S. Folestad, L. Johnson, F. Josefsson, and B. Galle, *Anal. Chem.*, **54**, 925 (1982).

148. S. Folestad, B. Galle, and B. Josefsson, *J. Chromatogr. Sci.*, **23**, 273 (1985).

149. E. J. Guthrie and J. W. Jorgenson, *Anal. Chem.*, **56**, 483 (1984).

150. J. H. Richardson, K. M. Larson, G. R. Haugen, D. C. Johnson, and J. E. Clarkson, *Anal. Chim. Acta*, **116**, 407 (1980).

151. K. Ishibashi, T. Ishibashi, and N. Ishibashi, *Anal. Chim. Acta*, **173**, 165 (1985).

152. D. J. Desilets, J. T. Coburn, D. A. Lantrip, P. T. Kissinger, and F. E. Lytle, *Anal. Chem.*, **58**, 1123 (1987); and subsequent private communication.

153. D. R. Van Hare, L. A. Carreira, L. B. Rogers, and L. Azarraga, *Appl. Spectrosc.*, **38**, 543 (1984).

154. D. R. Van Hare, L. A. Carreira, L. B. Rogers, and L. Azarraga, *Int. J. Environ. Anal. Chem.*, **22**, 85 (1985).

155. M. S. Klee, L. B. Rogers, L. A. Carreira, and L. Azarraga, in *Analytical Techniques in Environmental Chemistry 2*, J. Albaiges, Ed., Pergamon Press, Oxford, U.K., p. 177.

156. G. D. Boutilier, R. M. Irwin, R. R. Antcliff, L. B. Rogers, and L. A. Carreira, *Appl. Spectrosc.*, **35**, 576 (1981).

157. J. D. Winefordner and E. Voigtman, *ASTM Spec. Tech. Publ.*, 17 (1983); *Chem. Abstr.*, **101**, 203590 (1984).

158. G. K. C. Low, G. E. Batley, and C. I. Brockbank, *J. Chromatogr.*, **392**, 199 (1987).

159. P. Kucera, Ed., '*Microcolumn High-Performance Liquid Chromatography*, Journal of Chromatography Library, Vol. 28, Elsevier, Amsterdam (1984).

160. M. V. Novotny and D. Ishii, *Microcolumn Separations*, Journal of Chromatography Library, Vol. 30, Elsevier, Amsterdam (1985).

161. R. G. Christensen and E. V. White, *J. Chromatogr.*, **323**, 33 (1985).

162. M. Novotny, A. Hirose, and D. Weisler, *Anal. Chem.*, **56**, 1243 (1984).

163. M. Novotny, M. Konishi, A. Hirose, J. Gluckman, and D. Wiesler, *Fuel*, **64**, 523 (1985).

164. S. C. Beale, D. Wiesler, and M. Novotny, *J. Chromatogr.*, **393**, 391 (1987).

165. C. Borra, D. Wiesler, and M. Novotny, *Anal. Chem.*, **59**, 339 (1987).

166. J. F. Haw, T. E. Glass, and H. C. Dorn, *Anal. Chem.*, **53**, 2327 (1981).

167. P. G. Amatels and L. T. Taylor, *Anal. Chem.*, **56**, 966 (1984).

168. K. S. Seshadri and D. C. Cronauer, *Am. Chem. Soc. Div. Fuel Chem. Prepr.*, **27**, 64 (1982); *Chem. Abstr.* **99**, 161139 (1983).

169. M. Zander, J. Haase, and H. Dreeskamp, *Erdoel Kohle Erdgas Petrochem.*, **35**, 65 (1982).

170. W. Holstein and D. Severin, *Chromatographia*, **15**, 231 (1982).

171. J. F. McKay, M. M. Boduszynski, and D. R. Latham, *Liq. Fuels Technol.*, **1**, 35 (1983).

172. M. M. Boduszynski, R. J. Hurtubise, T. W. Allen, and H. F. Silver, *Anal. Chem.*, **55**, 225 (1983).

173. M. M. Boduszynski, R. J. Hurtubise, T.W. Allen, and H. F. Silver, *Anal. Chem.*, **55**, 232 (1983).

174. Y. Ohshima, J. Yen, and H. Ohnuma, *Kozangabuku Kenkyu Hokoku*, **5**, 71 (1984); *Chem. Abstr.*, **102**, 81349 (1985).

175. A. A. Grinberg, T. V. Bigdash, S. A. Leont'eva, and B. D. Kabulov, *Zh. Anal. Chim.*, **39**, 83 (1984); *Chem. Abstr.*, **100**, 88301 (1984).

176. S. Arsic, V. Butkovic, K. Humski, L. Kasinc, and R. Marcoc, *Nafta* (*Zagreb*), **34**, 701 (1983); *Chem. Abstr.*, **100**, 86301 (1984).

177. T. Ogawa, *Nenryo Kyokaishi*, **60**, 787 (1981); *Chem. Abstr.*, **96**, 54899 (1982).

178. J. F. McKay and D. R. Latham, *Anal. Chem.*, **45**, 1050 (1973).

179. P. Garrigues and M. Ewald, *Anal. Chem.*, **55**, 2155 (1983).

180. A. L. Colmsjo, C. E. Oestman, and Y. U. Zebuehr, in *Polynuclear Aromatic Hydrocarbons, 8th International Symposium*, M. Cooke and A. J. Dennis, Eds., Battelle Press, Columbus, Ohio, p. 273 (1985).

181. S. Scypinski, Doctoral dissertation, Seton Hall University, *Chem. Abstr.*, **102**, 231134 (1985).

182. T. Romanowski, W. Funcke, J. Koenig, and E. Balfanz, *Anal. Chem.*, **54**, 1285 (1982).

183. A. Obuchi, H. Aoyama, A. Ohi, and H. Ohuchi, *J. Chromatogr.*, **312**, 247 (1984).

184. T. Shiozaki, K. Tanabe, and H. Matsushita, *Taiki Osen Gakkaishi*, **19**, 300 (1984); *Chem. Abstr.*, **102**, 115198 (1985).

185. H. Stray, S. Manoe, A. Mikalsen, and M. Oehme, *J. High Resolut. Chromatogr. Chromatogr. Commun.*, **7**, 74 (1984).

186. L. Zoccolillo, A. Liberti, F. Coccioli, and M. Ronchetti, *J. Chromatogr.*, **288**, 347 (1984).

187. T. Neilsen, *Anal. Chem.*, **55**, 286 (1983).

188. K. Tanabe, H. Matsushita, C. T. Kuo, and S. Imamiya, *Taiki Osen Gakkaishi*, **21**, 535 (1986); *Chem. Abstr.*, **106**, 161813 (1987).

189. W. A. MacCrehan and W. E. May, in *Polynuclear Aromatic Hydrocarbons, 8th International Symposium*, M. Cooke and A. J. Dennis, Eds., Battelle Press, Columbus, Ohio, p. 857 (19).

190. S. M. Rappoport, Z. L. Jin, and X. B. Xu, *J. Chromatogr.*, **240**, 145 (1982).

191. G. Ginzer, R. Riggin, T. Bishop, M. A. Birts, and P. Strup, EPA report 600/4–84–063, *Chem. Abstr.*, **102**, 67098 (1985).

192. R. K. Symons and I. Crick, *Anal. Chim. Acta*, **151**, 237 (1983).

193. L. B. Yeats, G. B. Hurst, and J. E. Caton, *Anal. Chim. Acta*, **151**, 349 (1983).
194. A. Radecki, J. Grzybowski, J. R. Ochocka, and G. Rewkowska, *Bromatol. Chem. Toksykol.*, **17**, 199 (1984); *Chem. Abstr.* **102**, 154528 (1985).
195. S. K. Yang, M. Mushtaq, and P. F. Fu, *J. Chromatogr.*, **371**, 195 (1987).
196. F. J. Jongeneelen, R. B. M. Anzion, and P. T. Henderson, *J. Chromatogr.*, **413**, 227 (1987).
197. P. Jadaud, M. Caude, and R. Rosset, *Analusis (Paris)*, **14**, 491 (1986); *Chem. Abstr.*, **106**, 158927 (1987).
198. J. B. Green, J. S. Thomson, S. K. T. Yu, C. A. Treese, B. K. Stierwalt, and C. P. Renaudo, *Am. Chem. Soc., Div. Fuel Chem. Prepr.*, **31**, 198 (1986); *Chem. Abstr.*, **104**, 152031 (1986).
199. J. B. Green, J. S. Thomson, C. A. Treese, and S. K. T. Yu, *Energy Res. Abstr.*, **11** No. 35537 (1986); *Chem. Abstr.* **106**, 179433 (1987).
200. J. Chmielowiec, J. E. Beshai, and A. E. George, *Am. Chem. Soc. Div. Pet. Chem. Prepr.*, **25**, 532 (1980); *Chem. Abstr.*, **96**, 202139 (1982).
201. M. M. Boduszynski, J. F. McKay, D. R. Latham, *Am. Chem. Soc., Div. Pet. Chem. Prepr.*, **26**, 865 (1981); *Chem. Abstr.* **98**, 146135 (1981).
202. G. A. Odoerfer, L. R. Rudnick, and D. D. Whitehurst, *Am. Chem. Soc. Div. Fuel Chem. Prepr.*, **26**, 89 (1981); *Chem. Abstr.*, **97**, 147370 (1982).
203. A. Matsunaga and S. Kusayanagi, *Am. Chem. Soc., Div. Fuel Chem. Prepr.*, **26**, 59 (1981); *Chem. Abstr.*, **97**, 130388 (1982).
204. G. Guiochon, *Petroanal.*, **8**, 132 (1982); *Chem. Abstr.*, **99**, 73333 (1983).
205. S. Coulombe and H. Sawatsky, in *Polynuclear Aromatic Hydrocarbons, 8th International Symposium*, M. Cooke and A. J. Dennis, Eds., Battelle Press, Columbus, Ohio, p. 281.
206. J. Chmielwiec, *Anal. Chem.*, **55**, 2367 (1983).
207. M. Radke, H. Willsch, and H. Dietrich, *Anal. Chem.*, **56**, 2538 (1984).
208. H. Kadi and D. P. Lazar, *Rev. Roum. Chim.*, **29**, 497 (1984); *Chem. Abstr.*, **101**, 174124 (1984).
209. M. Novotny, M. Konishi, A. Hirose, J. Gluckman, and D. Wiesler, *Fuel*, **64**, 523 (1985).
210. G. Felix and C. Bertrand, *Analusis (Paris)*, **15**, 28 (1987); *Chem. Abstr.*, **106**, 159098 (1987).
211. L. R. Schronk, R. D. Grisby, and A. R. Hanks, *J. Chromatogr. Sci.*, **19**, 490 (1981).
212. W. Holstein and D. Severin, *Anal. Chem.*, **53**, 2356 (1981).
213. D. W. Grant, *Chem. Phys. Vapor. Coal*, **218** (1982); *Chem. Abstr.* **97**, 200500 (1982).
214. T. W. Allen, R. J. Hurtubise, and H. F. Silver, *Anal. Chim. Acta*, **141**, 411 (1982).
215. P. G. Amateis and L. T. Taylor, *Chromatographia*, **17**, 431 (1983).
216. K. G. Liphard, *Chem. Phys. Vapor. Coal*, **49** (1984); *Chem. Abstr.*, **102**, 64700 (1985).
217. K. G. Liphard, *Erdoel Kohle Erdgas Petrochem.*, **37**, 516 (1984).
218. S. Yokoyama, N. Tsuzuki, T. Katoh, and Y. Sanada, *Nenryo Kyokaishi*, **62**, 106 (1983); *Chem. Abstr.*, **98**, 218568 (1983).
219. S. Yokoyama, N. Tsuzuki, H. Uchino, T. Gatoh, and Y. Sanada, *Nippon Kagaku Kaishi*, **405** (1983); *Chem. Abstr.*, **98**, 146279 (1983).

220. M. K. Marsh, C. A. Smith, C. E. Snape, and B. J. Stokes, *J. Chromatogr.*, **283,** 173 (1984).
221. L. Szepesy, M. Horvath, and K. Lakszner, *Chromatographia*, **19,** 431 (1984).
222. R. J. Hurtubise and H. F. Silver, *Energy. Res. Abstr.*, **9,** No. 25207 (1984); *Chem. Abstr.* **101,** 232928 (1984).
223. S. Yu and R. Rujuan, *Sepu (Beijing)*, **1,** 77 (1984); *Chem. Abstr.* **102,** 223147 (1985).
224. R. J. Hurtubise and H. F. Silver, *Energy Res. Abstr.*, **10,** No. 6294 (1985); *Chem. Abstr.*, **102,** 169487 (1985).
225. S. C. Ruckmick and R. J. Hurtubise, *J. Chromatogr.*, **331,** 55 (1985).
226. T. W. Allen, R. J. Hurtubise, and H. F. Silver, *Anal. Chem.*, **57,** 666 (1985).
227. F. M. Lancas and H. M. McNair, *J. Liq. Chromatogr.*, **8,** 239 (1985).
228. M. Stefanova and L. Lazarov, *Izv. Khim.*, **19,** 534 (1986); *Chem. Abstr.*, **106,** 159266 (1987).
229. S. C. Ruckmick, R. J. Hurtubise, and H. F. Silver, *J. Chromatogr.*, **392,** 277 (1987).

CAPILLARY SUPERCRITICAL FLUID CHROMATOGRAPHY METHODS

BOB W. WRIGHT AND RICHARD D. SMITH

Chemical Methods and Separations Group
Chemical Sciences Department
Battelle, Pacific Northwest Laboratories
Richland, Washington

1. INTRODUCTION

The origin of supercritical fluid chromatography (SFC) using packed columns dates back over two decades, with a period of intense activity in the early and mid-1960s (1–8). Polycyclic aromatic compounds (PACs) were analyzed with this technique from near its inception. Sie and Rijnders (4, 5) used PAC standards ranging from naphthalene through coronene as model compounds to develop many of the theoretical aspects of this chromatographic method. SFC was also applied to real samples containing PACs, such as coal and polyphenyl tars (6) and automobile exhaust (7, 8). However, the lack of sufficiently selective and sensitive detectors, the complexity of the instrumentation, and the rapid growth of high-performance liquid chromatography (HPLC) led to a period where little SFC progress was reported. Fortunately, interest in other applications of supercritical fluids related to separations and chemical processing continued to grow during this period (9, 10).

A strong resurgence in SFC began early in the 1980s because of (a) a growing understanding of supercritical properties and related phenomena, (b) the introduction of capillary columns for SFC, and (c) the introduction of commercial packed-column SFC instrumentation. The development of non-extractable stationary phases for capillary gas chromatography (GC) and the potential compatibility of SFC with gas-phase detectors (e.g., flame ionization, mass spectrometry, etc.) also served to further increase interest in SFC. Since that time, there has been a large growth in the SFC literature, and the field has advanced rapidly. The general availability of commercial capillary (and packed-column) SFC instrumentation has helped promote this growth and will undoubtedly stimulate continued growth. Both the development of

111

improved SFC capabilities using PACs as model compound mixtures and the actual analysis of PAC have been the object of several of these studies.

The renewed interest in SFC can also be attributed to the limitations in both the chromatographic efficiency and detection methods for HPLC. The physical properties of supercritical fluids provide the potential for significantly enhanced chromatographic efficiency per unit time compared to HPLC and for an extended molecular-weight range compared to GC. These characteristics offer significant promise for the analysis of PACs since many mixtures are extremely complex and extend over a wide molecular-weight range. Although both GC and HPLC are used extensively for PAC analysis and provide entirely satisfactory results for distinct ranges of molecular weight and complexity, not all samples can be successfully analyzed by these methods. High-resolution GC provides the necessary separation power for the satisfactory analysis of complex mixtures of volatile compounds ranging in size up to about seven rings. HPLC provides the solvating power needed to separate many of the larger PACs, but it lacks the resolving power to adequately separate highly complex mixtures. The physical properties of supercritical fluids also provide the potential for successful coupling of SFC to mass spectrometry (MS). The advantages of coupling a chromatographic technique with MS are considerable, as evidenced by the major role of GC–MS in mixture analysis. This interest in SFC–MS has also driven the development of SFC and provides an important role in the analysis of PACs.

An overview of capillary SFC and related analytical techniques (fractionation and extraction) as they apply to the analysis of PAC will be given in this chapter. The properties of supercritical fluids which render them efficacious and form the basis of SFC will be discussed. Although this chapter will emphasize capillary SFC, the relative merits of both packed and capillary column operation will also be reviewed. Various instrumentation configurations and methods, including SFC–MS, will be described and illustrated with applications using a range of PAC samples. At present, most of the applications of SFC to PAC analysis have been aimed at instrumentation and methodology development. Only as SFC techniques mature and are seriously applied in actual problem-solving situations will their full potential be realized.

2. PROPERTIES OF SUPERCRITICAL FLUIDS

From a practical viewpoint, supercritical fluids may be defined as gases that are at temperatures above, but close to, their critical temperature and that are compressed to pressures (or densities) at which "liquidlike" interactions become significant. The combination of physical properties (viscosity and diffusion rates) with variable solvent properties provides the basis for the advantages of SFC and related analytical techniques. The physical properties of a supercritical fluid are variable between the limits of a normal gas and those of a liquid by control of pressure and temperature, with different solvent characteristics displayed at each density. Typically, supercritical fluids are used at densities ranging from 0.1 to 0.8 of their liquid density, and practical pressures for applications range from less than 50 bar to more than 500 bar. Under these conditions the diffusion coefficients of supercritical fluids are substantially greater than liquids. For example, the diffusivity of supercritical CO_2 varies between 10^{-4} and 10^{-3} cm^2/s over the range of conditions usually utilized, whereas liquids typically have diffusivities of less than 10^{-5} cm^2/s. Similarly, the viscosity of supercritical fluids mirrors the diffusivity and is typically 10–100 times less than for liquids (11). It is these more favorable physical properties that define the advantages of supercritical fluids in chromatography and extraction applications.

Figure 1 shows the pressure–density relationship for carbon dioxide in terms of reduced parameters (e.g., pressure, temperature, or density divided by the appropriate critical parameter), including the two-phase vapour–liquid region. This relationship is generally valid for most single-component systems. The isotherms at various reduced temperatures show the variations in density which can be expected with changes in pressure. Thus, the density of a supercritical fluid will be typically 10^2–10^3 times greater than that of the gas at ambient temperatures. Consequently, molecular interactions increase because of shorter intermolecular distances. The liquidlike density of a supercritical fluid results in enhanced solvating capabilities. The variable solvent properties as a function of density have been demonstrated for a number of fluids using the solvatochromic method (12). Table 1 gives the critical parameters for a number of common and potential supercritical solvents. In conjunction with the general pressure–temperature–density relationships shown in Figure 1, these data provide a vast amount of practical guidance for fluids and their gas–liquid equilibria.

For a pure supercritical fluid, as well as for infinitely dilute supercritical fluid solutions relevant to chromatographic separations, the relationships between pressure, temperature, and density can be estimated with reasonable accuracy from equations of state (except near the critical point), and they conform closely to the behavior shown in Figure 1. However, it is often advantageous to use a binary fluid mixture to obtain enhanced solvating power to increase the range of molecular species that (a) are soluble in a single supercritical fluid, (b) alter the critical temperature of the mobile phase, or (c) change the chromatographic selectivity of the mobile phase and/or stationary phase. The phase behavior of binary systems is highly varied and much more complex than single-component systems and has only been well described in

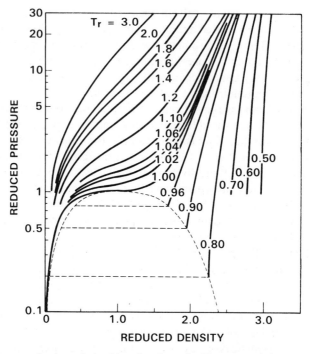

Figure 1. Pressure–density relationship of carbon dioxide expressed in terms of reduced parameters. The area below the dotted line represents the two-phase gas–liquid equilibrium region. T_r, reduced temperature.

Table 1. Common SFC Solvents

Compound	Boiling Point (°C)	Critical Temperature (°C)	Critical Pressure (bar)	Critical Density (g/cm^3)
CO_2	−78.5	31.3	72.9	0.448
NH_3	−33.4	132.4	112.5	0.235
H_2O	100	374.2	218.3	0.315
N_2O	−88.6	36.5	71.7	0.45
Ethane	−88.6	32.3	48.1	0.203
Ethylene	−102.7	9.2	49.7	0.218
Propane	−42.1	96.7	41.9	0.217
Pentane	36.1	196.6	33.3	0.232
Benzene	80.1	288.9	48.3	0.302
Methanol	64.7	240.5	78.9	0.276
Isopropanol	82.5	235.3	47.0	0.273

the literature for selected model systems (11). Five basic types of phase behavior have been identified, with the simplest type (Type I) characterized by a continuous mixture curve for pressure and temperature conditions over the composition range of the two components. Examples of binary fluid systems of this type are (a) carbon dioxide with isopropanol or methanol and (b) propane with isopropanol. A detailed discussion of the more complex types of binary fluid mixtures, as well as of the phase behavior of these systems, can be found elsewhere (13). The phase behavior of carbon dioxide–methanol is shown in Figure 2, which gives the pressure–composition vapor–liquid envelopes for three temperatures (14, 15). The single-phase supercritical regions are above the envelopes, and the regions within the envelopes correspond to two-phase subcritical mixtures at the respective temperatures.

To exploit the chromatographic advantages of SFC it is essential that the fluid mixtures used for mobile phases be selected so they can be mixed and pumped as a single phase and preferably at ambient temperatures. Proper operating conditions must also be chosen that will give a single-phase supercritical fluid; and care must be taken, when operating over a range of

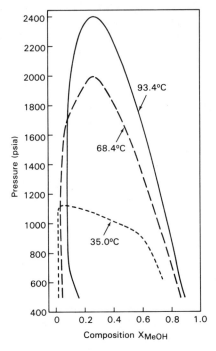

Figure 2. Pressure–composition diagrams for carbon dioxide–methanol fluid mixtures at three temperatures. Two-phase gas–liquid regions exist inside the curves at the given temperatures.

pressures (as is typical in capillary SFC), to avoid entering a two-phase region. In the absence of actual phase equilibria data, simple predictive methods used to obtain mixture critical parameters, such as taking linear mole-fraction averages of the pure-component critical parameters, can result in considerable error and lead to inadvertent operation in the vapor–liquid region of the phase diagram. More complex predictive methods utilizing equations of state (16, 17) or surface fraction functions (Chueh and Prausnitz method) (18) generally provide more accurate estimation of the true critical parameters. These considerations are important when pressure programming methods are used, but they are of lesser importance for packed column separations when relatively high isobaric pressures are used.

The chromatographic behavior observed from a fluid mixture can also be used to discern if a single-phase supercritical system or a subcritical two-phase system exists under a specified set of operating conditions. This is illustrated with the capillary SFC–MS chromatograms of coal tar shown in Figure 3. These separations were obtained at three different temperatures (but otherwise identical operating conditions), using a 10% (v/v) mixture of isopropanol in propane mobile phase. At 120°C (upper chromatogram), the PAC components were poorly resolved and they eluted close together. When the temperature was raised to 130°C (center chromatogram), the components were much better resolved and they eluted over a wider pressure range during the pressure ramp. The early eluting components were eluted at lower pressures, and the later eluting components were eluted at higher pressures. At 150°C (lower chromatogram), only slightly improved performance was obtained. This behavior is probably explained by the lower temperature mixture existing as a subcritical liquid with strong solvating properties that eluted the components close together. Under such conditions, pressure programming is of relatively little value since density does not change substantially. At the higher temperature of 130°C, the fluid mixture was likely in a single-phase supercritical state, and the expected higher chromatographic efficiency and the solvating power increase with increasing pressure were observed. The differences in the chromatograms obtained at 130°C and 150°C can be attributed to the different fluid properties (viscosity and densities at a given pressure) and analyte volatilities at the two temperatures.

The range of solvating power of practical supercritical fluids for SFC is of primary importance and ultimately defines the limits of application. The solubility of analytes typically increases with density; furthermore, a maximum rate of increase in solubility with pressure is generally observed near the critical pressure, where the rate of increase of density with pressure is greatest (19). There is often a linear relationship between log [solubility] and fluid density for dilute solutions of nonvolatile compounds (up to concentrations where solute–solute interactions become important). Where volatility is

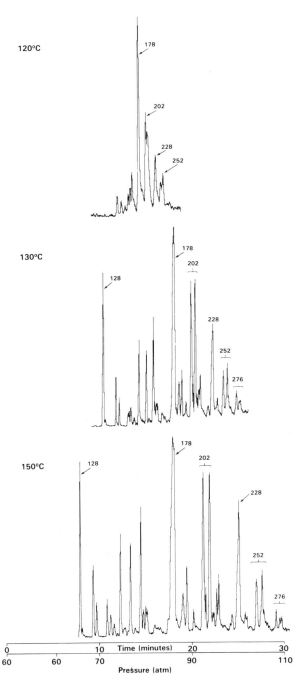

Figure 3. Capillary SFC–MS total-ion chromatograms of a coal-tar mixture obtained at 120°C (upper), 130°C (center), and 150°C (lower) using a 10% (v/v) mixture of isopropanol in propane mobile phase.

117

extremely low, and at densities less than or near the critical density, increasing temperature will typically decrease solubility (20). However, solubility may increase at sufficiently high temperatures, where the solute vapor pressure also becomes significant. Under conditions of constant density, solubility generally increases with temperature. Thus, while the highest supercritical fluid densities at a constant temperature are obtained near the critical temperature, the greatest solubilities and lowest chromatographic retention will often be obtained at somewhat lower densities but higher temperatures.

As with liquids, more polar solutes are most soluble in polar supercritical fluids, although nominally nonpolar fluids can be remarkably good solvents for many moderately polar compounds (11). Carbon dioxide, for example, can exhibit solvating properties at higher pressures intermediate between liquid pentane and methylene chloride. Many polar solvents that would be attractive fluids have critical temperatures that are excessively high to allow practical operation with current stationary phases and perhaps exceed the thermostability limits of the analytes themselves. This has generated interest in mixed or binary fluid mobile phases that will have enhanced solvating power and lower critical temperatures.

3. CHROMATOGRAPHY WITH SUPERCRITICAL MOBILE PHASES

In SFC the mobile phase is maintained at a temperature somewhat above its critical point (often at reduced temperatures of 1.02–1.5). It is generally advantageous to utilize the highest temperature compatible with the SFC system and the material being analyzed, since chromatographic separation efficiencies are improved with operation at higher temperatures and lower densities as a result of more favorable diffusion coefficients (21). Since low critical-temperature fluids can be chosen and low reduced temperatures can be utilized, mild thermal conditions can be achieved to allow application to labile compounds that cannot be addressed by GC. The density of the supercritical mobile phase is usually several hundred times that of the gas but is less than that of the liquid at typical SFC pressures.

The use of open tubular columns results in negligible pressure drops for normal mobile-phase linear velocities and column dimensions. Thus, large numbers of theoretical plates can be generated by using long columns. A low-pressure drop is also an important consideration if the pressure programming (or density programming) capability of SFC is to be fully exploited to control the mobile-phase solvating power. Pressure programming in SFC provides many of the advantages of gradient elution in HPLC while avoiding the complications impacting gradient reproducibility. Rapid-pressure programming methods can also be utilized to affect high-speed capillary SFC separations (22, 23). With the development of small-diameter (< 75 μm) fused silica columns coated with nonextractable stationary phases, high-resolution separations approaching those of conventional capillary GC have been obtained (24). Studies under isobaric conditions have demonstrated that more than 3000 and 12 000 effective theoretical plates per meter can be obtained with 50-μm- and 25-μm-i.d. columns, respectively (23). However, the use of such columns to obtain high efficiencies places great demands on detection sensitivity and limits the sample capacity. Studies have shown that increasing stationary-phase film thickness to allow increased sample capacity results in only minimal losses (10% loss for film increase from 0.25 to 1.0 μm in a 50-μm-i.d. column) in chromatographic efficiency. Since the ratios of the mobile-phase to stationary-phase diffusion coefficients in SFC are on the order of 10^2 compared to 10^4 for GC, the same increase in film thickness does not result in as large an increase in plate height for SFC as in GC (25).

3.1. Capillary vs. Packed-Column Considerations

Nearly all current SFC research utilizes either (a) columns packed with 5–10-μm particles prepared for HPLC or (b) wall-coated (0.1–1-μm film thickness) open tubular fused silica capillaries of 25–100-μm i.d. Simple plate-height arguments (26, 27) suggest that columns packed with 5-μm particles should have plate heights somewhat less than 50-μm-i.d. capillaries, resulting in the potential for either faster separations or generation of greater numbers of effective plates within a set time constraint. Packed columns of typical diameter (1–5-mm i.d.) provide much greater flow rates and generally allow relatively large sample loadings. The higher flow rates of packed columns make interfacing to gas-phase detectors more difficult, although the low flow rates of capillary columns requires rigorous elimination of dead-volume from the system. The lower stationary phase volume of capillary columns also limits the sample capacity that can be injected before overloading occurs. Limited sample capacity and the need to focus the solutes into a narrow band at the column head usually dictate that injection volumes of less than 50 nl (on-column) be used for small-diameter capillary columns. Small injection volumes are usually obtained by flow splitting, which can also lead to poor reproducibility and sample discrimination. Small injection volumes can also result in poor overall detection limits, even though high detector sensitivity is obtained. However, advances in capillary SFC injection techniques are being made (28), and these limitations will undoubtedly be overcome.

Capillary columns have much greater permeabilities and have demon-strated higher numbers of total effective plates than packed columns. The pressure drop is directly proportional to the linear velocity in both types of columns, and very fast separations with microparticle packed columns result in substantial pressure drops. Columns packed with 3–5-μm particles can generate 10–100 times as many plates per second as 50-μm-i.d. capillary columns (26, 27) but are more limited in the range of flow rates or column length because of the large pressure drop. The pressure drop for 5-μm packings at optimum linear velocities is $\sim 3 \times 10^4$ greater per plate than for 50-μm-i.d. capillary columns (26). This difference in column permeability limits effective high-speed application to very short packed columns. Separations requiring large numbers of plates using packed columns are ultimately limited by the rapid increase in plate height at lower-than-optimum linear velocities (26). The maximum efficiency in such separations depends upon the maximum pressure drop (directly proportional to linear velocity) which can be tolerated. For the same pressure drop, the maximum possible efficiency of a capillary column will be about 10^2–10^3 greater than for a packed column.

The reduced pressure drop in capillary columns allows pressure programming over a wider range of pressures than packed columns. Although 5-μm-particle packed columns generate more plates per second than 50-μm-i.d. capillary columns, the lower pressure drop of capillary columns allows greater resolution (proportional to number of plates) to be obtained in conjunction with pressure programming. A pressure drop does not intrinsically prevent pressure programming, but it does result in subtle difficulties. In a capillary column with a low pressure drop, the programmed pressure rise will take place across the entire column almost instantaneously; but as the pressure drop increases, the time to reach steady-state conditions also increases. The pressure drop also limits the minimum pressure at the column entrance and thus limits the range of pressure and density which can be used to vary retention.

The different retention characteristics of commercially available packed columns and SFC capillary columns are also an important consideration. For example, capacity ratios for phenanthrene in carbon dioxide with a 5-μm C_{18} microbore column are approximately 10^2 times greater than for an SE-54 capillary column (50-μm i.d., 0.25-μm film thickness) at the same density (29). Greater capacity ratios were observed for packed columns with carbon dioxide in all cases. The addition of solvent modifiers in packed column applications, however, can significantly reduce retention. Because one interest in SFC for PAC analysis is the analysis of higher-molecular-weight and more polar compounds, which have greater retention and require relatively higher fluid densities, capillary columns are usually more appropriate.

4. INSTRUMENTATION

The instrumentation for capillary SFC contains familiar elements of both HPLC and GC since both high-pressure mobile phases and greater-than-ambient operating temperatures are utilized. High pressure is maintained throughout the column, and detectors must either be designed to work at high pressures or be capable of being interfaced to the decompressed gas flow of the column effluent. The chromatographic columns must also be stable to the solvating influences of the mobile phase. A schematic diagram of a typical capillary SFC instrument is shown in Figure 4. The following sections describe the major elements of the instrumentation used for capillary SFC.

4.1. Pumping and Injection Systems

The mobile-phase flow rates needed for capillary SFC are very low, typically on the order of a few microliters per minute. Such low flow rates can most consistently be produced using a pulse-free syringe pump. Even with split-flow systems, reciprocating piston pumps are not generally suitable for the low flow rates needed for capillary operation (30). The pumping system must also be capable of providing pressure control of the mobile phase, rather than flow control, as is typically used in liquid chromatography. The pumping system should also allow for pressure or density programming of the mobile phase. Current systems (both home-made and commercial) typically utilize a microcomputer to provide wide flexibility in mobile-phase control.

The introduction of a narrow-band of sample into a capillary SFC column (50–100-μm i.d.) without any sample discrimination effects is not trivial. Similar problems exist in capillary GC and micro-LC, and varying degrees of success have been achieved in overcoming them. Sample introduction is best accomplished at the operating pressure so that minimal perturbations are imposed on the chromatographic process. To date, the most widely used injection procedure has involved sample splitting after ambient temperature injection using a commercial high-pressure valve with an internal sample loop volume of 60–200 nl. Split ratios can be as low as 1:2 or as high as 1:100, depending on the sample, column, and injection system. Sample splitting is necessary to obtain a narrow injection band on the column (31) and, in some cases, to prevent overloading the column. This injection method can suffer from poor reproducibility and nonlinear splitting of the sample.

Another approach to injecting small sample volumes is to connect the column directly to the valve and then, with fast pneumatic and electronic controls, time the travel of the rotor to allow only a small fraction of the loop volume to be transferred to the column (32). For instance, an electronically actuated 0.2-μl sample loop with a travel time of 70 ms was reported to inject a volume of 1–2 nl (33). Since no sample splitting is occurring, this method provides better reproducibility (28) and eliminates sample discrimination during injection. However, since only a few nanoliters of sample solution is introduced on the column, the sample must be relatively concentrated. This

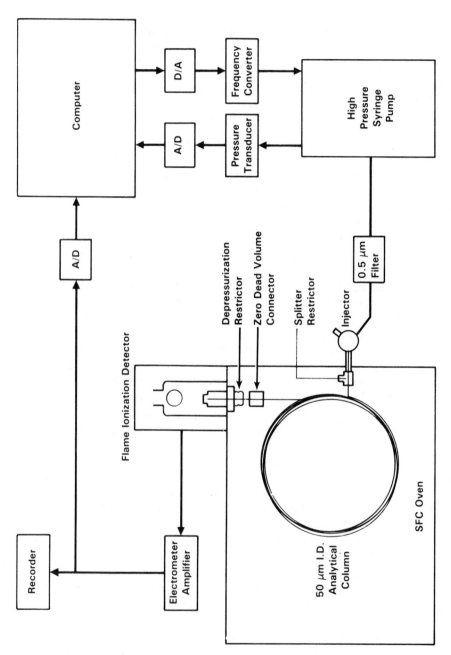

Figure 4. Schematic diagram of a typical capillary SFC instrument.

122

requirement restricts the use of capillary SFC for trace analysis and is presently a major weakness of the technique.

Methods for allowing the injection of larger and more dilute samples by concentrating and focusing the solutes at the head of the column are being studied. The use of an uncoated fused silica inlet tube, or retention gap, which allows partial separation of the sample solutes from the solvent was reported (34) to increase the amount of sample that could be introduced. Other studies using retention gaps (28, 35), solvent venting (35), and special manipulation of the density (28, 35) to focus the sample have also been described. New and improved methods for sample injection will undoubtedly be developed to allow larger sample sizes and improved solute focusing on the head of the column.

For the injection of materials that are relatively insoluble in organic solvents at room temperature, a supercritical fluid injection system has been described (36). High-molecular-weight PACs can be relatively insoluble in liquid pentane at room temperature (e.g., coronene) but can be easily chromatographed using supercritical pentane as the mobile phase. As the molecular weights of sample solutes increase, the possibility of complications in liquid injection increases. By utilizing the same supercritical fluid for injection as is used for the mobile phase, such problems are minimized. A schematic diagram (36) of the supercritical fluid injector is shown in Figure 5. More consistent split injections and better reproducibility of supercritical fluid solutions were obtained with the supercritical fluid injection system than with identical split injections of a liquid system. Furthermore, approximately 16 times more ovalene could be injected using supercritical pentane at 210°C and 122 bar than was possible with liquid dichloromethane at room temperature. Although use of such an injection system can be cumbersome (filling, cleaning, etc.), it provides distinct advantages for relatively insoluble samples. Injection systems based on supercritical fluid extraction are also being developed (37).

4.2. Capillary Columns

Theoretical calculations (38) have predicted that column diameters in capillary SFC must be less than 100 μm in inner diameter to maintain reasonable analysis times with high resolution. As a consequence, practical column dimensions of 50 μm in inner diameter and 10–30 m in length have generally been accepted as a good compromise between performance and ease of preparation and handling. Even so, the small column dimensions are a major consideration in the design of the instrumentation and in the preparation of suitable columns.

The column technology used for capillary SFC is similar to that used for

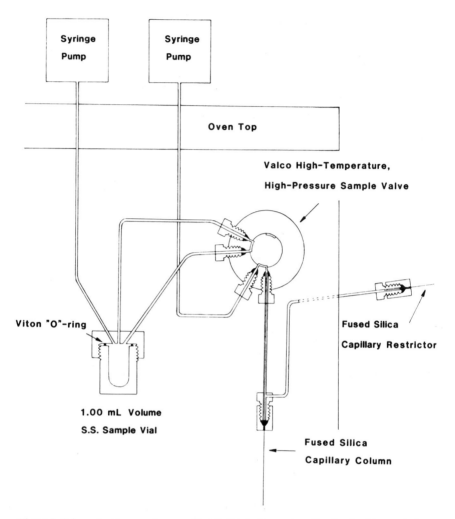

Figure 5. Schematic diagram of supercritical fluid injection system. Reprinted, with permission, from ref. 36. Copyright 1986, Huethig Verlag.

capillary GC, differing only in the smaller column diameters and the necessity of more firmly cross-linking and bonding the stationary phases to the column wall to resist the solvating influences of the mobile phase. In practice, however, it is much more difficult to prepare smaller-diameter columns. Essentially the same range of stationary phases that have been used for PAC analysis in capillary GC have also been used in capillary SFC and include 100% methyl- and 5% phenyl-, as well as 50% phenyl-, biphenyl-, and cyanopropyl-, methylpolysiloxanes. Polysiloxane phases are preferred since they can be stabilized by free-radical cross-linking. For good performance, small-diameter columns should possess the usual qualities of high efficiency, excellent deactivation, and lasting stability. Proper deactivation is especially important since only small sample amounts can be introduced. A deactivation layer that is bonded to the silica surface can also lead to improved stability since it can also be chemically attached to the stationary phase during cross-linking. Specific methods for the deactivation (39) and free-radical cross-linking (40) of small-diameter SFC columns are described elsewhere.

The choice of supercritical mobile phase has a significant effect on column stability. Columns coated with cross-linked polysiloxane stationary phases decrease in stability in order with the use of supercritical carbon dioxide, polar-modified (e.g., up to 10% methanol, isopropanol, acetic acid, acetonitrile, etc.) supercritical carbon dioxide, supercritical propane, supercritical pentane, and supercritical ammonia. Columns will generally last for months when used with carbon dioxide, will last for weeks when used with supercritical pentane, and will deteriorate within a few hours when subjected to supercritical ammonia. Increased stability of up to 2 days for ammonia has been reported with an octylpolysiloxane stationary phase (41). The decrease in column stability with supercritical pentane is due to its higher critical temperature and to the greater mutual solubility of the pentane and the stationary phases. Decreased stability can also be attributed to stationary-phase swelling. A methylpolysiloxane was reported to swell to approximately three times its original film thickness in supercritical hydrocarbons such as butane and pentane, while little change was observed in supercritical carbon dioxide (42).

An important feature of a stationary phase that governs the separation process is its selectivity for the sample solutes. In many cases, the high separation efficiencies that can be achieved with capillary chromatographic methods are insufficient to resolve isomeric compounds. Particular problems exist for the separation of many closely related isomeric PAC. Enhanced separations of many isomeric PACs can be achieved by choosing a more selective stationary phase. This is illustrated with the chromatograms of isomeric ethylene-linked naphtho-acridines shown in Figure 6 (43). Chromatogram A was obtained using an 8-m × 50-μm-i.d. column coated with a 50% liquid-crystalline polysiloxane phase, and chromatogram B was obtained using a column of the same dimensions coated with a 100% methylpolysiloxane phase. Carbon dioxide at 140°C with identical density programs was used to affect both separations. All six of the isomers were fully resolved on the liquid-crystal phase, whereas they essentially coeluted on the traditional methylpolysiloxane phase. Separation selectivity is based on shape and size for the liquid-crystal phase (44) rather than on polarity as is generally

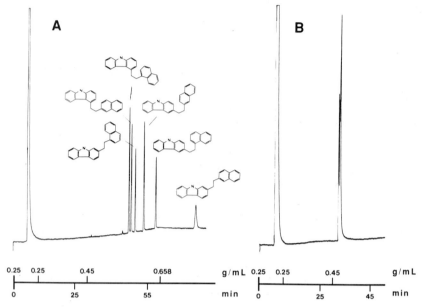

Figure 6. Capillary SFC chromatograms of ethylene-linkage PAC isomers obtained under identical operating conditions and column dimensions with (*A*) a 50% liquid-crystalline polysiloxane stationary phase and (*B*) a 100% methylpolysiloxane stationary phase.

the case for other stationary phases. Such a separation mechanism should be highly useful for the resolution of isomeric PACs.

4.3. Restrictors

In order to maintain pressurized conditions along the column and regulate the mobile-phase linear velocity, a flow restrictor is attached to the end of the capillary column. For optical detectors, the restrictor is placed after the detection region (on-column flow cell). Detection of the analytes takes place in the pressurized fluid (the temperature is usually cooled below the critical point to condense the fluid to a liquid), and few constraints are placed on the restrictor performance other than furnishing the appropriate linear velocity of the mobile phase and freedom from plugging.

For flame-based detectors (FID, NPD, etc.), mass spectrometry detection, and other gas-phase detectors, the chromatographic effluent must be depressurized before detection. Depressurization is generally achieved by expansion of the mobile phase through a restrictor into a region where detection can occur. Performance of this restrictor is crucial to allow expansion of the sample solutes to the gas phase for subsequent detection. The ideal restrictor for SFC would provide the following characteristics: (a) provide for uniform pulse-free flow, (b) offer immunity from plugging, (c) be easily replaced and provide for variation of flow rate, and (d) provide for complete transfer of labile or nonvolatile solutes to the detector without pyrolysis or formation of analyte particles. The performance of an SFC restrictor and the efficiency of transfer of less volatile solutes to the gas phase during expansion depend upon the state of the fluid prior to expansion, the dimensions and geometry of the restrictor, restrictor heating and heat-transfer properties of the restrictor, and the state of the fluid during and after the expansion. A detailed discussion of the thermodynamics of the expansion process from restrictors has been given elsewhere (45).

At present, most SFC utilizes some form of capillary restrictor with the dimensions empirically selected to provide the desired mobile-phase linear velocity for the chromatographic pressure and temperature range. The most commonly used restrictor types are schematically illustrated in Figure 7. Initially, short lengths (5–10 cm) of small-bore (5–10-μm i.d.) fused silica tubing (Figure 7A) were attached to the end of the column with a zero-dead-volume connector. With this restrictor design, depressurization occurs gradually over the restrictor length; and as the fluid solvating power drops, it is possible for less volatile solutes to precipitate and form particles. This behavior has been observed as detector "spiking" (34, 46) and can ultimately lead to plugging of the restrictor. By increasing the fluid temperature, the onset of the spiking phenomena can be delayed to later-eluting, typically less volatile compounds (47). Improved transfer of less volatile solutes can be

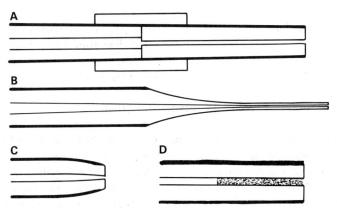

Figure 7. Schematic illustration of various restrictor geometries used in capillary SFC. (A) Small-bore fused silica tubing. (B) Tapered restrictor. (C) Integral restrictor. (D) Porous frit restrictor.

obtained with tapered (Figure 7B) or integral (Figure 7C) restrictors. With these designs, an incomplete pressure drop of the fluid occurs through the restrictor. The remaining decompression takes place outside the restrictor which maximizes fluid density and minimizes solute nucleation. The tapered restrictors are fabricated (34) by a controlled drawing of a heat region of fused silica tubing from which the polyimide coating (shown as dark lines in Figure 7) has been removed. Although these restrictors can be fabricated relatively easily, they are fragile and must be handled with care. The integral restrictors are fabricated by polishing the closed end of a piece of fused silica capillary obtained by careful heating until a small orifice giving the desired flow rate is obtained (48). Both the tapered and integral restrictors can be formed directly on the column end, thereby eliminating the necessity of a connector. The porous frit restrictor (Figure 7D) is quite different than the other restrictors (49). The frit resembles a column packed with submicron particles which creates multiple fluid paths. The tortuous fluid path, combined with the relatively long residence time, provides for more efficient heat transfer to the expanding gas. This results in improved transport for sufficiently volatile compounds but prevents detection of compounds with insufficient vapor pressures at the restrictor temperature. The frit restrictor is relatively rugged and is not easily plugged.

The design and implementation of efficient restrictors is crucial for the successful application of SFC and SFC–MS techniques to higher-molecular-weight PACs. Consequently, the improvement of restrictor technologies is an important and active area of research.

5. APPLICATIONS

Capillary SFC has been applied to a range of PAC samples using various supercritical fluid mobile-phase systems and detectors. Typical applications based on detection methods that illustrate the limitations and potential of SFC methods will be presented in the following sections.

5.1. Flame Detection

The same advantages of using FID in GC can also be realized in SFC. These advantages include near-universal response, high sensitivity, and a wide dynamic range. Unfortunately, many supercritical fluid mobile phases also respond to FID; and only a few systems, such as carbon dioxide, can be utilized. High-molecular-weight PACs (> 300 dalton) have limited solubility in carbon dioxide, thereby limiting this fluid system to the characterization of compounds in this molecular-weight range.

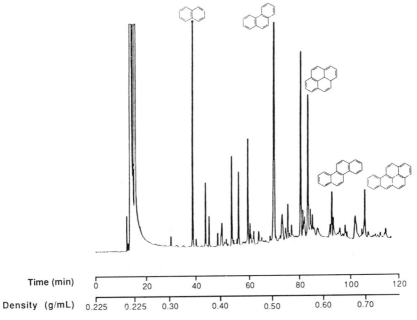

Figure 8. Capillary SFC chromatogram of coal tar.

An example of an SFC separation of coal tar using FID detection is shown in Figure 8 (46). The chromatogram was obtained with a 34-m × 50-μm-i.d. column coated with a cross-linked SE-54 stationary phase using carbon dioxide at 40°C as the mobile phase. This separation exhibited resolution similar to that obtained with capillary GC, but the analysis time was approximately twice as long. The density programming produced an elution order very similar to that of GC. The nitrogen–phosphorus detector (NPD) was also evaluated and was found to provide good selectivity for nitrogen-containing compounds over hydrocarbons when using carbon dioxide as the mobile phase (46). A dual-flame photometric detector (FPD) has also been demonstrated for sulfur-selective detection of sulfur heterocycles that were separated using supercritical carbon dioxide (50). However, nonlinear, non-exponential response and relatively poor detection limits (25 ng with an S/N ratio of 2 for benzo[b]thiophene) were obtained.

Another example of an SFC separation using carbon dioxide and FID detection is shown in Figure 9 (51). This separation of a highly complex alkylated two- and three-ringed mixture of neutral PAC from diesel fuel illustrates the high resolution that can be achieved using capillary SFC. The chromatogram was obtained using a 15-m × 50-μm-i.d. column coated with a

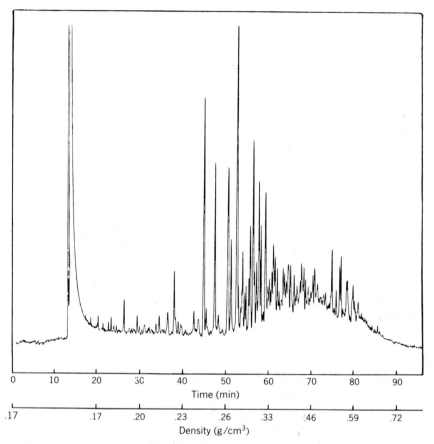

Figure 9. Capillary SFC chromatogram of the neutral PAC fraction from a diesel fuel.

cross-linked 50% phenyl polymethylphenylsiloxane stationary phase and a mobile-phase operating temperature of 60°C. A dual-ramp linear density program was used to slowly increase the solvating power of the mobile phase to achieve the best resolution. The ability to precisely control the solvating power of the mobile phase is an important feature of capillary SFC.

Capillary SFC using supercritical carbon dioxide is ideally suited for the analysis of thermally labile compounds. It has been reported that several nitro-PAC could not be eluted by capillary GC (52), even when using a cold on-column injector. These compounds were hydroxynitropyrenes, hydroxynitrofluorene, and nitropyrene quinones. Presumably these compounds thermally decomposed during analysis. The characterization of these compounds is important because of their high mutagenic activity (53). Capillary

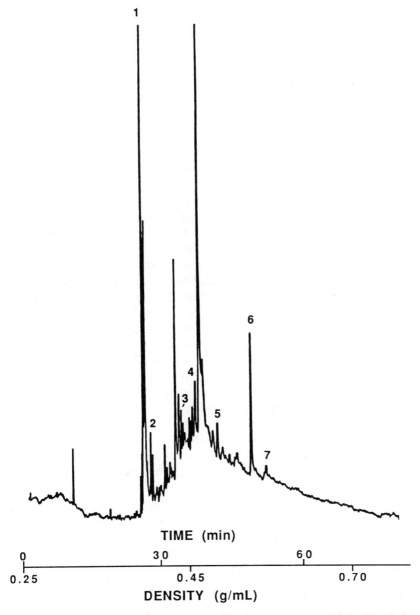

Figure 10. Capillary SFC chromatogram of a subfraction from the NBS 1650 diesel particulate extract. Peak identifications: (1) 5-nitroisoquinoline, (2) 1-hydroxy-2-nitronaphthalene, (3) 5-nitroindole, (4) 9-hydroxy-2-nitrofluorene, (5) 1-hydroxy-8-nitropyrene, (6) 1-hydroxy-6-nitropyrene, (7) 1-hydroxy-3-nitropyrene.

SFC successfully eluted these compounds, and sensitive and selective detection was obtained using a nitro-selective thermionic detector (54). A capillary SFC chromatogram of a subfraction containing the nitro-PAC of the NBS diesel particulate sample is shown in Figure 10 (54). A complex mixture of nitro-containing compounds was separated, and several of the compounds were identified by retention comparison with standards. This separation utilized a 15-m × 50-μm-i.d. column coated with cross-linked SE-54 stationary phase and a carbon dioxide mobile phase at 101°C. The TID-1-N$_2$ detector (55) was used to obtain selective detection of the nitro-PAC.

5.2.   UV and Fluorescence Detection

The UV absorbance detector has been used extensively in packed-column SFC because most mobile phases are transparent in the UV region. However, because of (a) stringent demands on the allowable UV-absorption cell volumes (38) in capillary SFC, (b) the high pressure requirements, and (c) the resulting lower sensitivities, UV detection has not been used extensively in capillary SFC. Highly compressible mobile phases, such as carbon dioxide, also present problems because of refractive index changes and resulting baseline drift which occur during density programming. By utilizing a flow cell constructed of 100–250-μm-i.d. fused silica tubing (polyimide removed to form a window), on-column detection (56) can be achieved to meet the low cell-volume requirements. By cooling the column effluent prior to detection, it is possible to reduce the baseline shifts created by refractive index changes to an acceptable level (57). An SFC chromatogram of coal tar obtained with UV detection at 254 nm using this approach is shown in Figure 11 (57). The narrow peaks and smooth baseline indicate that acceptable detector volume and stability were obtained. The separation was obtained with a 10-m × 50-μm-i.d column coated with a cross-linked 100% methylpolysiloxane phase (SE-33) stationary phase and with pressure-programmed carbon dioxide at 80°C.

UV detection provides the capability of using mobile-phase systems that offer higher solvating power than FID-compatible fluids (e.g., carbon dioxide). Retention studies using polar-modified carbon dioxide mixtures have shown that significant retention decreases can be obtained with modifier concentrations in the 5–20% range (58). Retention data (59) for several PACs as a function of 2-propanol, dichloromethane, and acetonitrile concentration in carbon dioxide are shown in Figure 12. All measurements were made at 100°C and 150 atm. These conditions were well above the critical boundaries for all the fluid mixtures that were employed. Retention shifts (k') of over an order of magnitude were observed for some of the standards. Programming the composition of polar-modified carbon dioxide fluid mixtures in con-

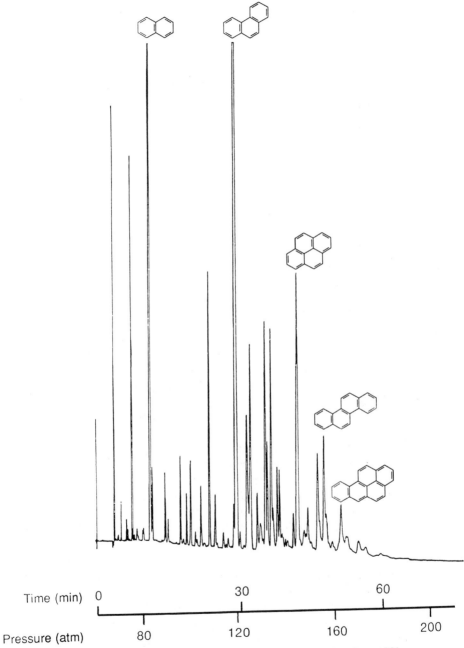

Figure 11. Capillary SFC chromatogram of coal tar with UV detection at 254 nm.

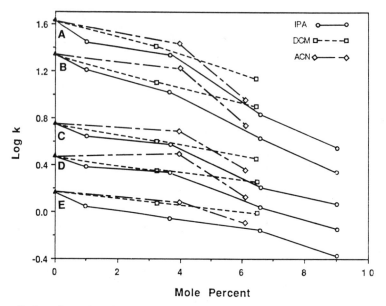

Figure 12. Log of retention of (*A*) coronene, (*B*) picene, (*C*) chrysene, (*D*) pyrene, (*E*) phenanthrene vs. modifier mole fraction. IPA = 2-propanol; DCM = dichloromethane; ACN = acetonitrile.

junction with pressure programming has also been reported (60). UV detection has been used with SFC analyses of PAC using a supercritical ammonia mobile phase (41). Ammonia provided significantly reduced retention as compared to that of carbon dioxide. These studies suggest that very high-molecular-weight compounds may be amenable to SFC analysis in reasonable analysis times with more polar and mixed mobile phases.

Fluorescence detection was utilized for the detection of PACs using supercritical pentane as a mobile phase in the first reports of capillary SFC (61, 62). In these early studies, the glass capillary column itself was straightened and carefully placed in the light path. Since column diameters of 100–300-i.d. were used, sufficient sensitivity and acceptably small detector volumes to minimize significant band broadening were obtained. With the use of smaller-diameter columns, flow cells similar to those described for UV detection (56) have been utilized. A flow cell using optical fibers to transmit the excitation and emission light has been described which allows on-the-fly excitation or emission fluorescence spectra to be obtained during elution of sample solutes (63). Fluorescence detection generally provides better sensitivities for PAC than for FID and is compatible with virtually any fluid phase. Highly selective detection is also possible by choosing the appropriate excitation energy and/or detection of selected emission wavelengths.

An example of a capillary SFC chromatogram utilizing fluorescence detection is shown in Figure 13 (64). This separation of the polar fraction of a diesel fuel was obtained with a 10 mol% isopropanol modified carbon dioxide fluid mixture at 110°C and with a 10-m × 50-μm-i.d. column coated with a cross-linked SE-54 stationary phase. A polar-modified fluid mixture was used to solvate the more polar components that potentially could not be eluted with carbon dioxide. A greater number of components appeared to elute in the SFC analysis of this fraction than were eluted during GC analysis. The pressure was held at 117 atm for 10 min after injection and then slowly ramped upwards. Detection excitation was at 254 nm, and all fluorescence greater than 270 nm was collected. The numbers on the labeled peaks refer to probable molecular weights of the main components which can tentatively be identified as alkyl carbazoles.

An example of an SFC analysis of a high-molecular-weight material is shown in Figure 14 (65). This carbon-black extract was eluted with pentane at 210°C on a 10-m × 50-μm-i.d. column coated with a cross-linked 50% phenyl

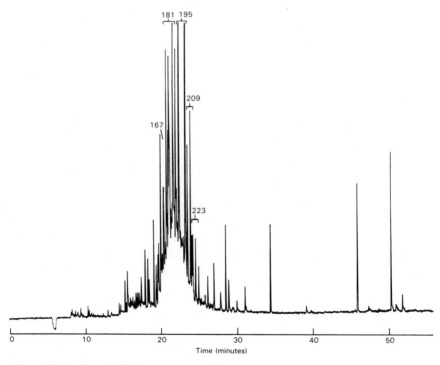

Figure 13. Capillary SFC chromatogram with fluorescence detection of the polar fraction of a diesel fuel. Peak labels indicate probable molecular weights of marked components.

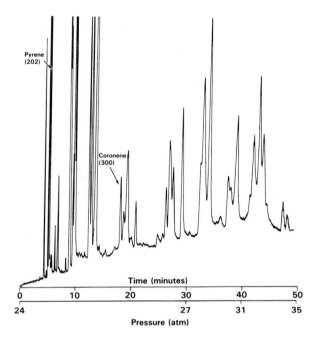

Figure 14. Capillary SFC chromatogram with fluorescence detection of a carbon-black extract. Reprinted, with permission, from ref. 65. Copyright 1988, Preston Publications, Inc.

polymethylphenylsiloxane stationary phase. Fluorescence detection excitation was at 272 nm, and emissions greater that 300 nm were collected. Since standards were not available for the high-molecular-weight PACs, peak groups relative to the pyrene and coronene peaks were counted, and it can be estimated that compounds up to 11 or 12 rings were eluted. Ovalene (10 rings) has been verified to elute from a similar sample using similar SFC conditions (40).

Another example of the applicability of SFC for the analysis of high-molecular-weight material is illustrated with the chromatograms shown in Figure 15 (66). These chromatograms are of a coal-liquid vacuum stillbottoms sample with detection obtained at an excitation wavelength at 340 nm (Figure 15, upper chromatogram) and 300 nm (Figure 15, lower chromatogram). Fluorescence emission was at 460 nm. Note the enhancement of pyrene and the alkylated pyrenes in the 340-nm chromatogram, whereas in the 300-nm chromatogram, the pyrenes disappear completely. At the same time, the 300-nm chromatogram strongly enhances the detection of coronene. These chromatograms were obtained on a 27-m × 50-μm-i.d. column coated with a

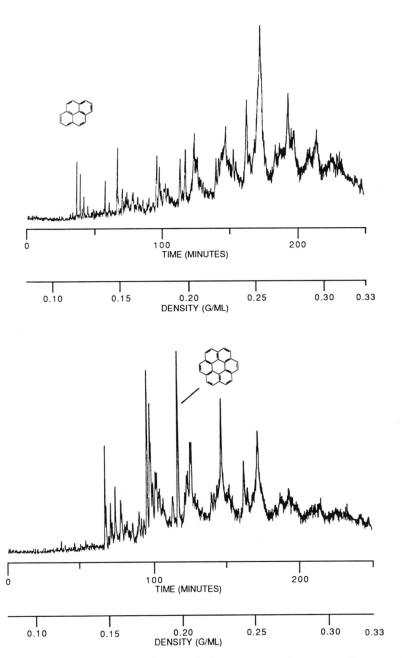

Figure 15. Selected wavelength capillary SFC chromatograms of vacuum stillbottoms at excitation wavelengths of (upper) 340 nm and (lower) 300 nm.

cross-linked 50% phenyl polymethylphenylsiloxane stationary phase and with a 9:1 pentane/benzene mobile-phase mixture at 223°C.

5.3. SFC–Mass Spectrometry

The advantages of coupling a chromatographic technique with mass spectrometry (MS) are considerable, as evidenced by the major role of GC–MS in organic analysis. For SFC, the mass spectrometer is an ideal detector, providing both sensitive and selective detection with universal applicability. In contrast to many detection methods, MS is compatible with a broad range of mobile phases. The added flexibility to choose ionization modes such as electron impact (EI) or chemical ionization (CI) with nearly any reagent gas, independent of the mobile phase, provides the basis for obtaining either structural information or high sensitivity and selectivity.

The instrumentation for capillary SFC with MS detection (SFC–MS) is described in detail elsewhere (67). Since mobile-phase flow rates are small, the chromatographic effluent can be directly introduced into the ionization region after expansion through a restrictor. Detection limits for CI, which are significantly better than with EI, typically range from 0.1 to 10 pg depending upon the compound, analysis time, separation efficiency, and CI reagent gas (67).

The analysis of a standard mixture of PACs ranging from naphthalene to cornene was described in the first report of capillary SFC–MS (68). This separation utilized a 20-m × 200-μm-i.d. column, supercritical pentane as the mobile phase, and a pinched piece of Pt–Ir tubing for a restrictor. An example of a separation of the PAC in coal tar using SFC–MS with a 50-μm-i.d. fused silica column is shown in Figure 16A (69). This separation used a 15-m-long column coated with cross-linked SE-54 stationary phase and a mobile phase of supercritical carbon dioxide at 60°C programmed from 75 bar to 175 bar at 1 bar/min following elution of the solvent front. Data were not stored until after the solvent peak eluted; hence there is no solvent peak or baseline prior to the elution of the compounds. The resolution of this separation was comparable to that obtained with FID (see Figure 8) illustrating the efficient interfacing between SFC and MS that was obtained.

Figure 16 also provides a comparison between the high-resolution separation just described and a rapid separation of the same mixture. The high-resolution separation required over 110 min, and the rapid separation was effected in less than 6 min. The rapid separation shown in Figure 16B (69) utilized a 1.8-m × 50-μm-i.d. column coated with cross-linked SE-54, carbon dioxide at 100°C as the mobile phase, and a rapid pressure ramp of 50 bar/min. Although significant resolution was lost, the rapid separation provided a level of characterization that would be adequate for many

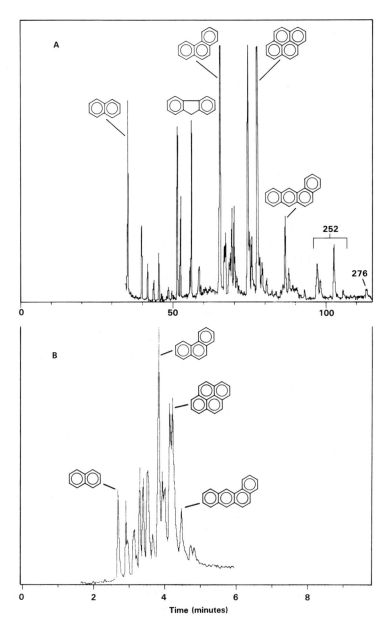

Figure 16. Capillary SFC–MS total-ion chromatograms comparing a (*A*) high-resolution separation of coal tar obtained using a 15-m × 50-μm-i.d. column with a (*B*) rapid separation of coal tar using a 1.8-m × 50-μm-i.d. column with a 50 bar/min pressure ramp. Reprinted, with permission, from ref. 69. Copyright 1986, American Chemical Society.

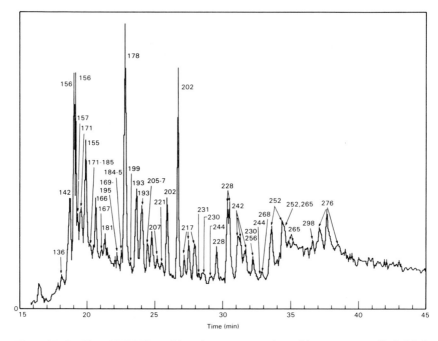

Figure 17. Capillary SFC–MS total-ion chromatogram of a solid-waste extract. Peak labels indicate probable molecular weights of the marked components.

purposes, especially with the inherent high selectivity that can be obtained with MS. Even greater chromatographic selectivity can be obtained by using smaller-diameter columns (23).

An example of an SFC–MS separation of a complex mixture is shown in Figure 17. This sample was an extract from a solid hazardous waste obtained by supercritical extraction (see next section) with an ethanol-modified pentane mixture. The analysis was obtained using a 12-m × 50-μm-i.d. column coated with cross-linked 50% phenyl polymethylphenylsiloxane stationary phase. Supercritical pentane at 230°C was used as the mobile phase, which was pressure programmed from 25 bar at rates of 0.5, 1.0, and 2.0 bar/min sequentially during the analysis. Pentane CI generated predominantly protonated molecular ions $(M+1)^+$. The molecular weights of many of the components in the mixture are noted on the total-ion chromatogram.

The total-ion chromatogram from an SFC–MS analysis of a complex polar fraction from a diesel fuel is shown in Figure 18 (65). This separation was obtained using a mobile phase consisting of a 6.5 mol% mixture of isopropanol in carbon dioxide at 100°C and a 15-m × 50-μm-i.d. column coated with

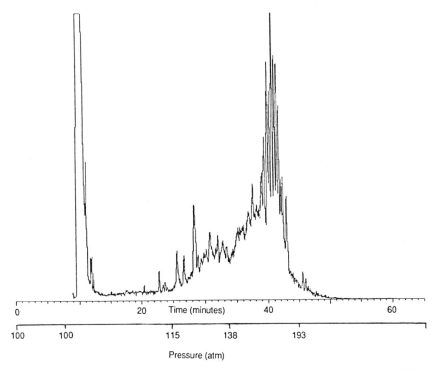

Figure 18. Capillary SFC–MS total-ion chromatogram of the polar fraction of a diesel fuel. Reprinted, with permission, from ref. 65. Copyright 1988, Preston Publications, Inc.

a cross-linked SE-54 stationary phase. Thirteen minutes after injection, the mobile phase was pressure ramped from 100 bar at 1.25/min, then at 2.5 bar/min at 25 min, then at 5 bar/min at 34 min, and finally at 10 bar/min at 45 min. Methane was used for the CI reagent gas and gave predominantly protonated molecular ions. Although numerous peaks of the complex mixture were resolved, it is clear that many additional components were not chromatographically resolved. By using the mass-selective detection capabilities available with MS analysis, numerous ion series (>60), each containing several peaks, could be resolved from the overlapping and coeluting compounds. Typical single-ion profiles at various m/z values are shown in Figure 19 (65). Probable identifications based on molecular weights and retention times are also listed for each specific ion series. High concentrations of alkylated indoles and carbazoles were the primary components in this sample matrix. It is possible that the m/z 157 compounds correspond to alkyl naphthalenes that were present as contaminants in the polar fuel fraction. This example illustrates the large amount of chemical characterization

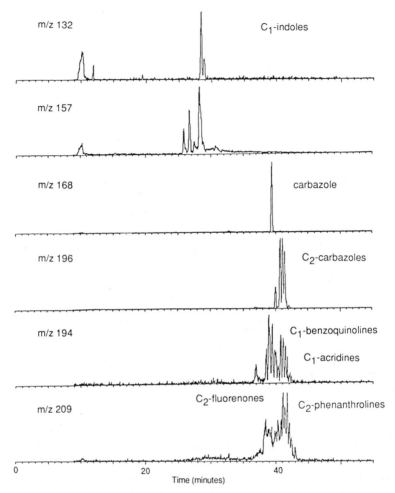

Figure 19. Typical single-ion plots obtained from the SFC–MS analysis of the sample described in Figure 18. The m/z values correspond to protonated molecular ions. Reprinted, with permission, from ref. 65. Copyright 1988, Preston Publications, Inc.

information that can be obtained with SFC–MS analysis of a complex mixture.

The real potential for SFC–MS lies in the analysis of high-molecular-weight materials that cannot be addressed by GC–MS. As SFC–MS techniques mature and improved interfacing methods are developed, it can be expected that this potential will be more fully realized.

6. RELATED TECHNIQUES

The physical properties of supercritical fluids also render them useful for sample preparation purposes. The liquidlike solvent power and high diffusivity provide the potential for more rapid extraction rates and more efficient extractions due to better penetration of the sample matrix than is feasible with liquids. Supercritical fluid extraction has been used extensively in chemical processing applications (9, 70) and more recently was demonstrated to be useful for analytical sample preparation (71–73). Supercritical fluids can also be used as mobile phases for semipreparative separations and, in conjunction with appropriate collection, can be used for sample fractionation. Both supercritical fluid extraction (SFE) and supercritical fluid fractionation methods have been applied to PAC matrices, and specific examples will be described in the following sections.

6.1. Supercritical Fluid Extraction

SFE has been successfully used for rapid and efficient complex mixture sample preparation (71). Typical examples using hazardous waste samples are described below. Sample A was a soil boring contaminated with coal gasification residuals, and sample B was from a waste stream from a treatment facility. Table 2 presents a comparison between (a) the extraction efficiencies (e.g., amount of material extracted) obtained using three different fluid systems and (b) the specific extraction conditions.

The extraction efficiencies for the three different fluid systems were very similar for sample A. However, the amount of material removed from sample B was approximately 1.8 times greater with methanol-modified carbon dioxide and ethanol-modified pentane than with pure carbon dioxide. The

Table 2. Supercritical Fluid Extraction Comparison of Hazardous Waste Samples

	Extraction Conditions			Percent Extracted	
Fluid System	Temperature ($^\circ$C)	Pressure (bar)	Fluid Volume (ml)	Sample A	Sample B
Carbon dioxide	150	415	500	27	28
Carbon dioxide–methanol (80:20)	150	400	500	26	49
Pentane–ethanol (93:7 mol%)	215	165	750	29	51

higher extraction temperature used for the pentane system could have contributed to the improved efficiency for this fluid. Identical temperatures were used for the carbon dioxide and carbon dioxide–methanol fluid systems, which suggests that the components were more soluble in the more polar fluid system. This is consistent with the sample composition, since subsequent analyses of sample B indicated that it contained phenolics, amines, esters, nitrated aromatics, and other polar compounds. Sample A, on the other hand,

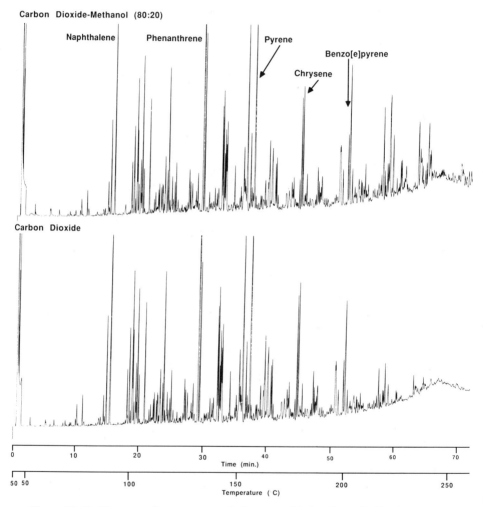

Figure 20. Capillary gas chromatograms of the supercritical carbon dioxide (lower) and methanol-modified carbon dioxide (upper) extracts of a hazardous waste sample. Reprinted, with permission, from ref. 71. Copyright 1988, American Chemical Society.

contained neutral polycyclic aromatic compounds that would not be expected to have significantly higher solubility in the more polar fluid systems. However, the GC profiles of the carbon dioxide extract and the carbon dioxide–methanol extract of sample A were slightly different. As can be observed from the capillary GC chromatograms shown in Figure 20 (71), there are greater relative quantities of the higher-molecular-weight compounds in the carbon dioxide–methanol extract than in the pure carbon dioxide extract. This may be accounted for by slightly greater solubility of these higher-molecular-weight compounds in the more polar fluid. These examples illustrate the potential of analytical SFE for rapid and efficient complex mixture sample preparation.

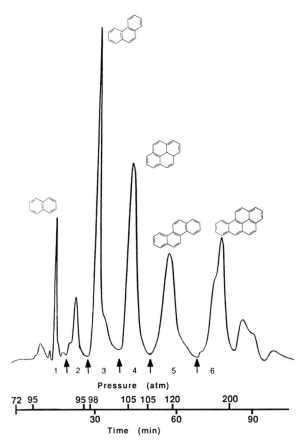

Figure 21. Supercritical fluid fractionation of a coal tar with UV detection. Numbers refer to the fractions that were collected.

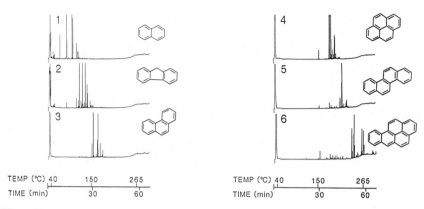

Figure 22. Capillary gas chromatograms of the fractions obtained during supercritical fractionation of a coal tar. The numbers refer to the fractions shown in Figure 21.

6.2. Supercritical Fluid Fractionation

Supercritical fluid fractionation of a variety of complex coal- and petroleum-derived PAC mixtures based on the number of aromatic rings has been reported (74, 75). These fractionations were obtained using supercritical carbon dioxide and columns packed with NH_2-modified silica particles. The stationary phase particles were of an intermediate size (30–70 μm) to minimize the pressure drop along the column so the full advantages of pressure programming could be used to increase the selectivity of the separation. The fractionation apparatus (74) incorporated a UV detector to monitor the peaks as they eluted and used a system of pressurized collection vessels to isolate each peak. An example (74) of the fractionation obtained on a coal-tar sample is shown in Figure 21. Carbon dioxide at 40°C was programmed in a step fashion to elute each of the different fractions. Capillary gas chromatograms (74) of each of the collected fractions are shown in Figure 22. Each fraction consisted essentially of parent ring structures and their alkylated homologues, indicating that the selectivity of the fractionation according to ring number was very good. The alkylated species eluted near the same time as their parent compounds, while compounds of different ring number were widely separated. This example illustrates the potential of supercritical fluid fractionation for selective separations with easy solvent removal.

ACKNOWLEDGMENT

This work was supported by the U.S. Department of Energy, Offices of Fossil Energy and Health and Environmental Research, under Contract DE-AC06-76RLO 1830.

REFERENCES

1. E. A. Klesper, H. Corwin and D. A. Turner, *J. Org. Chem.*, **27**, 700 (1962).
2. M. N. Meyers and J. C. Giddings, *Sep. Sci.*, **1**, 761 (1966).
3. S. T. Sie, W. van Beersum, and G. W. A. Rijnders, *Sep. Sci.*, **1**, 459 (1966).
4. S. T. Sie and G. W. A. Rijnders, *Sep. Sci.*, **2**, 729 (1967).
5. S. T. Sie and G. W. A. Rijnders, *Anal. Chim. Acta*, **38**, 31 (1967).
6. S. T. Sie and G. W. A. Rijnders, *Sep. Sci.*, **2**, 755 (1967).
7. R. E. Jentoft and T. H. Gouw, *J. Chromatogr. Sci.*, **8**, 138 (1970).
8. R. E. Jentoft and T. H. Gouw, *Anal. Chem.*, **48**, 2195 (1976).
9. G. M. Schneider, E. Stahl, and G. Wilke, *Extraction with Supercritical Gases*, Verlag Chemie, Deerfield Beach, Fla. (1980).
10. M. E. Paulaitis, V. J. Krukonis, R. T. Kurnick, and R. C. Reid, *Rev. Chem. Eng.*, **1**, 179 (1983).
11. M. A. McHugh and V. J. Krukonis, *Supercritical Fluid Extraction*, Butterworth, Boston (1986).
12. C. R. Yonker, S. L. Frye, D. R. Kalkwarf, and R. D. Smith, *J. Phys. Chem.*, **90**, 3022 (1986).
13. W. B. Street, in *Chemical Engineering at Supercritical Fluid Conditions*, M. E. Paulaitis, J. M. L. Penninger, R. D. Gray, and P. Davidson, Eds., Ann Arbor Science Publishers, Ann Arbor, Mich (1983).
14. E. Brunner, *J. Chem. Thermodyn.* **17**, 671 (1985).
15. A. J. Seckner, Ms Thesis, Johns Hopkins University, 1987.
16. D. Peng and D. B. Robinson, *AlChE J.*, **23**, 137 (1977).
17. R. A. Heidemann and A. H. Khalil, *AlChE J.*, **26**, 769 (1980).
18. P. L. Chueh and J. M. Prausnitz, *AlChE J.*, **13**, 1099 (1967).
19. M. Gitterman and I. Procaccia, *J. Chem. Phys.*, **78**, 2648 (1983).
20. R. D. Smith and H. R. Udseth, *Anal. Chem.*, **55**, 2266 (1983).
21. S. M. Fields and M. L. Lee, *J. Chromatogr.*, **349**, 305 (1985).
22. R. D. Smith, E. G. Chapman, and B. W. Wright, *Anal. Chem.*, **57**, 2829 (1985).
23. B. W. Wright and R. D. Smith, *J. High Resolut. Chromatogr.*, **9**, 73 (1986).
24. S. M. Fields, R. C. Kong, J. C. Fjeldsted, M. L. Lee, and P. A. Peaden, *J. High Resolut. Chromatogr.*, **7**, 312 (1984).
25. S. M. Fields, R. C. Kong, M. L. Lee, and P. A. Peaden, *J. High Resolut. Chromatogr.*, **7**, 423 (1984).
26. G. Guiochon and H. Colin, *J. Chromatogr. Lib.*, **28**, 1 (1984).
27. L. G. Randall, in *Ultrahigh Resolution Chromatography*, ACS Symposium Series, Vol. 252, American Chemical Society, Washington, D.C., p. 135 (1984).
28. B. E. Richter, D. E. Knowles, M. R. Andersen, N. L. Porter, E. R. Campbell, and D. W. Later, *J. High Resolut. Chromatogr.*, **11**, 29 (1988).
29. R. D. Smith and B. W. Wright, in *Microbore Column Chromatography: A Unified Approach to Chromatography*, F. J. Yang, Ed., Marcel Dekker, New York, Chapter 10 (1988).
30. T. Greibrokk, B. E. Berg, A. L. Blilie, J. Doehl, A. Farbrot, and E. Lundanes, *J. Chromatogr.*, **394**, 429 (1987).
31. P. A. Peaden, J. C. Fjeldsted, M. L. Lee, S. R. Springston, and M. Novotny, *Anal. Chem.*, **54**, 1090 (1982).

32. M. C. Harvey, S. D. Stearns, and J. P. Avarette, *LC*, **3**, 434 (1985).
33. M. L. Lee and K. E. Markides, *J. High Resolut. Chromatogr.*, **9**, 652 (1986).
34. T. L. Chester, D. P. Innis, and G. D. Owens, *Anal. Chem.*, **57**, 2243 (1985).
35. A. Farbrot Buskhe, B. E. Berg, O. Gyllenhaal, and T. Greibrokk, *J. High Resolut. Chromatogr.*, **11**, 16 (1988).
36. W. P. Jackson, K. E. Markides, and M. L. Lee, *J. High Resolut. Chromatogr.*, **9**, 213 (1986).
37. B. E. Richter, N. L. Porter, D. E. Knowles, M. R. Andersen, E. R. Campbell, and D. W. Later, *Proceedings of the 1988 Pittsburgh Conference*, Abstract No. 424 (1988).
38. P. A. Peaden and M. L. Lee, *J. Chromatogr.*, **259**, 1 (1983).
39. R. C. Kong, C. L. Wooley, S. M. Fields, and M. L. Lee, *Chromatographia*, **18**, 362 (1984).
40. R. C. Kong, S. M. Fields, W. P. Jackson, and M. L. Lee, *J. Chromatogr.*, **289**, 105 (1984).
41. J. C. Kuei, K. E. Markides, and M. L. Lee, *J. High Resolut. Chromatogr.*, **10**, 257 (1987).
42. S. R. Springston, P. David, J. Steger, and M. Novotny, *Anal. Chem.*, **58**, 997 (1986).
43. H-C. K. Chang, K. Markides, and M. L. Lee, unpublished data.
44. K. E. Markides, M. Nishioka, B. J. Tarbet, J. S. Bradshaw, and M. L. Lee, *Anal. Chem.*, **57**, 1296 (1985).
45. R. D. Smith, J. L. Fulton, R. C. Petersen, A. J. Kopriva, and B. W. Wright, *Anal. Chem.*, **58**, 2057 (1986).
46. J. C. Fjeldsted, R. C. Kong, and M. L. Lee, *J. Chromatogr.*, **279** 449 (1983).
47. T. L. Chester, *J. Chromatogr.*, **299**, 423 (1984).
48. E. J. Guthrie and H. E. Schwartz, *J. Chromatogr. Sci.*, **24**, 236 (1986).
49. K. E. Markides, S. M. Fields, and M. L. Lee, *J. Chromatogr. Sci.*, **24**, 254 (1986).
50. K. E. Markides, E. D. Lee, R. Bolick, and M. L. Lee, *Anal. Chem.*, **58**, 740 (1986).
51. B. W. Wright, H. R. Udseth, R D. Smith, and R. N. Hazlett, *J. Chromatogr.*, **314**, 253 (1984).
52. M. C. Paputa-Peck, R. S. Marano, D. Schuetzle, T. Riley, C. V. Hampton, T. J. Prater, L. M. Skewes, T. E. Jensen, P. H. Ruehle, L. Bosch, and W. P. Duncan, *Anal. Chem.*, **55**, 1946 (1983).
53. M. Moller, I. Hagen, and T. Ramdahl, *Mutat. Res.*, **157**, 149 (1985).
54. W. R. West and M. L. Lee, *J. High Resolut. Chromatogr.*, **9**, 161 (1986).
55. P. A. Patterson, *Chromatographia*, **16**, 107 (1982).
56. J. C. Fjeldsted, W. P. Jackson, P. A. Peaden, and M. L. Lee, *J. Chromatogr. Sci.*, **21**, 222 (1983).
57. S. M. Fields, K. E. Markides, and M. L. Lee, *Anal. Chem.*, **60**, 802 (1988).
58. C. R. Yonker and R. D. Smith, *J. Chromatogr.*, **361**, 25 (1986).
59. S. M. Fields, K. E. Markides, and M. L. Lee, *J. Chromatogr.*, **406**, 223 (1987).
60. C. R. Yonker and R. D. Smith, *Anal. Chem.*, **59**, 727 (1987).
61. M. Novotny, S. R. Springston, P. A. Peaden, J. C. Fjeldsted, and M. L. Lee, *Anal. Chem.*, **53**, 407A (1981).
62. P. A. Peaden, J. C. Fjeldsted, M. L. Lee, S. R. Springston, and M. Novotny, *Anal. Chem.*, **54**, 1090 (1982).

63. J. C. Fjeldsted, B. E. Richter, W. P. Jackson, and M. L. Lee, *J. Chromatogr.*, **279,** 423 (1983).

64. B. W. Wright, H. T. Kalinoski, H. R. Udseth, and R. D. Smith, in *2nd International Conference on Long Term Storage Stabilities of Liqid Fuels*, L. L. Stavinoha, Ed., Southwest Research Institute, San Antonio, Texas, p. 526 (1986).

65. B. W. Wright, H. R. Udseth, E. K. Chess, and R. D. Smith, *J. Chromatogr. Sci.*, **26,** 228 (1988).

66. W. P. Jackson, R. C. Kong, and M. L. Lee, in *Polynuclear Aromatic Hydrocarbons: Mechanisms, Methods, and Metabolism*, M. Cooke and A. J. Dennis, Eds., Battelle Press. Columbus, Ohio, p. 609 (1985).

67. R. D. Smith, H. T. Kalinoski, and H. R. Udseth, *Mass Spectrom. Rev.* **6,** 445 (1987).

68. R. D. Smith, J. C. Fjeldsted, and M. L. Lee, *J. Chromatogr.*, **247,** 231 (1982).

69. R. D. Smith, B. W. Wright, and H. R. Udseth, in *Chromatography and Separation Chemistry, Advances and Developments*, S. Ahuja, Ed., ACS Symposium Series, Vol. 297, American Chemical Society, Washington, D.C. p. 260 (1986).

70. L. G. Randall, *Sep. Sci. Technol.*, **17,** 1 (1982).

71. B. W. Wright, J. L. Fulton, A. J. Kopriva, and R. D. Smith, in *Supercritical Fluid Extraction and Chromatography, Techniques and Applications*, B. A. Charpentier and M. A. Sevenants, Eds., ACS Symposium Series, Vol. 366, American Chemical Society, Washington, D.C. Chapter 3 (1988).

72. B. W. Wright, C. W. Wright, R. W. Gale, and R. D. Smith, *Anal. Chem.*, **59,** 38 (1987).

73. B. W. Wright, S. R. Frye, D. G. McMinn, and R. D. Smith, *Anal. Chem.*, **59,** 640 (1987).

74. R. M. Campbell and M. L. Lee, *Anal. Chem.*, **58,** 2247 (1986).

75. M. Nishioka, D. G. Whiting, R. M. Campbell, and M. L. Lee, *Anal. Chem.*, **58,** 2251 (1986).

CHAPTER

5

MICELLE-MEDIATED METHODOLOGIES FOR THE PRECONCENTRATION AND SEPARATION OF POLYCYCLIC AROMATIC COMPOUNDS

WILLIE L. HINZE, H. N. SINGH,* ZHENG-SHENG FU,[†]
RONALD W. WILLIAMS, DONALD J. KIPPENBERGER,[‡]
MARTIN D. MORRIS,[‡] AND FAWZY S. SADEK,[§]

Department of Chemistry
Laboratory of Analytical Micellar Chemistry
Wake Forest University
Winston-Salem, North Carolina

1. INTRODUCTION

Polycyclic aromatic compounds and their derivatives (PACs) are ubiquitous environmental pollutants, and their concentrations must be closely monitored because many are known to be toxic and/or carcinogenic (1, 2). Reviews of the occurrence of PACs in the environment (e.g., in water, in soil, and in the atmosphere) have been published (2–5). In most of these complex environmental samples, the individual concentrations of the PACs are in the parts-per-billion (ppb) or sub-ppb levels (3–5). Therefore, the analyses of PACs in these situations dictate that a preconcentration extraction step and subsequent separation procedures be performed so that the PAC concentration is adequate for their detection and quantitation. The literature contains numerous protocols for the sampling, preconcentration, separation, and detection of PACs in environmental samples (2–6). The purpose of this chapter is to briefly outline and illustrate the potential use of novel micelle-mediated procedures as alternative approaches to the analysis of PACs, with emphasis given to

*Present address: Department of Chemistry, Aligarh Muslim University, Aligarh, India.

†Present address: Chemistry Department, Northwestern Teachers' College, Lanzhou City, Gansu Province, People's Republic of China.

‡Present address: U.S. Army Environmental Hygiene Agency, Aberdeen Proving Grounds, Maryland.

§Present address: Department of Physical Science, Winston-Salem State University, Columbia Heights, Winston-Salem, North Carolina.

151

their analyses in water-borne sample matrices. As will be discussed in later sections of this chapter, it is thought that use of micellar-mediated methodologies will offer the analyst several important advantages over current procedures for the separation and analysis of PACs.

Aqueous surfactant micelles are colloidal-sized organized structures that can form in aqueous media containing surfactant (surface active agents, i.e., soap or detergent) molecules at a concentration above their critical micelle concentration (CMC). Such aqueous normal micellar aggregates are typically composed of 40–140 individual surfactant molecules. The aggregates are roughly spherical in shape such that the surfactant hydrophilic headgroups are in contact with the bulk water while their long-chain alkyl hydrocarbon tails are oriented inward, forming a nonpolar central core region (for reviews, refer to refs. 7–10). There are four charge-type micellar systems possible based upon the nature of the surfactant molecule's hydrophilic headgroup. Table 1 summarizes some of the typical micelle-forming surfactants of each charge-type along with data on their micellar parameters in water at 25°C. More detailed information of such micellar systems, their aggregation behavior, and critical parameters are given in several excellent monographs (7–11).

Aqueous micellar systems are stable, optically transparent, photochemically inactive, inexpensive, and relatively nontoxic. In addition, they exhibit several unique properties that can be used advantageously in chemical separation and detection schemes. These include their ability to (a) solubilize, compartmentalize, and concentrate (or separate) solutes, (b) alter the local environment (i.e., fluidity, polarity, acidity, etc.) about associated solutes, (c) alter the position of equilibrium systems, and (d) alter photophysical and chemical pathways and rates (7–10). Although all of these micellar features can be exploited to aid the separation scientist in specific instances, the main basis for the successful utilization of aqueous micellar media in separations stems from the fact that they can differentially solubilize and incorporate a variety of solutes. The general use of micellar and related media in separation science has been the topic of several recent reviews (see refs. 12–14).

The purpose of this overview is to focus on the utilization of aqueous micellar media to facilitate analysis of polycyclic aromatic compounds. As will be detailed, the use of such micellar media seems to be ideal for the sampling, preconcentration, extraction, chromatographic separation, and detection of PACs found in aqueous sample matrices. Additionally, a brief section will be devoted to speculating on future developments in this field.

2. MICELLAR-FACILITATED SAMPLING CONSIDERATIONS

The first promising area of application of aqueous micellar media to PAC analyses concerns sampling. There are numerous literature references which

Table 1. Summary of Micelle-Forming Surfactants and Their Critical Micelle Parameters[a]

Surfactant Structure, Name, and Abbreviation	CMC (mM)	N^b
Nonionic		
$C_{12}H_{25}(OCH_2CH_2)_4OH$, polyoxyethylene(4)dodecanol $[C_{12}E_4]$	0.06	—
$C_{12}H_{25}(OCH_2CH_2)_{23}OH$, polyoxyethylene(23)dodecanol $[C_{12}E_{23}$ or Brij-35]	0.1	40
$(CH_3)_3CCH_2C(CH_3)_2C_6H_4(OCH_2CH_2)_{9.5}OH$, polyoxyethylene(9.5)-*p-t*-octylphenol [Triton X-100 or TX-100]	0.2	143
$C_9H_{19}C_6H_4(OCH_2CH_2)_{7.5}OH$, polyoxyethylene(7.5)-*p*-nonylphenol [PONPE-7.5]	0.12	—
$C_9H_{19}C_6H_4(OCH_2CH_2)_9OH$, polyoxyethylene(9)-*p*-nonylphenol [Igepal CO-630]	—	—
Zwitterionic		
$C_{12}H_{25}(CH_3)_2N^+CH_2CH_2CH_2SO_3^-$, *N*-dodecylsultaine [SB-12]	1.2	55
$C_{12}H_{25}(CH_3)_2N^+CH_2CO_2^-$, *N*-dodecylbetaine [DoDAA]	1.5	73
Cationic		
$C_{16}H_{33}N^+C_5H_5Cl^-$, cetylpyridinium chloride [CPC]	0.9	95
$C_{16}H_{33}N^+(CH_3)_3Br^-$, hexadecyltrimethylammonium bromide [CTAB]	0.9	61
$C_{16}H_{33}N^+(CH_3)_3Cl^-$, hexadecyltrimethylammonium chloride [CTAC]	1.3	78
Anionic		
$C_{10}H_{21}CO_2^-Na^+$, sodium laurate [NaL]	24.0	56
$C_{12}H_{25}OSO_3^-Na^+$, sodium dodecylsulfate [NaLS]	8.1	62

, sodium deoxycholate [NaDC]   2.8–6.4[c]   14[c]

[a] Micellar parameters given for aqueous medium at 25.0°C. Data taken from ref. 7–14.
[b] Aggregation number, that is, number of monomer surfactant molecules per micellar aggregate in water at 25°C with surfactant concentrations slightly above CMC.
[c] CMC and N are pH dependent (see ref. 15).

indicate that serious problems can occur when handling PAC solutions in which the PAC concentration is in a range of less than 1 ppb (3–5). That is, erroneous results can arise from losses of PAC in the sampling step (or subsequent steps in the assay) or during sample storage because the PAC can adsorb onto most surfaces with which it comes into contact. In addition, any particulate matter suspended in the sample solution may contain appreciable amounts of sorbed PAC (3–5).

Our preliminary data indicate that the rate of loss of PAC due to its adsorption onto container surfaces can be significantly decreased in the presence of aqueous micellar media (16). Figure 1 shows the relative fluorescence signal vs. time of an aqueous solution of benzo[*a*]pyrene (BaP) housed in polyethylene and pyrex containers. As can be seen, in aqueous solution alone, the fluorescence intensity decreases with time in both containers examined. After 90 min, the signal has decreased by over 83% because of the loss of BaP by adsorption onto the surface of the polyethylene container. However, addition of surfactant micellar species to the BaP sample slows or eliminates such losses caused by adsorption. For example, if the BaP solution contained micellar CTAC (2 g/liter), the rate of loss of BaPs fluorescence signal was much less, that is, only about 0.8% loss in 90 min if in a pyrex

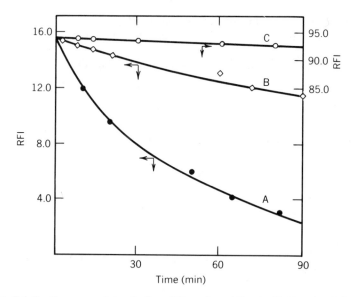

Figure 1. Relative fluorescence intensity ($\lambda_{ex} = 300$ nm, $\lambda_{em} = 410$ nm) of BaP vs. time (minutes) for BaP solutions in water alone stored in (*A*) a polyethylene bottle and (*B*) a pyrex beaker or for (*C*) BaP in an aqueous CTAC (2 g/liter) solution in a polyethylene bottle. The solutions were stirred at a constant rate throughout the experiment, and the temperature was $25 \pm 2°C$.

container (not shown in Figure 1) and only 5% if in a polyethylene container. Similar results have been observed for pyrene as analyte in CTAC micelles as well as in micellar NaLS and Triton X-100 media (16).

Additionally, the presence of surfactant micellar media has the potential to desorb PAC species from any suspended particulate matter found in the aqueous or water-related samples. This stems from the fact that the solubility of PACs is dramatically increased in aqueous micellar media because of the PACs' ability to partition, as well as to bind, the hydrophobic regions of the micellar entity (8). For instance, the molar solubility of 5, 9-dimethyldibenz[c, g]carbazole and 5, 7, 9-trimethyldibenz[c, g]carbazole is enhanced by more than four orders of magnitude in the presence of aqueous micellar NaLS, SB-12, or CTAB media compared to that in bulk water alone [water solubility $\sim (5.3$–$9.8) \times 10^{-9}$ M (see ref. 17)]. Although much more study is required, it appears that the incorporation of micellar systems is effective in preventing loss of PAC due to adsorption to the surface of sample containers or to any suspended particulate matter present in the original PAC-containing sample. If this proves to be the case, then current difficulties with sampling and storage of samples can be greatly reduced, if not entirely eliminated.

3. MICELLE-MEDIATED EXTRACTION/PRECONCENTRATION OF PACs

As previously mentioned, another problem with PAC analyses stems from the fact that in many groundwater samples, the individual PAC concentrations are in the range of 1 ppb or less. Thus, these low initial concentrations necessitate use of some sort of preconcentration/extraction step in order to raise the PAC concentrations to levels appropriate for the detection system being employed. Previously, liquid–liquid extraction procedures or use of solid adsorbents have been employed in order to preconcentrate PAC samples prior to further separation and analysis (4, 5). In this chapter, we discuss simpler alternative procedures based on either the critical behavior (cloud point) exhibited by neutral micellar solutions (i.e., nonionic or zwitterionic) or the selective solubilizing ability of all charge-type micelles for the extraction and preconcentration of PACs.

3.1. Cloud-Point Extractions

Aqueous solutions of neutral surfactant micellar compositions can exhibit critical phenomena upon temperature alteration. Nonionic surfactants such as the alkyl (or aryl) polyoxyethylene ethers (Brij-35, TX-100, $C_{12}E_4$, etc.; see

Table 1), *n*-alkylsulfinyl alcohols, dimethylalkylphosphine oxides, and so on, can exhibit so-called cloud-point behavior (8, 18). That is, upon heating the isotropic micellar solution, a critical temperature is eventually reached at which the solution suddenly becomes turbid (cloud point) because of the diminished solubility of the surfactant in water. After some time interval (which can be speeded up by centrifugation), demixing into two transparent liquid phases occurs (i.e., a surfactant micellar-rich phase in equilibrium with almost pure water). A plot of the clouding temperature vs. surfactant concentration data exhibits a minimum in the coexistence curve, with the temperature and concentration at which the minimum occurs referred to as the critical temperature and concentration of the surfactant, respectively. Table 2 lists the cloud-point temperatures for some nonionic surfactant micellar systems (8, 18). It should be noted that the presence of impurities/additives in water (such as salt, organic species, etc.) can significantly alter (i.e., raise or lower) the cloud-point temperature compared to that observed in pure water (Table 2) (8).

Table 2. Summary of the Cloud-Point Temperatures of Some Nonionic Surfactants[a]

Surfactant	Cloud Point (°C)
Polyoxyethylene(23)dodecanol, Brij-35	100
Polyoxyethylene(4)dodecanol, $C_{12}E_4$	4
Polyoxyethylene(7.5)-*p*-nonylphenol, PONPE-7.5	1–4
Polyoxyethylene(9.5)-*p*-*t*-octylphenol, TX-100	64
Polyoxyethylene(9)-*p*-nonylphenol, Igepal CO-630	48–50
Polyoxyethylene(4)decanol, $C_{10}E_4$	21

[a] Cloud-point temperatures for aqueous 1% surfactant solutions.

Based upon the use of such nonionic surfactant systems and their cloud-point phase separation behavior, it should be possible to design simple, practical, and efficient extraction/concentration schemes for PACs. Previously, a variety of metal ions have been extracted and preconcentrated from aqueous media using such an approach (12, 14, 19, 20). The reported general procedure for such metal-ion extractions consisted of adding an appropriate chelating agent to the aqueous solution which was specific for the metal ion to be separated so that a relatively insoluble metal–chelate complex formed. Next, the nonionic surfactant was added to solubilize the metal complex, the solution temperature was adjusted to above the cloud point, and the two phases were separated after centrifugation of the solution. Using this technique, extraction efficiencies for the metal ion in excess of 98% were achieved (19). Since only a very small amount of the nonionic surfactant is required to form micelles (due to their very low CMC values; see Table 1), high analyte preconcentration factors (i.e., in range of 20–100) were attainable (19).

A similar approach should also be applicable for the extraction and preconcentration of organic species that are sparingly soluble in water. We have preliminary data which demonstrate the potential of using such procedure for the extraction and preconcentration of PACs (16). For example, we have reported that a pesticide test mix containing 10 ppm each of endrin aldehyde, endrin, chloradane, endosultan, lindane, aldrin, BHC, DDT, methoxychlor, and chloropyrifos in water could be essentially quantitatively extracted into the surfactant-rich phase using PONPE-7.5. Analysis of the original aqueous medium after the nonionic micellar phase separation extraction revealed the presence of no pesticides (at least at the detection limit of an electron-capture detector using GC analysis). Similar results were obtained for the extraction and preconcentration of a PAC test mix containing benzo[a]pyrene, fluorene, fluoranthene, pyrene, and benzo[e]pyrene using Igepal CO-630 as the nonionic surfactant (16). Thus, this general approach shows considerable promise for the preconcentration/extraction of PACs prior to gas or liquid chromatographic analysis.

Potential advantages of such nonionic extractions/preconcentrations include simplicity, convenience, relatively small sample volumes required, safety (the aqueous nonionic surfactant solutions are less toxic, less volatile, and less flammable compared to the usual organic solvents employed in liquid–liquid extractions), low cost, and the ability to attain large concentration factors. For instance, in the two mentioned examples in the previous paragraph, the concentration factor attained was about 50–100 (i.e. 1.0 ml of the nonionic surfactant was added to 50 or 100 ml of the sample test mix). Additionally, as will be detailed in a later section, nonionic micellar solutions can also function as chromatographic mobile phases (12, 13, 21), thus ensuring compatibility of the extraction "solvent" with the mobile phase.

Obviously, much more work concerning the extraction efficiency, solute distribution, and effect of sample matrix is required before one can meaningfully compare and contrast this proposed cloud-point extraction procedure with the liquid extraction and/or solid adsorbant techniques currently being utilized for the preconcentration/extraction of PACs. Work in these areas is in progress in our laboratory.

3.2. Extractions Based on the Differential Solubilizing Ability of Micelles

Aqueous micellar media have previously been employed to extract a variety of biological components from cells, bacterial cultures, and solid matrices, among others (14, 22). The basis for these extractions stems from differences in the rates and degree to which different components associate with the micellar aggregate. To date, there have been no reports of using this approach to extract or concentrate PACs. However, the general method may be ideal for extracting PACs from a variety of solid matrices (soils, sludges, sediments, suspended particulate matter in water, etc.). In one preliminary study, we achieved average percent recoveries in the range of 75% for the extraction of anthracene, acenaphthene, phenanthrene, and chrysene from spiked soil samples using CTAC or Brij-35 micellar media as the extracting solvent (16). If the recovery efficiencies can be improved, then this approach to extract PACs from solid matrices may prove beneficial because of relatively short processing time and ease of use.

Although not the central focus of this chapter, it is worth mentioning that aqueous micellar systems have also been employed to extract specific organic species from hydrocarbon matrices (13, 14). The experimental procedure is simple in that a volume of an aqueous micellar solution is mixed with the organic sample containing components to be separated. Because different solutes in the organic mixture exhibit different rates of solubilization into the aqueous micellar systems, the micelle will become enriched in one component at the expense of the other. In many instances, the incorporation of the organic component(s) into the aqueous micelle may lead to microemulsion formation (10, 14, 23).

Most of the work to date in this area has concerned model studies employing simple binary mixtures of PACs (24–26). Results from such studies indicate that the technique may prove to be useful in separating aromatic or cyclic hydrocarbons from noncyclic hydrocarbons (14). Selectivity factors approaching 10 were reported for the extraction of benzene from hexane using aqueous micellar CPC or NaLS (24, 25). In a more practical application, the technique has proven to be effective in recovering phenolic compounds from carbolic oils (27, 28). Most recently, the technique has been successfully employed to separate isomeric and nonisomeric PACs in close boiling solid mixtures (29). For instance, aqueous solutions of CPC, CTAC, and CTAB were reportedly employed to separate o-nitrochlorobenzene from p-nitrochlorobenzene, 2,4-dichlorophenol from 2,4,6-trichlorophenol, and o-nitrotoluene from p-nitrotoluene, among others.

In a related application, aqueous solutions of micelles can potentially be employed as a scrubbing extraction solvent for the removal of PACs in gas, vapor, or aerosol matrices. There is one recent report concerning the continuous absorption of toluene vapor by use of micellar Aerosol OT [i.e., sodium bis(2-ethylhexyl)sulfosuccinate] or microemulsions involving Aerosol OT (30). Again, as in most of this area concerning the use of micellar media for the extraction/separation of PACs, much more work needs to be done to refine and demonstrate the advantages and limitations of the particular approach mentioned. However, the use of aqueous micellar media in this area appears very promising for some specific separation problems involving PACs.

4.   MICELLAR-ENHANCED ULTRAFILTRATION

Whereas traditional ultrafiltration is ineffective at removing dissolved low-molecular-weight PACs from aqueous streams, a recently developed technique, called micelle-enhanced ultrafiltration (MEUF) (31, 32), has been demonstrated to be effective. This modified ultrafiltration technique is based on the fact that once PACs partition and bind to the micellar aggregates present in the aqueous solution, they (along with the unoccupied micelles) cannot pass through an ultrafiltration membrane of appropriate pore size. Membranes of pore size $\leq 10,000$ daltons prevent the passage of micellar assemblies. Consequently, the retentate contains the micellar aggregates plus the micellar-bound PACs, whereas the permeate contains only a very small concentration of the unsolubilized PAC as well as surfactant monomers (surfactant present at a concentration equal to its CMC value). An excellent description of the theory, salient features, and details of this MEUF technique is given elsewhere (32). The general technique has been shown to be successful in removing such PACs as cresols, benzene, and phenols as well as alcohols and toxic ions from aqueous streams via use of NaLS or CPC as the micellar systems (31–34). Rejection efficiencies of $> 99\%$ have been reported using MEUF for the removal of 4-*tert*-butylphenol from wastewater (35). Work to date indicates that this method can be utilized to perform the large-scale removal of PAC molecules from aqueous streams (32, 35).

5.   MICELLAR SYSTEMS IN CHROMATOGRAPHIC SEPARATION AND DETECTION SCHEMES

A number of analytical techniques are available for the analysis of PACs in water samples. The most commonly encountered procedures consist of some sort of chromatographic separation (TLC, HPLC, etc.) followed with spectroscopic detection. A number of review articles summarize the current status of such conventional methods with respect to the separation and detection of PACs (3–6). This section will outline and give some examples of the use of micellar media and their potential in this area. Specifically, a description of micellar electrokinetic capillary chromatography, micellar liquid chromatography, and micellar-mediated chromatographic detection, along with a summary of their utilization in the separation and detection of PACs will be presented.

5.1. Micellar Electrokinetic Capillary Chromatography

This technique was first reported by Terabe et al. (36) and is the most recent and fastest developing micellar-mediated separation procedure (14). In our opinion, this technique holds the greatest potential with respect to being able to resolve complex mixtures of PACs. Micellar electrokinetic capillary chromatography (MECC) is an extremely efficient separation technique in which solutes in a mixture are separated based on their differential distribution between an electroosmotically pumped bulk aqueous mobile phase and a slower moving, electrophoretically retarded, ionic micellar phase (which acts as a "pseudostationary phase"). Recent reports have described in detail the fundamental characteristics, experimental considerations, detection modes, and factors that influence retention in MECC (36–45). In addition, the topic has been recently reviewed (12, 41).

MECC has been utilized to separate a variety of classes of PACs. Table 3 summarizes the PAC class types and micellar systems that have been utilized in successful MECC separations. In all of these reported separations, the efficiencies obtained have been spectacular, exceeding 100,000 theoretical plates per meter in all cases. Current research with MECC has focused on means to extend the elution range possible (44) and on expanding the number of detection modes compatible with MECC (45). Based on literature reports to date, MECC seems to be ideally suited for the efficient separation of many

Table 3. Summary of Successful Separations of Different Classes of PACs with MECC

Class-Type PACs Separated	Micellar Phase Employed	Ref.
Aromatic sulfides	Aq. 0.05 M NaLS or Aq. 0.03 M NaLS containing 20% MeOH, pH 7	42
Aromatic test mix	Aq. 0.50 M NaLS, Aq. 0.50 DTAC[a], or Aq. 0.05 M NaTS[b]	38
Isomeric chlorophenols	Aq. 0.10 M NaLS, pH 7	43
Nitroaromatic compounds	Aq. 0.01 M NaLS, pH 7	41
Phenols	Aq. 0.05 M NaLS, pH 7	37
Xylenols	Aq. 0.05 M NaLS, pH 7	37

[a] DTAC represents dodecyltrimethylammonium chloride.
[b] NaTS represents sodium decylsulfate.

classes of PACs. In addition, the sampling and extraction micellar systems previously discussed are compatible with this separation technique.

5.2. Micellar Liquid Chromatography

The technique of micellar liquid chromatography (MLC), in which surfactants are added to aqueous mobile-phase solutions at concentrations at or above their CMC values, was first introduced by Armstrong and co-workers (12, 46). Since then, the technique has gained popularity and is being increasingly employed in TLC, HPLC, gel filtration, and column chromatographic separations. This general area has been the subject of several recent review articles (12–14, 47–49); thus the present account will be brief and will focus on separations of PACs done to date using MLC.

In MLC, solutes present in a mixture can partition between the bulk aqueous and surfactant-modified stationary phase (P_{ws}) and between the bulk aqueous and micellar aggregate present in the mobile phase (P_{wm}). In addition, for water-insoluble solutes, partitioning can occur directly between the micelle–solute species and the surfactant-coated stationary phase (P_{ms}) (see ref. 62). The theory of MLC has been worked out for the different LC modes (12, 50). The fundamental relationship (for HPLC) relating retention behavior (k') and the surfactant micelle concentration (C_m) is given in equation (1):

$$\frac{1}{k'} = \frac{1}{\theta} \left[\frac{(K_b) C_m}{P_{ws}} + \frac{1}{P_{ws}} \right] \tag{1}$$

where k' is the capacity factor, θ is the phase ratio, K_b is the micelle–solute binding constant, and C_m is the micelle concentration which, in turn is equal to the total surfactant concentration, C_T, minus the CMC value, all divided by the micelle aggregation number, N (i.e., $C_m = [C_T - CMC]/N$) (see ref. 50). As can be seen, unlike conventional HPLC using hydroorganic mobile phases, there are two partition processes that govern the separation. Consequently, micellar media can be viewed as a sort of "super" secondary equilibria system for LC. However, it is more than just that since micellar systems can bind a variety of solutes as a result of favorable electrostatic, pi, H-bonding, hydrophobic, or combination of such interactions, between the solute and the micellar entity (12). Thus, such mobile phases are much more versatile than any of the conventional mobile-phase systems.

Many different classes of PAC as well as other compounds have been successfully separated using MLC (12–14). Table 4 summarizes some of the types of PACs that have been reportedly separated using this technique, and it also gives information on the chromatographic systems employed (51–63).

Table 4. Summary of PAC Separations Reported Using MLC

Class of PACs Separated (Ref.)	Micellar Mobile Phase Employed	Stationary Phase	Mode[a]
Phenols (51–53)	Aq. NaLS (gradient), pH 2.5	Ultrasphere C_8	HPLC
	Aq. NaLS	MicroPak MCH-10	HPLC
	Aq. NaLS or Aq. CTAB, with 8% added PrOH	Whatman No. 3 paper strips	PC
Hydroxybenzenes (i.e., phenols, quinols, catechols) (54)	Aq. NaLS	C-18 RP	HPLC
Polycyclic aromatic hydrocarbons (55–57)	Aq. NaLS or Aq. CTAB	Alumina or polyamide	TLC
	Aq. 0.20 M NaLS	C-8 silica	HPLC
	Aq. NaLS	MicroPak MCH-10	HPLC
Alkylbenzenes (58)	Aq. NaLS with 2% added BuOH	Hypersil silica	GF
Pesticides (59–60)	Aq. NaLS or Aq. CTAB	Polyamide	TLC
	Aq. NaLS with 10% added PrOH	Cyano spheri-10	HPLC
s-Triazine herbicides (60)	Aq. 0.10 M NaLS with 10% added PrOH	C_8 spheri-10	HPLC
Thiols, nitrosoamines, aromatic quinones (16)	Aq. NaLS, Aq. NaDC, or Aq. CTAC with 2–5% added amyl alcohol	C-18 or C-8 RP polyamide	HPLC TLC
Mycotoxins (61)	Aq. NaLS	Polyamide, RP, or alumina	TLC
Benzene derivatives (63)	Aq. SB-12, phosphate buffer	RP-18	HPLC

[a] PC = paper chromatography, GF = gel filtration, TLC = thin-layer chromatography.

Although good separations were reported, one major problem initially encountered using MLC, especially when separating relatively hydrophobic PACs, is that of decreased chromatographic efficiency (52, 64, 65). Subsequent work showed that some improvement in efficiency could be achieved if small amounts (i.e., $\leq 10\%$) of short-chain alcohols were added to the micellar solutions (64) and/or if the separation was performed at elevated temperature

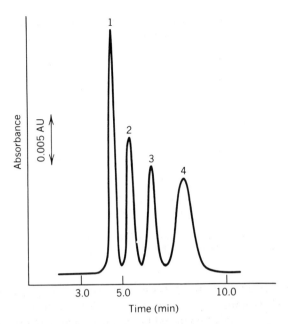

Figure 2. Trace of a chromatogram showing the separation of an aromatic quinone test mix containing: (1) anthraquinone, (2) 2-methylanthraquinone, (3) 2-ethylanthraquinone, and (4) 2-*t*-butylanthraquinine with a 0.285 M NaLS micellar mobile phase containing 5% added amyl alcohol on a 10-cm C-18 column. Conditions: mobile-phase flow rate = 1 ml/min; temperature 23°C; solute concentrations were 10 mg/ml in acetone; and absorbance was monitored at 254 nm.

(~ 40°C) (65). Although this recipe works well for PACs of intermediate polarity, the efficiency is still much too poor for the resolution of individual PACs in complex mixtures (16). In our own work, we have found that the addition of 1–5% of a longer-chain alcohol, such as amyl alcohol, is more effective in increasing the efficiency (refer to data presented in Table 5). Figure 2 shows the baseline resolution of an aromatic quinone mixture utilizing a micellar NaLS mobile phase containing 5% added amyl alcohol. Consequently, we strongly recommend that amyl alcohol be added to micellar mobile phases in order to enhance the efficiency. This is particularly important when chromatographing very hydrophobic PACs (16).

With respect to the separation of PACs using MLC, we have also found that bile salt micelles appear to be superior to that of the conventional aqueous micellar systems that were described in the introduction. Compared to the normal, long-chain surfactants employed in most previous analytical micellar work, bile salt amphiphilic molecules are different in that they are chiral, possess a hydrophobic as well as hydrophilic face, and undergo a

Table 5. Effect of Added Alcohols upon Chromatographic Efficiency[a]

Additive to Mobile Phase	Efficiency
None[b]	150
5% added methanol	180
5% added ethanol	320
5% added n-propanol	970
5% added n-butanol	2175
5% added n-pentanol	2500
5% added triethanolamine	1870

[a] Solute was phenanthrene; temperature was 30°C; data were taken from ref. 16.
[b] Micellar mobile phase consisted of aqueous 0.305 M NaLS; flow rate was 0.8 ml/min; 10-cm C-18 column.

different type of aggregational process. That is, their aggregational process can be viewed as consisting of the stepwise formation of initial primary micelles composed of only two to eight monomers held together by hydrophobic interactions between the bile salt molecule's nonpolar faces. Next, at higher bile salt concentration, these primary micelles can further aggregate to form larger, rod-shaped secondary micellar structures as a result of intermicellar hydrogen bonding between their hydroxyl groups (14). Since these micellar systems possess planar, hydrophobic regions, they appear to be ideal candidates for use as mobile phases in the MLC separation of PAC molecules, which are also planar. Indeed, our preliminary data support this expectation. For example, an aqueous 0.15 M NaDC bile salt micellar mobile phase containing 2.5% amyl alcohol in conjunction with a C-18 RP column nicely afforded the separation of a series of polyaromatic compounds [HPLC retention times of 6.4, 7.8, 9.0, and 10.0 min for pyrene, anthracene, benzene, and naphthalene, respectively (16)].

Techniques currently under development in this area which may eventually prove useful in the separation of PACs include the use of appropriate micellar media in field flow fractionation as well as in supercritical fluid chromatography (66, 67). Apart from the use of micellar chromatography to separate PACs, recent work indicates that octanol/water partition coefficients for PACs correlate well with hydrophobicity measurements obtained by MLC (68, 69). Consequently, this has the potential of offering a more convenient means to collect hydrophobicity data and establish linear quantitative PAC structure–activity relationships (i.e., correlations to predict toxicity, carcinogenicity, etc.) (69).

5.3. Micellar-Enhanced Detection

The success of any separation process ultimately depends upon the detection or location of the separated PAC components. Different surfactant micellar media have been successfully utilized to develop new detection modes as well as to greatly enhance existing schemes (13, 14). Most applications involving surfactant micelles have concerned spectroscopic detection following the HPLC or TLC separation of PACs.

The presence of appropriate aqueous micellar media can lead to enhanced luminescence (fluorescence, room-temperature liquid phosphorescence, chemiluminescence) (70–73). These micelle-mediated detection techniques have been successfully utilized in the HPLC detection of some PACs (73–75). In these applications, the required micellar assembly can be added post column (74) or can be present as part of the chromatographic mobile phase (73). The detection of PACs following TLC separation has also been shown to be greatly enhanced if the TLC plates are sprayed with micellar reagents prior to location and quantitation of the components using fluorescence (76). Table 6 presents some data on this latter application, which illustrates the enhancements possible depending upon the stationary-phase–micellar-spray-reagent combination employed.

A micellar spray reagent that functions to quench the native luminescence of PACs has been described for use in Raman spectroscopic detection modes (77). In the past, such Raman detection was difficult because the more intense fluorescence signals typically obscured the weaker Raman bands.

Table 6. Relative Fluorescence Enhancements Observed for PACs on Different TLC Plates Sprayed with Micellar Reagents

Stationary Phase	Enhancements in Fluorescence of PAC in the Presence of the Following Micellar Spray Reagents[a]:			
	CTAC	NaLS	NaC[b]	SB-12
Silica gel	3–20	2.3–27	3–47	4.7–47
Alumina	2.5–7.9	2–39	3.3–30	3.7–27
Cellulose	1.2–2	1.4–2.2	1–1.1	1–1.1
C-18 RP	0.8–1.1	1–1.2	1.1–1.3	1–1.1
Polyamide	0.9–1.2	1.2–2.1	1.1	0.9–1.0

[a] PAC test solutes included: pyrene, benzanthracene, benzo[a]pyrene, and benzo[e]pyrene. Data are taken from ref. 77.
[b] NaC refers to the bile salt micelle, sodium cholate (14).

Another application has been in the area of amperometric electrochemical detection (78). Previously, this form of electrochemical detection had been very difficult to employ in gradient elution liquid chromatography as a result of the baseline shifts caused by the changing mobile-phase composition. However, when utilizing micellar mobile phases in MLC, the ionic strength can be easily kept constant, which facilitates this particular detection mode. The approach has been successfully demonstrated for the detection of phenolic compounds following their separation in MLC using an NaLS gradient (78).

The most recent potential application in this area of micellar-enhanced chromatographic detection concerns thermal lensing detection (79). The continued successful development of this micellar-enhanced detection system would prove to be ideal for the detection of PACs separated via use of supercritical micellar mobile phases in supercritical fluid chromatography (67).

6. CONCLUSIONS

As has been identified in this overview, there are various areas in separation science where the unique properties and features of aqueous micellar media can be potentially advantageously employed to facilitate analysis of PACs. These include sampling, preconcentration, extraction, and chromatographic separation and detection of these compounds. However, the field is in its infancy, and much more work is required before the full extent of their usefulness for analysis of PACs is known. Future work should concentrate on the continued development and optimization of the techniques mentioned as well as on systematic evaluation of their potential for the analysis of PAC in real environmental and energy-related samples. It is thought that the successful development of micellar-mediated procedures will yield techniques that are superior to many existing procedures currently employed in the separation and analysis of different classes of PAC.

ACKNOWLEDGMENTS

The support of the National Science Foundation (CHE-8215508), The Petroleum Research Fund, The Department of Energy (DE-FG22-85PC80542), and Wake Forest University through a Research and Publication Fund Grant is gratefully acknowledged. This chapter was presented, in part, at the 191st National Meeting of the American Chemical Society, New York City, April 17, 1986 (Abstract No. ANYL 117).

REFERENCES

1. R. G. Harvey, Ed., *Polycyclic Hydrocarbons and Carcinogenesis*, American Chemical Society, New York (1985), and references therein.
2. M. Cooke and A. J. Dennis, Eds., *Polynuclear Aromatic Hydrocarbons: Mechanisms, Methods and Metabolism*, Battelle Press, Columbus, Ohio (1985).
3. A. Bjorseth, Ed., *Handbook of Polycyclic Aromatic Hydrocarbons*, Marcel Dekker, New York (1983), and references therein.
4. D. J. Futoma, S. R. Smith, T. E. Smith, and J. Tanaka, *Polycyclic Aromatic Hydrocarbons in Water Systems*, CRC Press, Boca Raton, Fla. (1981).
5. M. L. Lee, M. V. Novotny, and K. D. Bartle, *Analytical Chemistry of Polycyclic Aromatic Compounds*, Academic Press, New York (1981), and references therein.
6. Committee D-19 on Water, *Annual Book of ASTM Standards, Part 31*, Water, American Society for Testing and Materials, Philadelphia (1981).
7. J. H. Fendler and E. J. Fendler, *Catalysis in Micellar and Macromolecular Systems*, Academic Press, New York (1975).
8. D. Attwood and A. T. Florence, *Surfactant Systems*, Chapman and Hall, London (1983), and references therein.
9. M. J. Rosen, *Surfactants and Interfacial Phenomena*, Wiley-Interscience, New York (1978).
10. J. H. Fendler, *Membrane Mimetic Chemistry*, Wiley-Interscience, New York (1982), and references therein.
11. P. Mukerjee and K. Mysels, *Critical Micelle Concentrations of Aqueous Surfactant Systems*, National Standards Reference Data Series, Vol. 36, National Bureau of Standards, Washington, D.C. (1971), and references therein.
12. D. W. Armstrong, *Sep. Purif. Methods*, **14,** 212 (1985).
13. W. L. Hinze, in *Colloids and Surfactants: Fundamentals and Applications*, E. Barni and E. Pelizzetti, Eds., Society Chimica Italiana, Rome, pp. 167–207 (1987).
14. W. L. Hinze, in *Ordered Media in Chemical Separations*, W. L. Hinze and D. W. Armstrong, Eds., American Chemical Society, Washington D.C., pp. 2–82 (1987), and references therein.
15. D. M. Small, in *The Bile Acids*, P. P. Nair and D. Kritchevsky, Eds., Plenum Press, New York, Chapter 8 (1971), and references therein.
16. W. L. Hinze, F. S. Sadek, Z. S. Fu, R. W. Williams, L. Guiney, T. E. Riehl, K. N. Thimmaiah, and D. Y. Pharr, Unpublished data, 1979–1988.
17. R. Drasnoshekova, M. Gubergrits, P. Jacquignon, and F. Perin, in *Polynuclear Aromatic Hydrocarbons: Mechanisms, Methods and Metabolism*, M. Cooke and A. J. Dennis, Eds., Battelle Press, Columbus, Ohio, p. 763 (1985).
18. V. DeGiorgio, in *Physics of Amphiphiles: Micelles, Vesicles and Microemulsions*, V. DeGiorgio and M. Coti, Eds., Soc. Italiana di Fisica, Bologna, Italy, pp. 303–335 (1985).
19. H. Watanabe, in *Solution Behavior of Surfactants*, Vol. 2, K. L. Mittal and E. J. Fendler, Eds., Plenum Press, New York, p. 1305 (1982).
20. E. Pramauro, C. Minero, and E. Pelizzetti, in *Ordered Media in Chemical Separations*, W. L. Hinze and D. W. Armstrong, Eds., American Chemical Society, Washington, D.C., pp. 152–161 (1987).

21. M. F. Borgerding and W. L. Hinze, *Anal. Chem.*, **57**, 2183 (1985).
22. A. Helenius and K. Simons, *Biochim. Biophys. Acta*, **415**, 29–79 (1975).
23. K. Shinoda and S. Friberg, *Emulsions and Solubilization*, Wiley-Interscience, New York (1986).
24. M. A. Chaiko, R. Nagarajan, and E. Ruckenstein, *J. Colloid Interface Sci.*, **99**, 168 (1984).
25. R. Nagarajan and E. Ruckenstein, *Sep. Sci. Technol.*, **20**, 285 (1985).
26. B. G. C. O'Rourke, A. J. I. Ward, and B. J. Carroll, *J. Pharm. Pharmacol.*, **39**, 865 (1987).
27. P. Plucinski, *J. Heat Mass Transfer*, **28**, 451 (1985).
28. P. K. Plucinski, *Comun. Joun. Com. Esp. Deterg.*, **16**, 481 (1985).
29. A. Mahapatra and M. M. Sharma, *Solvent Extr. Ion Exch.*, **5**, 781 (1987).
30. L. M. Gan, C. H. Chew, M. K. Wong, L. L. Koh, and K. H. Ng, *J. Dispersion Sci. Technol.*, **8**, 385 (1987).
31. G. A. Smith, S. D. Christian, E. E. Tucker, and J. F. Scamehorn, in *Ordered Media in Chemical Separations*, W. L. Hinze and D. W. Armstrong, Eds., American Chemical Society, Washington, D.C., pp. 184–198 (1987).
32. R. O. Dunn, J. F. Scamehorn, and S. D. Christian, *Sep. Sci. Technol.*, **20**, 257–284 (1985).
33. L. L. Gibbs, J. F. Scamehorn, and S. D. Christian, *J. Membrane Sci.*, **30**, 67 (1987).
34. J. F. Scamehorn, R. T. Ellington, S. D. Christian, B. W. Penney, R. O. Dunn, and S. N. Bhat, *AIChE Symp. Ser.*, **82**, 48 (1986).
35. R. O. Dunn, J. F. Scamehorn, and S. D. Christian, *Sep. Sci. Technol.*, **22**, 763 (1987).
36. S. Terabe, K. Otsuka, K. Ichikawa, A. Tsuchiya, and T. Ando, *Anal. Chem.*, **56**, 111 (1984).
37. S. Terabe, K. Otsuka, and T. Ando, *Anal. Chem.*, **57**, 834 (1985).
38. D. E. Burton, M. J. Sepaniak, and M. P. Maskarinec, *J. Chromatogr. Sci.*, **25**, 514 (1987).
39. M. J. Sepaniak and R. O. Cole, *Anal. Chem.*, **59**, 472 (1987).
40. Z. Prusik, V. Kasicka, S. Stanek, G. Kuncova, M. Hayer, and J. Vrkoc, *J. Chromatogr.*, **390**, 87 (1987).
41. M. J. Sepaniak, D. E. Burton, and M. P. Maskarinec, in *Ordered Media in Chemical Separations*, W. L. Hinze and D. W. Armstrong, Eds., American Chemical Society, Washington, D.C., pp. 142–151 (1987).
42. K. Otsuka, S. Terabe, and T. Ando, *Nippon Kagaku Kaishi*, **7**, 950 (1986).
43. K. Otsuka, S. Terabe, and T. ando, *J. Chromatogr.*, **348**, 39 (1985).
44. A. T. Balchunas and M. J. Sepaniak, *Anal. Chem.*, **59**, 1470 (1987).
45. R. A. Wallingford and A. G. Ewing, *Anal. Chem.*, **60**, 258 (1988).
46. D. W. Armstrong and J. H. Fendler, *Biochim. Biophys. Acta*, **478**, 75 (1977).
47. D. W. Armstrong, *Am. Lab.*, **13**, 14 (1981).
48. J. G. Dorsey, *Chromatogr.*, **2**, 13 (1987).
49. E. Pelizzetti and E. Pramauro, *Anal. Chim. Acta*, **169**, 1 (1985).
50. D. W. Armstrong and F. Nome, *Anal. Chem.*, **53**, 1662 (1981).
51. M. G. Khaledi and J. G. Dorsey, *Anal. Chem.*, **57**, 2190 (1985).
52. D. W. Armstrong and S. J. Henry, *J. Liq. Chromatogr.*, **3**, 657 (1980).
53. J. P. Rawat and O. Singh, *J. Indian Chem. Soc.*, **63**, 248 (1986).

54. E. Pramauro and E. Pelizzetti, *Anal. Chim. Acta*, **154,** 153 (1983).

55. D. W. Armstrong and R. Q. Terrill, *Anal. Chem.*, **51,** 2160 (1979).

56. D. W. Armstrong, *J. Liq. Chromatogr.*, **3,** 895 (1980).

57. S. Ji, *Fenxi Huaxue*, **13,** 660 (1985).

58. J. N. LePage and E. M. Rocha, *Abstracts of Papers*, 1982 Pittsburgh Conference, Abstract No. 458 (1982).

59. H. Liu, J. Ling, and X. Yan, *Zhejiang Gongxueyuan Xuebao*, **29,** 39 (1985).

60. N. P. Cege, *Micellar Chromatography, Optimization and Use*, M. S. Thesis, Michigan State University, 1985.

61. H. M. Stahr and M. Domoto, in *Planar Chromatography*, R. Kaiser, Ed., Springer-Verlag, New York (1985).

62. M. F. Borgerding, W. L. Hinze, F. H. Quina, H. McNair, and J. Bowermaster, *Anal. Chem.* (1988), submitted for publication.

63. J. P. Berry and S. G. Weber, *J. Chromatogr. Sci.*, **25,** 307 (1987).

64. J. G. Dorsey, M. T. DeEchegaray, and J. S. Landy, *Anal. Chem.*, **55,** 924 (1983).

65. P. Yarmchuk, R. Weinberger, R. F. Hirsch, and L. J. Cline Love, *J. Chromatogr.*, **283,** 47 (1984).

66. A. Berthod, D. W. Armstrong, M. N. Myers, and J. C. Giddings, *Anal. Chem.*, **60,** (1988), submitted for publication.

67. R. D. Smith, R. W. Gale, and J. L. Fulton, *LC/GC Magazine*, **6,** 134 (1988).

68. R. Graglia, E. Pramauro, and E. Pelizzetti, *Ann. Chim. (Rome)*, **74,** 41 (1984).

69. F. Gago, J. Alvarez-Builla, J. Elguero, and J. C. Diez-Masa, *Anal. Chem.*, **59,** 921 (1987).

70. W. L. Hinze, H. N. Singh, Y. Baba, and N. G. Harvey, *Trends Anal. Chem.*, **3,** 193 (1984).

71. T. E. Riehl, C. L. Malehorn, and W. L. Hinze, *Analyst*, **111,** 931 (1986).

72. L. J. Cline Love, M. Skrilec, and J. G. Habarta, *Anal. Chem.*, **52,** 754 (1980).

73. D. W. Armstrong, W. L. Hinze, K. H. Bui, and H. N. Singh, *Anal. Lett.*, **14,** 1659 (1981).

74. R. Weinberger, P. Yarmchuk, and L. J. Cline Love, *Anal. Chem.*, **54,** 1552 (1982).

75. P. E. Nelson, *J. Chromatogr.*, **59,** 59 (1984).

76. A. Alak, E. Heilweil, W. L. Hinze, and D. W. Armstrong, *J. Liq. Chromatogr.*, **7,** 1273 (1984).

77. D. W. Armstrong, L. A. Spino, T. Vo-Dinh, and A. Alak, *Spectroscopy*, **2,** 54 (1987).

78. M. G. Khaledi and J. D. Dorsey, *Anal. Chem.*, **57,** 2190 (1985).

79. C. D. Tran, *Anal. Chem.*, **60,** 182 (1988).

CHAPTER

6

ULTRAVIOLET ABSORPTION AND LUMINESCENCE SPECTROMETRY: AN OVERVIEW OF RECENT DEVELOPMENTS

M. ZANDER

Rütgerswerke AG
Castrop-Rauxel
Federal Republic of Germany

1. INTRODUCTION

It is not the aim of this chapter to give a detailed description of all more recently developed ultraviolet (UV) absorption and luminescence spectroscopical techniques as proposed for, or already applied in, polycyclic aromatic compound (PAC) analysis. Instead, some general features of these developments will be discussed, and methods will be compared with regard to their usefulness in PAC analysis.

The most efficient strategies in analysis depend strongly on the particular problems to be solved. Thus the specific role of electronic spectroscopical methods in PAC analysis will be discussed in some detail (Section 4).

Throughout the chapter, nomenclature as recommended by IUPAC for molecular absorption (1) and luminescence spectroscopy (2) will be used.

2. UV ABSORPTION SPECTROMETRY IN PAC ANALYSIS

2.1. Identification and Quantitative Determination

It has long been known that polycyclic aromatic hydrocarbons (PAHs) and structurally related heteroaromatic systems have characteristic, in most cases well-structured, UV/visible absorption spectra which are particularly useful for identification purposes. A simple method to further enhance the information content (i.e., number of absorption bands) of these spectra, and hence to improve their usefulness for PAC identification, is low-temperature UV/visible absorption spectrometry in solid matrices (3,4). Particularly useful is the application of Shpol'skii-type solvents at very low temperatures (e.g., 4 K). An

171

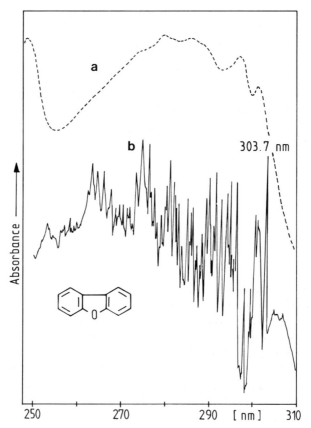

Figure 1. UV absorption spectra of dibenzo[b, d]furan in n-hexane at room temperature (curve a) and 4 K (curve b) (according to ref. 5).

example is given in Figure 1. The curves are the UV/visible absorption spectra of dibenzo[b,d]furan measured in n-hexane at room temperature (curve a) and in the frozen solvent at 4 K (curve b) (5). The half-width of the band marked in spectrum b is 1.7 cm^{-1}.

Derivative spectroscopy is another method for obtaining spectra that are characterized by a large number of signals and hence are useful for identification purposes (6). The first (second, third, etc.) derivative absorption spectrum of a molecule is defined as the first (second, third, etc.) derivative, $dA (\tilde{v})/d\tilde{v}$ $(d^2 A(\tilde{v})/d\tilde{v}^2, \ldots)$ of the absorbance A as a function of wavenumber or wavelength. Although the use of derivative spectroscopy has the disadvantage of decreasing the signal/noise ratio, it improves the detection of spectral bands in a broad background, or of shoulders, on a broad main band.

By combining the low-temperature and derivative spectroscopy technique, it is possible to generate very characteristic curves (fingerprints) that can be used for the recognition of complex PAC mixtures (e.g., in environmental samples). Figure 2 gives an example. Curve *a* is the short-wavelength region of the absorption spectrum of a coal-tar fraction measured in fluid solution (ethanol) at room temperature, and curve *b* is the long-wavelength region. Curves *e* and *f*, respectively, are the corresponding spectra measured at 95 K, while curves *c* and *d* (room temperature) and *g* and *h* (95 K) are the second-derivative spectra of the sample (7). As can be clearly seen, the number of signals exploitable for product characterization increases strongly in this order.

Although UV absorption spectra are useful for PAC identification purposes, absorption spectroscopy suffers from rather low sensitivity. This disadvantage can be surmounted by measuring luminescence (fluorescence or phosphorescence) excitation instead of absorption spectra. Since most commercially available luminescence spectrometers allow the measurement of

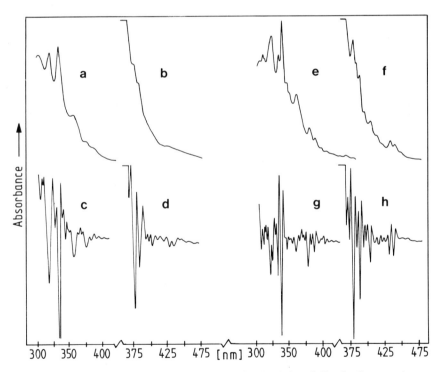

Figure 2. UV absorption spectra of a coal-tar fraction (ethanol) (for details see text).

energy-corrected excitation spectra, these can be regarded as an easily accessible and powerful means for the identification of PACs (8).

Probably the most important role of UV absorption spectrometry in modern PAC analysis is its usage as a detection and identification method in liquid chromatography. On-line coupling of high-pressure liquid chromatography (HPLC) and UV absorption spectrometry was introduced in 1976 (9). It proved to be a powerful method for the analysis of complex PAC mixtures (e.g., coal-tar pitch) (10). In both the original and subsequent versions, scanning spectrometers are used for spectra registration. However, since registration times are rather long with this type of equipment, the flow has to be interrupted for spectra measurement ("stop-and-go" method); hence the method is rather time-consuming. This disadvantage can be surmounted by using photodiode array detectors (PDA) for spectra registration. The very fast PDA detectors, some of which can acquire a complete spectrum in as little as 0.01 s, have no trouble obtaining spectra in real time, without any disturbance of the eluent flow. By means of the optical system of a PDA detector, source radiation is focused through a flow cell at the end of the LC column, and the emerging radiation is dispersed by a grating onto the linear PDA. A PDA is a series of light-sensitive elements etched onto a silicon chip. The elements work in parallel to simultaneously monitor a range of wavelengths spread across the face of the chip. Incident photons generate a charge that is stored on individual diodes. The accumulated charges are then switched sequentially by shift registers to form the detector output (11, 12).

Recently, supercritical fluid chromatography (SFC) coupled with a PDA detector was used for the analysis of PAHs in extracts of diesel particulate matter (13). Thirteen different PAHs could be easily identified by absorption spectra comparison. Since SFC has several advantages compared to gas chromatography and HPLC, the approach has a high potential for analysis of various kinds of complex PAC mixtures.

Although UV absorption spectrometry can be used for multicomponent analysis of mixtures (6), the method is clearly inferior to other techniques used in PAC analysis. Wehry (14) concluded that UV absorption spectrometry is "a mature 'workhouse' technique especially suited for quantitative analysis in samples of limited complexity." However, if a compound to be determined in a PAC mixture absorbs at longer wavelengths than all other components present, UV absorption provides an easily applicable means for quantitative analysis. An example of practical significance is the determination of anthracene in fractions derived from coal tar (15). The anthracene content of such a fraction was determined by UV absorption spectrometry, HPLC, gas chromatography, and a chemical method (Diels–Alder reaction with maleic anhydride) (16). The results obtained are listed in Table 1; agreement is rather satisfactory.

Table 1. Quantitative Determination of Anthracene (and Phenanthrene) in a Technical Sample using Different Analytical Methods

Method	Anthracene (% per weight)	Phenanthrene (% per weight)
UV	24.3	—
HPLC	25.5	18.6
GC	26.9	18.1
Maleic anhydride	25.0	—

2.2. Structure Elucidation

Chemical analysis of PACs does not only mean identification and quantitative determination of known PACs in complex mixtures but also includes structure evaluation of PACs that have not been described in the literature. This type of problem often occurs in organic chemistry but also occurs in coal chemistry and environmental science. Although X-ray structure analysis, nuclear magnetic resonance, and photoelectron spectroscopy (17) are the most efficient methods for structure elucidation of unknown PAC, the UV absorption spectrum in many cases can provide useful information.

Since excellent books on the subject are available (18), only two points of interest will be discussed. In many cases the UV absorption spectra of heteroaromatic systems are very similar to that of topologically related hydrocarbons. These relationships have been studied in detail in, for example, the carbazole series (19–21). As an example the UV absorption spectra of the anthracenocarbazole (1) and the corresponding hydrocarbon (benzo[a]pentacene (2)) are shown in Figure 3. Both the transition energies and the vibrational structures observed are very similar in this pair of topologically related compounds. In Figure 4 the longest-wavelength absorption band of the phenes (phenanthrene, tetraphene, pentaphene, . . . , heptaphene) and of the corresponding carbazoles are plotted against the number of rings as a measure of molecular size. As can be easily recognized, the band position and its dependence on anellation is very similar in these two series of compounds. The same applies to the benzoacenes and their carbazole analogs (Figure 4). Thus in many cases the comparison of the spectrum of a carbazole whose structure is unknown with the spectra of known hydrocarbons provides a simple and fairly reliable means for structure elucidation.

Another approach of structure determination of unknown PAC consists of calculating the transition energies (and oscillator strengths) by SCF–MO theory in a given case of possible structures and comparing these data with the

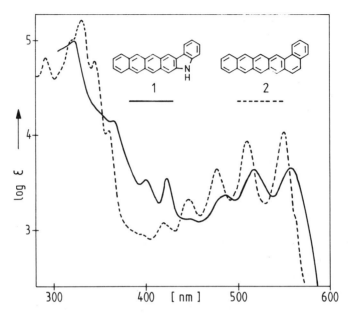

Figure 3. UV absorption spectra of anthraceno[2, 3-*b*]carbazole (**1**) (trichlorobenzene) and benzo[*a*]pentacene (**2**) (benzene) at room temperature (according to ref. 19).

experimental spectrum (22). At least some of the structures suspected (e.g., on the basis of the mass spectrum) can thus be excluded. However, since the energies of the MOs that are occupied in the ground state of a molecule can be more precisely calculated than transition energies by SCF–MO theory, the approach is much more useful when applied to photoelectron spectra (17).

3.   LUMINESCENCE SPECTROMETRY IN PAC ANALYSIS

PAC analysis requires techniques that are highly sensitive and selective because typical analytical samples consist of a large number of different PAC where each of them is present in low concentrations (see also Section 4). Because of their high sensitivity and selectivity, luminescence analytical methods are well suited for PAC analysis. During the last two decades, many interesting high-performance luminescence analytical methods have been developed that are characterized by extremely high sensitivity and selectivity, but it should not be overlooked that chromatographic methods are often superior and also have a better benefit/cost ratio.

In the following sections, some more recent developments in the field will be discussed.

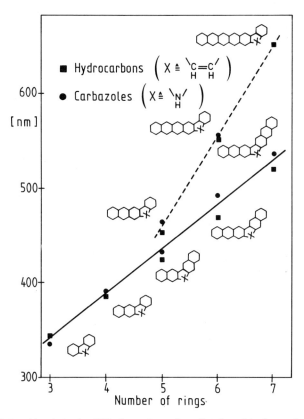

Figure 4. Relationships between UV absorption of anellated carbazoles and topologically related PAHs (for details see text) (according to ref. 20).

3.1. Sensitivity

The sensitivity of an analytical method can be described by (i) the gradient dS/dc of the linear part of the analytical function, where S is the signal observed and c is the concentration of the analyte, (ii) the lower limit of analyte detection, and (iii) the signal/noise ratio.

Since these characteristics depend strongly on the inherent fluorescence or phosphorescence quantum yield of the analyte, quenching effects must be avoided (8). On the other hand, with many compounds, particularly PACs, the inherent phosphorescence quantum yield can be enhanced by external heavy-atom perturbers (e.g., jodoalkanes). The method was originally developed to improve sensitivity (and selectivity) of low-temperature phosphorimetry (23–26) and was found to be rather useful. However, application of

external heavy-atom effects proved indispensible in room-temperature phosphorimetry (27).

3.1.1. External Heavy-Atom Effects

Two types of external heavy-atom perturbers suitable for phosphorimetry can
be distinguished: Type-A perturbers form phosphorescent ground-state complexes with the aromatic compound, whereas associative forces between type
B perturbers and the aromatic compound are negligible. Methyl iodide is an
often used type-B perturber, whereas silver perchlorate is a type-A perturber.
The phosphorescent ground-state complexes formed with type-A perturbers
are of the charge-transfer type (28), where the π-electronic system acts as the
electron donor and the metal ion acts as the electron acceptor; in the case of,
for example, the silver ion, both charge transfer from the highest-occupied π-
MO of the organic molecule to the vacant $5s$ AO of Ag^+ and back-donation
from the filled $4d$ AO of Ag^+ to the lowest-unoccupied π-MO of the organic
molecule have to be assumed (29).

   Though there are other differences regarding the effect of type-A and type-B
perturbers on a phosphorescent molecule (30), they can be best discriminated
by their different influence on the emission decay behavior of a phosphorescent compound. In the presence of a type-B perturber (e.g., methyl iodide),
the decay curve is multiexponential (30, 31) even at very high perturber
concentrations, that is, when the perturber compound (in low-temperature
phosphorimetry) is used as the solvent. On the contrary, the decay curve is
biexponential at low to medium type-A perturber concentrations containing
only the time constant of the unperturbed aromatic molecule and that of the
phosphorescent ground-state complex (which normally is in the millisecond
range); however, at type-A perturber concentrations sufficiently large for
complete complexation, the decay curve is monoexponential with the time
constant of the phosphorescent complex (30, 32).

   At a given type-A perturber concentration, phosphorescence enhancement
factors (i.e., ratio of phosphorescence intensity in the presence or absence of
perturber under otherwise identical conditions) increase with the π-donor
strength of the aromatic molecule. Thus the effect on aza-arenes as the
stronger π-donors is much larger than on aromatic hydrocarbons (33). This
has been used for the specific detection of nitrogen-containing compounds in
the presence of arenes by room-temperature phosphorimetry (RTP) (34).
Mercury(II) chloride is the selective heavy-atom perturber. The utility of
Hg(II) as a heavy-atom perturber in RTP is readily seen in the analyses of
complex mixtures. The RTP spectrum of a sample of phenanthrene, pyrene,
isoquinoline, and acridine with lead(II) acetate/thallium(I) acetate as heavy-
atom perturbers is given in Figure 5. The more intense phenanthrene and

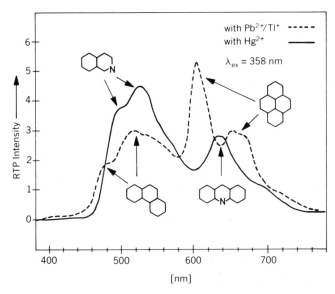

Figure 5. Room-temperature phosphorescence spectra of a four-component mixture on filter paper using different heavy-atom perturbers (for details see text) (according to ref. 34).

pyrene RTP signals completely mask the weaker emissions of isoquinoline and acridine that occur in the same spectral region. However, the RTP spectrum of the sample in the presence of mercury(II) chloride (Figure 5) shows that the isoquinoline and acridine signals are enhanced, while the phenanthrene and pyrene signals are quenched.

Even compounds that are virtually nonphosphorescent in the absence of external heavy-atom perturbers form ground-state complexes with type-A perturbers which are phosphorescent. Thus N-vinyl carbazole that exhibits fluorescence but no phosphorescence in rigid glasses at 77 K (35) shows intense phosphorescence in the presence of silver perchlorate (36).

Since the type-A perturber-induced phosphorescences do not stem from the aromatic donor molecules but from their ground-state complexes with the metal ion, spectral changes are often observed in the presence of type-A perturbers. A striking example is given in Figure 6. Curve a is the phosphorescence spectrum of 2-napthylamine (ethanol, 77 K) in the absence of a perturber, while curve b is the spectrum in the presence of 9 $\times 10^{-2}$ M $AgNO_3$. Since spectrum b is very similar to the phosphorescence spectrum of 2-naphthylamine hydrochloride (curve c), it is assumed that 3 is the emitting species in the presence of $AgNO_3$ (37).

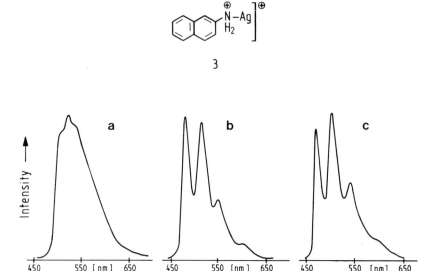

Figure 6. Phosphorescence spectra (ethanol, 77 K) of 2-naphthylamine in the absence (curve *a*) and presence (curve *b*) of $AgNO_3$ and of 2-naphthylamine hydrochloride (curve *c*) (according to ref. 37).

3.1.2. Instrumental Factors

Sensitivity (i.e., detection limits) in luminescence spectrometry in solution is limited by large background signals. The predominant sources of background are scattered excitation light from windows, Raman scatter from the solvent, and fluorescence from solvent impurities. A method was developed that uses a modulated laser beam with phase-sensitive detection electronics, as well as careful spatial and spectral filtering of the scattered light. Using, in addition, a flow cell with an ultrasmall detection volume (11 pl), 18 ag (18×10^{-18} g)—corresponding to 22,000 molecules of a highly fluorescent compound (rhodamine G)—could be detected (38). Reduction of Raman background in laser-induced fluorescence was also possible by second harmonic detection. Fluorescence spectra from a steady-state concentration of 210 molecules in the probed volume were obtained (39). However, applications of these ultrasensitive techniques in PAC analysis have not yet been reported.

In addition to excitation source intensity and signal/noise ratio of the system applied, absolute detection limits in luminescence spectrometry depend on the probed sample volume. Various microfluorescent systems have been introduced (40–49), most of which are flow cells designed for use in chromatography, particularly HPLC. However, a reloadable nanoliter cell useful for discrete samples has also been described (50). The cell has an optical volume of 0.5 nl and is designed for convenient sample introduction and easy cleaning. For anthracene, a detection limit of 85 fg $(8.5 \times 10^{-14}\,\text{g})$ was achieved using laser-excited fluorescence spectrometry.

3.2. Selectivity

In contrast to "sensitivity" (see above), the term "selectivity" in luminescence spectrometry seems not to be so clearly defined; it is differently used by different authors, and its meaning depends on the context in which it is used. In the following, the term "selectivity" will be used as a measure of the ability of an analytical method to respond to different compounds with (concentration-independent) different "signals" (spectra, lifetimes, etc.). The larger the number of compounds that can be distinguished by the method and the smaller the differences between the photophysical properties (energies of electronically excited states, rate constants of radiative/radiationless deactivation paths, etc.) of these compounds, the higher the selectivity of an analytical method. If the application of an analytical method is restricted to one or few compounds but enables us to detect these compounds in very complex mixtures, the method is highly "specific."

3.2.1. Selective Fluorescence Quenching

Because of the different HOMO/LUMO situations in alternant and non-alternant PAHs (51), the fluorescence of alternants is quenched by suitable electron acceptors (e.g., nitromethane), but the fluorescence of nonalternants is not (52–54). Conversely, electron donors (e.g., trimethoxybenzene) quench the fluorescence of nonalternants, but they do not quench the fluorescence of alternants (55, 56). This behaviour is known to apply to all alternants and nonalternants, provided that their fluorescence transition lies at wavelengths longer than approximately 400 nm.

Examples are shown in Figure 7. Curve *a* is the fluorescence spectrum of a mixture of four alternant and one nonalternant PAHs (NA) in the absence of a quencher. The spectrum observed is the superposition of the spectra of the five compounds present in the sample. Curve *b* is obtained after adding nitromethane as a fluorescence quencher to the solution. The spectrum is identical with that of the unperturbed nonalternant hydrocarbon (NA). Curve *c* is the fluorescence spectrum of a mixture consisting of four nonalternants and one alternant (A). By addition of trimethoxybenzene as a quencher, the fluorescence of all nonalternants present is completely quenched, and the spectrum (curve *d*) that remains is that of the alternant PAH (A).

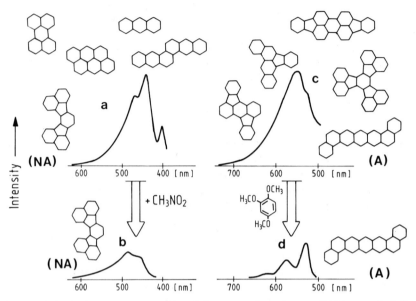

Figure 7. Selective fluorescence quenching of alternant and nonalternant PAH, respectively (for details see text) (according to ref. 56).

The method is particularly useful in connection with HPLC (57, 58). Two examples are given in Figure 8. The curves are HPLC chromatograms; fluorescence was used for detection. The empty peaks correspond to alternants, while the blackened peaks correspond to nonalternants. Curve *a* is the chromatogram of a synthetic mixture. After addition of a small amount of nitromethane to the eluent, curve *b* is obtained. Because of the fluorescence quenching effect, only the nonalternants give signals. Curve *c* is the chromatogram of a "real sample" (a low-molecular-weight fraction obtained from coaltar pitch). Chromatogram *d* is obtained after adding nitromethane to the system. Comparison of chromatograms *c* and *d* renders possible a straightforward distinction between alternants and nonalternants present in the sample (as shown in chromatogram *c*). The important advantage of the method lies in the fact that reference substances are not needed for the assignment of peaks to alternants and nonalternants.

Fluorescence quenching by electron acceptors also proved useful in structure elucidation of unknown PAH (see Section 3.3).

A highly specific method that is based on fluorescence quenching by an external heavy-atom perturber (methyl iodide) has been proposed for the detection and quantitative determination of perylene in complex PAC mixtures (26, 59–61).

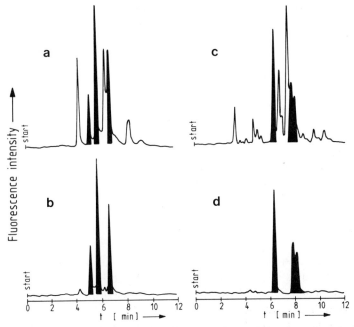

Figure 8. HPLC chromatograms of a synthetic mixture of PAH (*a* and *b*) and a low-molecular-weight fraction from coal-tar pitch (*c* and *d*) in the absence and presence of nitromethane in the eluent (for details see text) (according to ref. 57).

3.2.2. Selective Phosphorescence Enhancement

The application of external heavy-atom effects in phosphorimetry is an analytically useful method because the heavy-atom perturber generally not only increases the phosphorescence quantum yields of PACs (see Section 3.1.1) but it does so with greatly different enhancement factors, making it a rather specific method (23–26, 61). The technique of selective heavy-atom perturbation (SEHAP) was originally applied to low-temperature phosphorimetry but also was useful in room-temperature phosphorimetry (RTP) (62). The phosphorescence enhancement factors (RTP, filter paper) of various PACs using five different heavy-atom perturbers are given in Table 2. It is noteworthy to mention that the data in Table 2 have been obtained under the sample and perturber concentrations listed; studies have shown that a variation in those conditions can induce significant variations in the enhancement factor values (62). However, this does not apply to enhancement factors observed in low-temperature phosphorimetry in solid glasses (23). The data listed in Table 2 clearly show that different perturbers influence the

Table 2. Phosphorescence Enhancement Factors (62)

Compound	Concn. [M]	$AgNO_3$ (0.5 M)	CsJ (0.2 M)	NaJ (2 M)	NaBr (2 M)	$Pb(OAc)_2$ (0.5 M)
Quinoline	10^{-4}	40	15	12	75	210
Acridine	10^{-4}	10	2	1.5	8	40
Carbazole	10^{-4}	<1	72	75	18	35
Phenanthrene	10^{-4}	<1	30	5	20	110
Dibenz[a, h]-anthracene	5×10^{-5}	50	350	150	220	250
Benzo[a]pyrene	10^{-4}	<1	24	27	30	240
Benzo[e]pyrene	5×10^{-5}	10	95	10	28	380
1-Naphthol	10^{-4}	<1	140	140	50	60

phosphorescence signal of PACs to different degrees. It was concluded that one of the features offered by the method is the ability to "pick out" specific target PAC compounds of interest (e.g., in an environmental sample using appropriate heavy atoms).

It is well known from low-temperature experiments that external heavy-atom perturbers can also affect significantly the spectral structure of phosphorescence (23, 24, 63–70). The effect was also observed in RTP (71). Spectral changes induced by the heavy-atom effect lead to improved spectral assignment and more accurate identification of individual components in complex mixtures because the components can be characterized by two different spectra (one is measured in the presence of the perturber, while the other is measured in its absence).

3.2.3. Excitation-Energy Selection Techniques

Excitation-energy selection techniques are the most selective methods presently known in luminescence spectrometry. They are best suited for PAC analysis and have been widely used in the field.

The theory of these techniques has been discussed in detail (72–74), so only a short description of the underlying principles will be given. Organic molecules embedded in either glassy matrices or host crystals at low temperature are situated in different individual microenvironments (sites). In amorphous glasses or polymers the solute molecules adopt essentially an infinite number of sites, with a broad distribution of energies for each vibronic transition. In crystalline (Shpol'skii) matrices the situation is somewhat different. The solute molecules are not randomly dispersed among the laminar planes of the solvent microcrystallites but, instead, substitute a limited number of crystallographically different sites. Using narrow-line laser excitation it is possible to selectively excite the different sites present in Shpol'skii matrices, whereas in amorphous glasses a comparably narrow subset of sites (an "isochromat") can be excited to luminesce. As a result, luminescence band broadening as observed with classical broad-band excitation of luminescence is strongly reduced. The resulting emission spectra display very sharp lines, with bandwidth determined either by the bandwidth of the light source or the bandwidth of the transition. Fluorescence and phosphorescence line-narrowing (FLN and PLN, respectively) is the generic principle because it can be applied to either amorphous, polymeric, or crystalline matrices, while techniques such as Shpol'skii or matrix isolation spectrometry can be regarded as special variants of LN.

Site-selective excited Shpol'skii fluorescence spectrometry (*n*-octane, 15 K) was used to analyze a mixture of benz[*a*]anthracene (BaA) and 11 alkylated BaA's (75). By site-selective excitation it proved possible to obtain spectra, each of which is dominated by the spectral lines of one component only. Some of the spectra obtained are shown in Figure 9.

A somewhat different approach to analyze mixtures of structurally (and spectroscopically) closely related PAHs consists in preparing the sample in a host crystal with the same molecular dimensions as the analytes (76). The

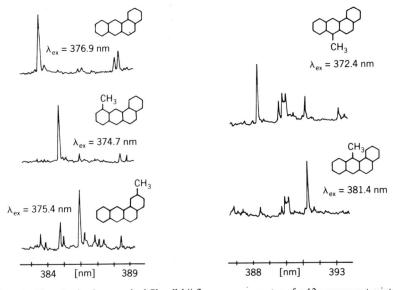

Figure 9. Site-selective laser-excited Shpol'skii fluorescence spectra of a 12-component mixture (λ_{ex} = excitation wavelength) (for details see text) (according to ref. 75).

sample molecules will then be incorporated into nearly identical sites within the crystalline lattice. In order to demonstrate the utility of the technique, methyl derivatives of naphthalene were used as prototypes. Since durene has approximately the same molecular dimensions as the methylnaphthalenes, it was used as the crystalline host. Direct 0–0 band excitation was shown to be the best way to selectively excite the individual components of the mixture. S_1 origins of the methylnaphthalenes studied are listed in Table 3. Using dye laser excitation, 1-methylnaphthalene and 1,4-, 1,5-, and 2,3-dimethylnaphthalene were spectroscopically isolated (10 K).

Table 3. S_1 Origins (0–0 Bands) for Methylnaphthalenes (76)

Compound	0–0 Band (nm)
Naphthalene	316.7[a]
1-Methylnaphthalene	319.5; 320.0[b]
2-Methylnaphthalene	319.5; 320.5[b]
1,3-Dimethylnaphthalene	321.1; 322.4[b]
1,4-Dimethylnaphthalene	320.9
1,5-Dimethylnaphthalene	322.5[a]
2,3-Dimethylnaphthalene	321.1

[a] 0–0 band was broad, and measurement is subject to uncertainty.
[b] Doublet structure in the fluorescence spectra due to molecules occupying two distinctively different sites in the durene lattice. Consequently, two 0–0 band origins are listed in these cases.

A 10-K durene crystal as the sampling medium and methylnaphthalenes as the analytes were also used to test the analytical utility of combining low-temperature sample preparation procedures with two-photon excitation (77). The two-photon excitation spectrum of fluorescence of a nine-component methylnaphthalene mixture was measured; all compounds present were spectroscopically isolated.

Fluorescence line-narrowing spectrometry in glasses at 4.2 K was used for direct determination of PAHs in solvent-refined coal (78). In an SRC II sample it proved possible to quantitatively determine pyrene, benzo[e]pyrene, and benzo[a]pyrene without sample pre-separation. Results obtained by direct-insertion capillary column gas chromatography–mass spectrometry (GC–MS) were found to agree with the FLNS values within estimated experimental uncertainty of $\sim \pm 20\%$ for both techniques.

A very interesting application of FLN (4.2 K) involves the direct identification of PAH metabolites and their DNA adducts. Solid-state fluorescence techniques (e.g., laser-excited Shpol'skii or matrix isolation spectroscopy) cannot readily be applied to large biomolecules like DNA–PAH adducts because of the limitations imposed by either solubility or vapor pressure, but the same is not true for FLN using water-containing solvents. The direct identification of metabolites 4–9 and DNA adducts 10–14 in (laboratory) mixtures by comparison to standard spectra of the pure compounds was accomplished (79, 80) (Scheme 1). In addition, mixtures of metabolites and DNA adducts were also resolved. The sensitivity of the technique is adequate for the identification of PAH metabolite-adducts at a level of approximately five adducts per 10^6 bases. Typically, using cells in culture, DNA damage

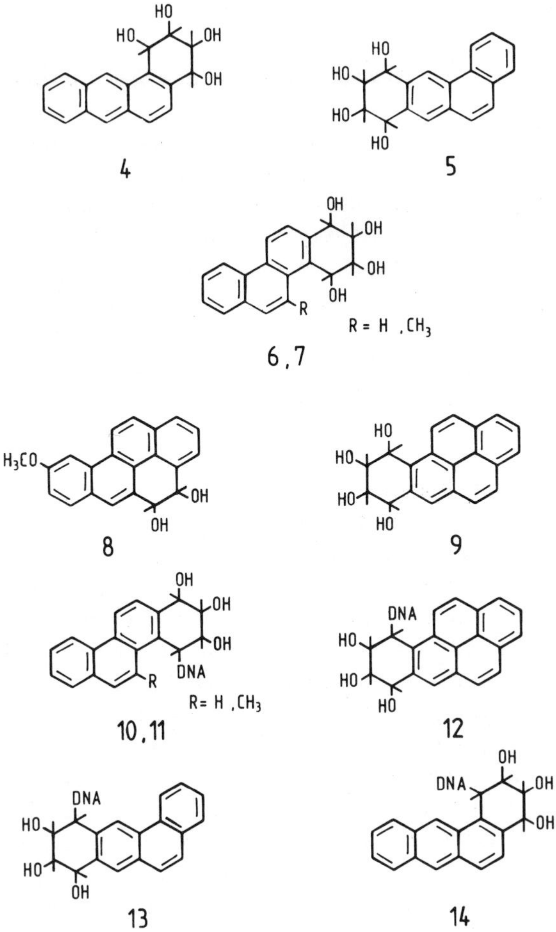

Scheme 1

levels from PAH carcinogens are approximately one adduct per 10^5–10^6 base pairs.

In quite another approach to FLN spectroscopy, the principle of free jet expansion was applied (81). In a free jet expansion the analyte molecules are present as isolated ultracold gas phase species. As a result, very characteristic spectra consisting of narrow lines can be obtained by laser excitation. As an example, the fluorescence excitation spectra of isomeric monomethylanthracenes obtained with the method have been reported. Compared to other low-temperature luminescence techniques (e.g., matrix isolation), the new method has the great advantage that because of the fast cooling rate, high-resolution fluorescence real-time analysis of chromatographic (e.g., GC) effluents becomes possible.

3.2.4. Synchronous Luminescence Spectrometry

A well-established approach to improve selectivity of luminescence spectrometry is to use simultaneously the information that is contained in the emission and excitation spectrum of a compound (or compound mixture). It is interesting to note that from its very beginning, "total luminescence spectrometry" has been used for rapid identification and monitoring of aromatic compounds in environmental samples (82, 83) (for a detailed review of total luminescence spectrometry, see ref. 84). Synchronous luminescence spectrometry has evolved from the basic principles of total luminescence spectrometry and constitutes a rather easily applicable method to utilize simultaneously part of the information content of the emission (fluorescence or phosphorescence) and excitation spectrum of a luminescent compound.

Since a rigorous mathematical description of the basic principles underlying synchronous excitation techniques has been given (85), the practical procedure will just be recalled. In conventional luminescence spectrometry, an emission spectrum can be monitored by scanning the emission wavelength λ_{em} while the luminescent compound is excited at fixed excitation wavelength λ_{ex}. On the other hand, an excitation spectrum can be obtained by scanning λ_{ex} while the emission is monitored at a given λ_{em}. In synchronous luminescence spectrometry, λ_{ex} and λ_{em} are simultaneously varied while the wavelength interval $\Delta(\lambda_{em} - \lambda_{ex})$ is kept constant (86). An important advantage of synchronous luminescence spectrometry compared to more sophisticated variants of total luminescence spectrometry lies in the fact that most commercially available high-standard luminescence spectrometers can be used for its application (at least in the constant *wavelength* mode).

The exceptional quality of synchronous luminescence spectrometry can be visualized by using an analogy. The excellent suitability of chromatographic techniques for the analysis of multicomponent mixtures resides in the fact that each component of the mixture provides one signal only (one "peak"). Synchronous luminescence excitation can be regarded as an interesting attempt to apply this property of chromatographic methods to spectrometry. This can be best demonstrated by Vo-Dinh's famous, and often cited, synchronous fluorescence spectrum of a mixture consisting of naphthalene, phenanthrene, anthracene, perylene, and tetracene (Fig. 10) (85). Each compound gives practically one signal only and apart from the order of "peaks" a chromatogram of this mixture would not look much different.

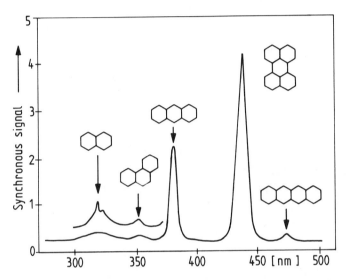

Figure 10. Synchronous fluorescence spectrum of a five-component mixture (ethanol, room temperature; $\Delta(\lambda_{em} - \lambda_{ex}) = 3$ nm) (according to ref. 85).

In the original version of synchronous fluorescence spectrometry, a *wavelength* interval between excitation and emission is kept constant (86). This version, however, does not account for the fact that neither the Stokes shift between fluorescence and absorption nor the vibrational spacing of the spectra of related compounds (e.g., PAHs) is constant in wavelength units if the absorption spectra lie in different spectral regions. On the other hand, these characteristics are approximately constant in wavenumber units. Consequently, constant-*energy* synchronous fluorescence spectrometry has been developed to improve the utility of the method. In this version the excitation and emission monochromators are scanned simultaneously, synchronized so that a constant-*energy* difference is maintained between the monochromators (87). The method has been applied to PAHs.

Although many applications of synchronous luminescence spectrometry in PAC analysis have been described in the literature (e.g., ref. 88) and hyphenated methods have been developed that include synchronous excitation techniques (89, 90), it is still controversial whether this method is really useful for the practicing analyst (see, e.g., ref. 91). It has been suspected that only a fortuitous combination of substances, concentrations, and fluorescence quantum yields will provide reliable qualitative and quantitative data. On the other hand, the method may be useful for routine analyses in special cases.

Various other techniques in luminescence spectrometry have been developed that will not been dealt with in this chapter [e.g., wavelength modulation spectrometry (92) or time-resolved fluorimetry (93)], though these techniques have found applications in PAC analysis. However, as stated in the introduction, it is not the aim of this chapter to give a full account of all developments made in this field. Regarding these methods, the reader is referred to refs. (8 and 94–96).

3.3. Structure Elucidation

Although luminescence (fluorescence and phosphorescence) spectrometry has found much fewer applications in structure elucidation of PAC than UV absorption spectrometry (see Section 2.2), some results have been obtained, and some methods have been developed, that are worth mentioning.

In Section 3.2.1 an easily applicable method (selective fluorescence quenching by electron donors and acceptors, respectively) has been described that enables us to determine whether a given PAH of unknown structure is an alternant or nonalternant system. In fact, this information can be very useful in connection with the structure elucidation of unknown PAHs. There are some other methods that can provide the information, but they are either laborious [photoelectron spectroscopy (17)] or less reliable and also require more substance material than fluorimetry [^{13}C NMR (97)]. In several cases, selective fluorescence quenching with electron acceptors has been used in structure elucidation of unknown PAHs (obtained by synthesis). Two examples will be discussed in some detail.

The aluminum-chloride-catalyzed reaction of benzo[b]chrysene (15) with benzene leads to a complex mixture of compounds that has been separated into pure substances in a preparative scale. Two of the substances obtained have molecular formula $C_{28}H_{16}$. Structures 16 and 17 were suggested. Fluorescence quenching experiments using nitromethane as the quencher molecule easily revealed that the higher-melting-point isomer is an alternant system, whereas the lower-melting-point one is a nonalternant. Structures 16 and 17 have been confirmed by independent syntheses (98).

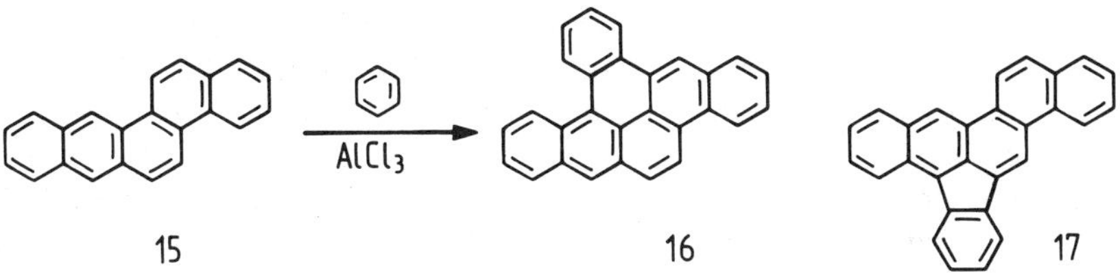

Scheme 2

Under the influence of aluminum chloride plus tin tetrachloride in boiling benzene (contrary to the solvent-free system), 1-(1-phenanthryl)pyrene (**18**) does not form the expected product (**19**) of dehydrogenating cyclization of **18**. Instead, complete rearrangement of **18** to isomeric biaryls occurs. Besides biaryls **20** and **21**, isolated products include fluoranthenes **22–24** derived from unphenylated **20** by dehydrogenating cyclizations and phenylations. Fluorescence quenching experiments (quencher: nitromethane) gave clear indication that the ring-closure products formed in this reaction are nonalternant hydrocarbons. With this information in hand, complete structure determination became possible by[1]H-NMR spectroscopy (99).

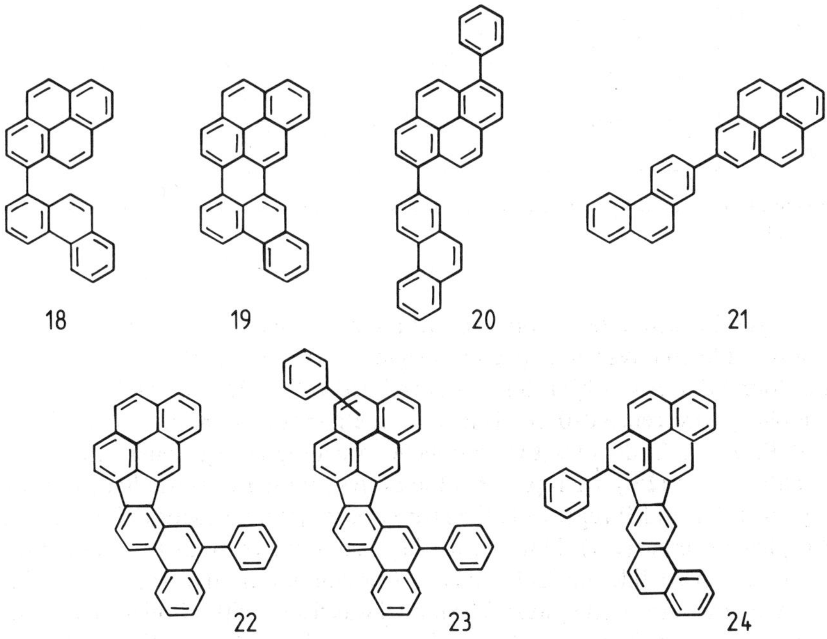

Scheme 3

In many cases, the dependence of the energy of the lowest triplet state (T_1) on compound structure is very similar for PAHs and their heteroaromatic analogues. In Figure 11 the T_1 state energies of PAHs and topologically related carbazoles are compared. As can be clearly seen, structure dependence of T_1 state energy is very similar in both series of compounds (100). This relation would be useful in the structure determination of unknown carbazoles.

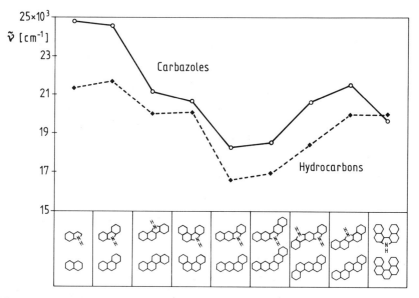

Figure 11. T_1 state energies of anellated carbazoles and topologically related PAHs. (Taken from ref. 100).

Polycyclic aromatic thiophenes (thia-arenes) exhibit vibronically well-resolved phosphorescence spectra in rigid matrices (e.g., ethanol at 77 K). It was found that $\Delta(\tilde{v}_p^T - \tilde{v}_p^P)$ is linearly correlated with $\Delta(\tilde{v}_p^{Ar} - \tilde{v}_p^P)$ and that $\tilde{v}_p$ are the phosphorescence 0–0 bands (cm^{-1}) in ethanol at 77 K of systems T, Ar, and P, where T denotes the thiophene benzologue (e.g., phenanthro[4,5-*bcd*]thiophene (**25**)), Ar represents the corresponding isosteric hydrocarbon (pyrene (**26**)), and P represents the largest part system present in both T and Ar (phenanthrene (**27**)). Thus the model is constructed so as to compare the influence of two different "substituents" (the sulphur atom or the C–C double bond) on the same part system. The model was tested with 17 thia-arenes. The result is shown in Figure 12. A good linear correlation (correlation coefficient $r = 0.9779$) was obtained for 12 of the 17 thia-arenes examined. Three

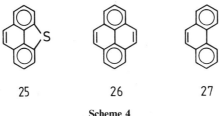

| 25 | 26 | 27 |

Scheme 4

additional systems fit into another linear correlation (Figure 12). These compounds are topologically clearly related in that the corresponding part systems contain only benzene rings connected by single bonds (biaryl type). On the other hand, no systems of that kind are contained in the group of compounds that fit the other relationship. Thus, a straightforward assignment of compounds to the correlations can be made on structural grounds. The relationships observed are expected to be useful for the structure determination of unknown thiophene benzologues (101). Since thiophene benzologues, some of which are known to be mutagenic, are widely distributed in the environment and, in addition, the number of possible isomers is much larger than in the PAH series, methods suitable for structure determination of thiophene benzologues are of great interest.

4. ON THE ROLE OF ELECTRONIC SPECTROSCOPIC METHODS IN PAC ANALYSIS

The most important objective of PAC analysis is the qualitative and quantitative characterization of PAC mixtures with regard to sample composition. Complex PAC mixtures in varying matrices (e.g., air dust, water, food, etc.) occur in the human environment but also constitute technically important materials (e.g., liquids derived from coal).

The enormous compositional complexity of PAC mixtures occurring, for example, in environmental samples results from the fact that PAC formation during pyrolysis or incomplete combustion of organic material is a rather unspecific process; PAC formation occurs via free radical reactions and, at least in the first steps of reaction, is mainly controlled by diffusion, that is, formation pathways leading to different PAC are not discriminated by different activation energies (102, 103). On the other hand, the number of isomeric PACs increases tremendously with the enlargement of molecular size. For convenience we may define isomeric PAHs as those having the same number of hexagons. In Table 4 the number of possible isomers of kata- and pericondensed PAHs, with two to eight rings are listed (104). For PAHs

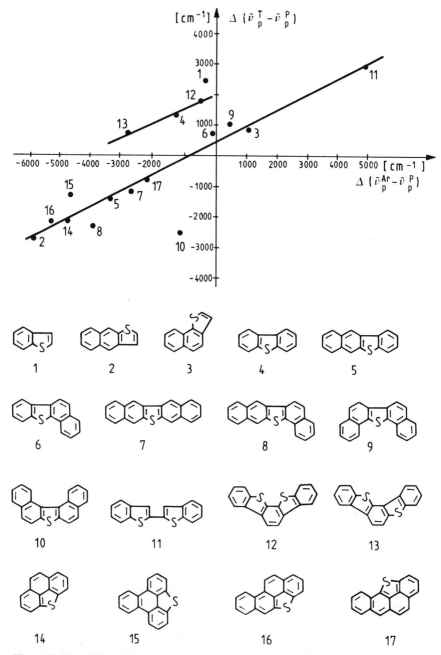

Figure 12. Correlation between T_1 state energy of thia-arenes and PAHs (for details see text) (according to ref. 101).

Table 4. Number of PAHs (Ring Isomers) Consisting of n Hexagons (104)[a]

n	Katacondensed PAH	Pericondensed PAH	Σ
2	1	0	1
3	2	1	3
4	5	2	7
5	12	10	22
6	37	45	82
7	123	210	333
8	446	1002	1448

[a] Nonalternant systems are not included.

containing 12 hexagons, 683, 101 isomers are possible (105). However, if we take into account that these large numbers of isomers contain also all possible pericondensed openshell free radical systems, most of which are extremely instable, the number of relatively stable systems that can occur in real samples will still be very large. In addition, a CH or CH=CH group may be replaced by N, S, or NH, which, in turn, leads to still higher figures for the possible isomeric compounds, since the N or S atom can appear in many different positions in the aromatic ring. The number of compounds that occur in environmental or technical material samples is multiplied further by substitutions of an H atom on the aromatic ring by, for example, alkyl, NO_2, or NH_2 groups.

The difficulties in analyzing complex ("real") PAC samples arise further from the fact that concentrations of individual components are normally extremely low and decrease with increasing size of molecules as a result of increasing number of individual isomers. Thus in conclusion, PAC analysis requires extremely selective and sensitive analytical methods.

In most cases, high-resolution chromatographic techniques, coupled with spectroscopic methods for compound identification, are best suited for the analysis of PAC mixtures (106). GC–MS can be easily applied to the analysis of PAHs with molecular weigths of up to approximately 300 (see, e.g., ref. 107) or even higher (108), while HPLC renders possible the detection of PAHs with molecular weigths of up to approximately 600 (9). High-sensitivity detection has been reported. For example, limits of detection as low as 200 fg (2×10^{-13}g) have been achieved in GC–MS of PAHs, using laser-induced multiphoton ionization for ion generation (109). Thus, the question may occur as to whether there is any demand at all for UV absorption and luminescence spectroscopical methods in PAC analysis.

One of the most important advantages of spectroscopical methods, as compared to GC, resides in the fact that these methods can be applied to thermally instable compounds (or compounds present in thermally instable matrices). Although the same applies to HPLC, it is inferior to high-performance luminescence methods with regard to selectivity. A wide range of important applications of luminescence spectrometry can be foreseen in the field of biochemical research on PAC. Line-narrowing luminescence spectrometry in glassy or polymeric matrices (and also hyphenated methods, including LN) are expected to be particularly useful. The same applies to uses in biological monitoring (i.e., determination of PAC metabolites in body fluids) (see also Section 3.2.3).

Frequently in PAC analysis the rapid and highly specific determination of one or few components of complex samples is necessary. In those cases, chromatographic methods often can provide the required information but at the same time can produce a large amount of redundant information. For example, 90% of the time needed for the determination of perylene in complex PAC mixtures by glass capillary GC is wasted for the production of redundant information. On the other hand, determination of perylene in complex PAC mixtures by fluorimetry using heavy-atom perturber containing solvents requires approximately one-seventh of the time needed for GC analysis (26, 59–61). Also, room-temperature phosphorimetry is expected to provide rapid and highly specific methods for routine analyses (27).

It is quite a general experience that research on analytical methods frequently leads to rather unexpected results which prove important in other areas of science. For example, room-temperature phosphorimetry, originally developed as an analytical method, has been shown to be of great value in surface analysis of filter papers (110). Further, a deeper insight has been obtained regarding the nature of adsorbate–substrate interactions (111). Such "side effects" of analytical research are most welcomed (as in all areas of research) and are typical for the complex mechanisms governing the development of scientific knowledge.

REFERENCES

1. IUPAC recommendation on Molecular Absorption Spectroscopy (UV/VIS), *Pure Appl. Chem.*, in preparation.
2. IUPAC recommendation on Molecular Luminescence Spectroscopy, *Pure Appl. Chem.*, **56**, 231 (1984).
3. W. V. Mayneord and E. M. F. Roe, *Proc. R. Soc. London Ser. A.*, **158**, 643 (1937).
4. E. Clar, *Spectrochim. Acta* **4**, 116 (1950).
5. L. A. Nakimovsky and M. Lamotte, Unpublished data, cited in A. P. D'Silva and V. A. Fassel, *Anal. Chem.*, **56**, 985A (1984).

6. H.-H. Perkampus, *UV-VIS-Spektroskopie und ihre Anwendungen*, Springer, Berlin, pp. 75–81 (1986).

7. H. Schindlbauer and M. Zander, *Erdoel Kohle*, **37**, 206 (1984).

8. M. Zander, *Fluorimetrie*, Springer, Berlin, pp. 72–73 (1981).

9. R. Thoms and M. Zander, *Fresenius Z. Anal. Chem.*, **282**, 443 (1976).

10. G.-P. Blümer, R. Thoms, and M. Zander, *Erdoel Kohle*, **31**, 197 (1978).

11. Y. Talmi, *Multichannel Image Detectors*, American Chemical Society, Washington, D.C. (1979).

12. S. A. Borman, *Anal. Chem.*, **55**, 836A (1983).

13. K. Jinno, T. Hoshino, T. Hondo, M. Saito, and M. Senda, *Anal. Chem.*, **58**, 2696 (1986).

14. E. L. Wehry, Optical Spectrometric Techniques for Determination of Polycyclic Aromatic Hydrocarbons, in *Handbook of Polycyclic Aromatic Hydrocarbons*, A. Bjorseth, Ed., Marcel Dekker, New York, p. 334 (1983).

15. H.-G. Franck and G. Collin, *Steinkohlenteer*, Springer, Berlin (1968).

16. G.-P. Blümer and M. Zander, *DGMK-Compendium 78/79* (supplement to *Erdoel Kohle*), pp. 1472–1483.

17. E. Clar, J. M. Robertson, R. Schlögl, and W. Schmidt, *J. Am. Chem. Soc.*, **103**, 1320 (1981).

18. E. Clar, *Polycyclic Hydrocarbons*, Vols. 1 and 2, Academic Press, London, and Springer, Berlin (1964).

19. M. Zander and W. H. Franke, *Chem. Ber.*, **97**, 212 (1964).

20. M. Zander and W. H. Franke, *Chem. Ber.*, **98**, 588 (1965).

21. M. Zander and W. H. Franke, *Chem. Ber.*, **98**, 2814 (1965).

22. See, for example, A. Streitwieser, Jr., *Molecular Orbital Theory for Organic Chemists*, Wiley, New York (1961).

23. M. Zander, *Fresenius Z. Anal. Chem.*, **226**, 251 (1967).

24. M. Zander, *Fresenius Z. Anal. Chem.*, **227**, 331 (1967).

25. L. V. S. Hood and J. D. Winefordner, *Anal. Chem.*, **38**, 1922 (1966).

26. M. Zander, *Erdoel Kohle*, **22**, 81 (1969).

27. T. Vo-Dinh, *Room Temperature Phosphorimetry for Chemical Analysis*, Wiley, New York (1984).

28. R. S. Mulliken and W. B. Person, *Molecular Complexes*, Wiley, New York (1969).

29. S. Fukuzumi and J. K. Kochi, *J. Org. Chem.*, **46**, 4116 (1981).

30. M. Zander, *Z. Naturforsch.*, **39a**, 1145 (1984).

31. S. P. McGlynn, M. J. Reynolds, G. W. Daigre, and W. D. Christodoulas, *J. Phys. Chem.*, **66**, 2499 (1962).

32. H. Hopf and M. Zander, *Z. Naturforsch.*, **40a**, 1045 (1985).

33. M. Zander, *Z. Naturforsch.* **33a**, 998 (1978).

34. D. W. Abbott and T. Vo-Dinh, *Anal. Chem.*, **57**, 41 (1985).

35. G. E. Johnson, *J. Phys. Chem.*, **78**, 1512 (1974).

36. M. Zander, *Chem. Phys. Lett.*, **133**, 445 (1987).

37. M. Zander, *Z. Naturforsch.*, **37a**, 1348 (1982).

38. N. J. Dovichi, J. C. Martin, J. H. Jett, M. Trkula, and R. A. Keller, *Anal. Chem.*, **56**, 348 (1984).

39. M. Trkula and R. A. Keller, *Anal. Chem.*, **57**, 1663 (1985).

40. G. G. Vurek and S. E. Pegran, *Anal. Biochem.*, **16**, 409 (1966).
41. G. G. Vurek and R. L. Bowman, *Anal. Biochem.*, **29**, 238 (1969).
42. L. H. J. Thacker, *J. Chromatogr.*, **136**, 213 (1977).
43. R. M. Smith, K. W. Jackson, and K. M. Aldous, *Anal. Chem.*, **49**, 2051 (1977).
44. G. J. Diebold and R. N. Zare, *Science*, **196**, 1439 (1977).
45. L. W. Hershberger, J. B. Callis, and G. D. Christian, *Anal. Chem.*, **51**, 1444 (1979).
46. M. J. Sepaniak and E. S. Young, *J. Chromatogr.* **190**, 337 (1980).
47. E. Voigtman, A. Jurgenssen, and J. D. Winefordner, *Anal. Chem.*, **53**, 1921 (1981).
48. S. Folestad, L. Johnson, B. Josefsson, and B. Galle, *Anal. Chem.*, **54**, 925 (1982).
49. G. G. Vurek, *Anal. Chem.*, **54**, 840 (1982).
50. L. Hirschy, B. Smith, E. Voigtman, and J. D. Winefordner, *Anal. Chem.*, **54**, 2387 (1982).
51. E. Heilbronner and E. Straub, *Hückel Molecular Orbitals*, Springer, Berlin (1966).
52. E. Sawicki, T. W. Stanley, and W. C. Elbert, *Talanta*, **11**, 1433 (1964).
53. H. Dreeskamp, E. Koch, and M. Zander, *Z. Naturforsch.*, **30a**, 1311 (1975).
54. M. Zander, U. Breymann, H. Dreeskamp, and E. Koch, *Z. Naturforsch.*, **32a**, 1561 (1977).
55. U. Breymann, H. Dreeskamp, E. Koch, and M. Zander, *Chem. Phys. Lett.*, **59**, 68 (1978).
56. U. Breymann, H. Dreeskamp, E. Koch, and M. Zander, *Fresenius Z. Anal. Chem.*, **293**, 208 (1978).
57. G.-P. Blümer and M. Zander, *Fresenius Z. Anal. Chem.*, **296**, 409 (1979).
58. P. L. Konash, S. A. Wise, and W. E. May, *J. Liq. Chromatogr.*, **4**, 1339 (1981).
59. M. Zander, *Fresenius Z. Anal. Chem.*, **229**, 352 (1967).
60. M. Zander, *Fresenius Z. Anal. Chem.*, **263**, 19 (1973).
61. W. J. McCarthy, Phosphorescence Spectrometry, in *Spectrochemical Methods of Analysis*, J. D. Winefordner, Ed., Wiley, New York, pp. 467–488 (1971).
62. T. Vo-Dinh and J. R. Hooyman, *Anal. Chem.*, **51**, 1915 (1979).
63. M. Zander, *Naturwissenschaften*, **52**, 559 (1965).
64. K. B. Eisenthal, *J. Chem. Phys.*, **45**, 1850 (1966).
65. V. A. Butlar and D. M. Grebenschikov, *Opt. Spectroscopy (USSR)*, **22**, 413 (1967).
66. M. Zander, *Z. Naturforsch.*, **22a**, 1289 (1967).
67. G. G. Giachino and D. R. Kearns, *J. Chem. Phys.*, **53**, 3886 (1970).
68. L. G. Thompson and S. E. Webber, *J. Phys. Chem.*, **76**, 221 (1972).
69. M. Zander, *Z. Naturforsch.*, **31a**, 677 (1976).
70. M. Zander, *Z. Naturforsch.*, **34a**, 1143 (1979).
71. G. W. Suter, A. J. Kallir, U. P. Wild, and T. Vo-Dinh, *Anal. Chem.*, **59**, 1644 (1987).
72. C. Pfister, *Chem. Phys.*, **2**, 171 (1973).
73. V. B. Conrad, R. R. Gore, J. L. Hammons, J. R. Maple, M. P. Perry, and E. L. Wehry, Selective Spectrometric Analysis by Laser-induced Matrix Isolation Fluorimetry, in *New Direction in Molecular Luminescence* D. Eastwood, Ed., ASTM Special Technical Publication 822, ASTM, Philadelphia (1983).
74. R. J. Personov, in *Spectroscopy and Excitation of Condensed Molecular Systems*, V. M. Agranovich and R. M. Hochstrasser, Ed., North-Holland, New York (1983).
75. Y. Yang, A. P. D'Silva, and V. A. Fassel, *Anal. Chem.*, **53**, 894 (1981).

76. S. M. Thornberg and J. R. Maple, *Anal. Chem.*, **56**, 1542 (1984).
77. S. M. Thornberg and J. R. Maple, *Anal. Chem.*, **57**, 436 (1985).
78. J. C. Brown, J. A. Duncanson, Jr., and G. J. Small, *Anal. Chem.*, **52**, 1711 (1980).
79. M. J. Sanders, R. S. Cooper, and G. J. Small, *Anal. Chem.*, **57**, 1148 (1985).
80. M. J. Sanders, R. S. Cooper, R. Jankowiak, and G. S. Small, *Anal. Chem.*, **58**, 816 (1986).
81. B. V. Pepich, J. B. Callis, D. H. Burns, and M. Gouterman, *Anal. Chem.*, **58**, 2825 (1986).
82. A. W. Horning, H. G. Eldering, and H. J. Coleman, Presented at the 27th Pittsburgh Conference, 1976.
83. A. W. Horning, in *Proceedings of the International Congress on Analytical Techniques in Environmental Chemistry*, I. Albaiges, Ed., Pergamon Press, Oxford, pp. 127–134 (1980).
84. D. W. Johnson, J. B. Callis, and G. D. Christian, *Anal. Chem.*, **49**, 747A (1977).
85. T. Vo-Dinh, *Anal. Chem.*, **50**, 396 (1978).
86. J. B. F. Lloyd, *Nature*, **231**, 64 (1971).
87. E. L. Jnman, Jr., and J. D. Winefordner, *Anal. Chem.*, **54**, 2018 (1982).
88. P. John and I. Soutar, *Anal. Chem.*, **48**, 520 (1976).
89. T. Vo-Dinh and R. B. Gammage, *Anal. Chem.*, **50**, 2054 (1978).
90. R. A. Femia and L. J. Cline Love, *Anal. Chem.*, **56**, 327 (1984).
91. H. W. Latz, A. H. Ullman, and J. D. Winefordner, *Anal. Chem.*, **50**, 2148 (1978).
92. T. C. O'Haver and W. M. Parks, *Anal. Chem.*, **46**, 1886 (1974).
93. F. E. Lythe, *Anal. Chem.*, **46**, 545A, 817A (1974).
94. E. L. Wehry, Ed., *Modern Fluorescence Spectroscopy*, Vol. 4, Plenum Press, New York (1981).
95. D. Eastwood, Ed., *New Directions in Molecular Luminescence*, ASTM Special Technical Publication 822, ASTM, Philadelphia (1983).
96. S. G. Schulman, Ed., *Modern Luminescence Spectroscopy*, Wiley, New York (1985).
97. A. J. Jones, T. D. Alger, D. M. Grant, and W. M. Litchman, *J. Am. Chem. Soc.*, **92**, 2386 (1970).
98. G.-P. Blümer, K.-D. Gundermann, and M. Zander, *Chem. Ber.*, **109**, 1991 (1976).
99. R. Bunte, K.-D. Gundermann, J. Leitich, O. E. Polansky, and M. Zander, *Chem. Ber.*, **120**, 247 (1987).
100. M. Zander, *Phosphorimetry*, Academic Press, New York (1968).
101. M. Zander, J. Jacob, and M. L. Lee, *Z. Naturforsch.*, **42a**, 735 (1987).
102. M. Zander in Proceedings des Kolloquiums "Polycyclische aromatische Kohlenwasserstoffe", VDI-Verlag, Düsseldorf 1980, pp. 11–21.
103. H. W. Kleffner, J. Talbiersky, and M. Zander, *Fuel*, **60**, 361 (1981).
104. A. T. Balaban, Ed., in *Chemical Applications of Graph Theory*, Academic Press, London, pp. 64–105 (1976).
105. W. F. Lunnon, in *Graph Theory and Computing*, Academic Press, London, p. 87 (1972).
106. K. D. Bartle, Recent Advances in the Analysis of Polycyclic Aromatic Compounds by Gas Chromatography, in *Handbook of Polycyclic Aromatic Hydrocarbons*, Vol. 2, A. Bjorseth and T. Ramdahl, Eds., Marcel Dekker, New York (1985).
107. H. Borwitzky and G. Schomburg, *J. Chromatogr.*, **170**, 99 (1979).

108. T. Romanowski, W. Funcke, J. Grossmann, J. König, and E. Balfanz, *Anal. Chem.*, **55**, 1030 (1983).
109. G. Rhodes, R. B. Opsal, J. T. Meek, and J. P. Reilly, *Anal. Chem.*, **55**, 280 (1983).
110. M. M. Andino, M. A. Kosinski, and J. D. Winefordner, *Anal. Chem.*, **58**, 1730 (1986).
111. G. W. Sutter, A. J. Kallir, U. P. Wild, and T. Vo-Dinh, *J. Phys. Chem.*, **90**, 4941 (1986).

PHASE-RESOLVED FLUORESCENCE SPECTROSCOPY

LINDA B. McGOWN AND KASEM NITHIPATIKOM

Department of Chemistry
Duke University
Durham; North Carolina

1. INTRODUCTION

Fluorescence spectroscopy is a natural choice for the selective and sensitive analysis of polycyclic aromatic compounds (PACs). Because the fluorescence experiment is based on the measurement of an emission signal that is directly proportional both to the concentration(s) of fluorescent component(s) and to the intensity of the exciting light, detection limits are often at the sub-nanomolar level and the linear dynamic range often spans five orders of magnitude. Selectivity is a result of the native fluorescence of many PACs and the nonfluorescence of most other groups of compounds, both organic and inorganic. Within the PACs, selectivity is accomplished through the exploitation of the diverse dimensions of information inherent in the fluorescence process. These include emission and excitation spectral characteristics, fluorescence lifetime, collisional quenching, and both linear and circular polarization effects. Further selectivity can often be achieved with reagents that selectively quench or enhance fluorescence, as well as with organized media such as micelles and cyclodextrins.

In addition to being a selective and sensitive technique for chemical analysis, fluorescence spectroscopy is also nondestructive. Also, the fluorescence characteristics of a molecule depend not only on the molecule itself but also on the microenvironment of the molecule. Therefore, through the combination of all of the above-mentioned features, fluorescence spectroscopy is uniquely suited to the characterization of samples in terms of both (a) identification and quantitation of PACs at trace levels and (b) dynamic intermolecular interactions of PACs with each other and with sample matrix constituents. For example, fluorescence spectroscopy could be used not only to determine carcinogenic PACs in human serum samples but also to study the distributions of the PACs between the aqueous phase and the serum proteins. Interactions between the PACs themselves such as aggregation and energy transfer could also be studied.

Sample characterization by fluorescence spectroscopy is best achieved through the combination of all of the multiple dimensions of information available in a fluorescence experiment. For example, fluorescent compounds with highly overlapping emission and excitation spectra often have very different fluorescent lifetimes, so that their intensity contributions can be resolved in the time domain. Therefore, the combination of fluorescence spectral and lifetime dimensions can greatly strengthen the ability of fluorescence spectroscopy to analyze multicomponent samples. Both the wavelength-dependent and time-dependent aspects of fluorescence emission can be incorporated into a single intensity function by means of phase-resolved fluorescence spectroscopy (PRFS). Recent articles have reviewed advances in PRFS (1–8). In this chapter, we will discuss the theory, instrumentation, and techniques of PRFS, with emphasis on the various ways in which PRFS can be used for the determination of PACs.

2. THEORY

PRFS is based on the phase-modulation technique for the determination of fluorescence lifetimes. The sample is excited with light that has been modulated at a high frequency. For example, the exciting light may have a sinusoidal waveform:

$$E(t) = A(1 + m_{ex} \sin \omega t) \qquad (1)$$

where $E(t)$ is the total excitation intensity at time t, A is the steady-state (DC) component of the exciting light, m_{ex} is the modulation depth of the exciting light (i.e., the ratio of the amplitude of the modulated, or AC, component to the steady-state, or DC, component), and ω is the angular modulation frequency. The fluorescence response function will have the same frequency ω but will be demodulated and phase-shifted to an extent determined by the fluorescence lifetime of the emission:

$$F(t) = A'[1 + m_{ex}m \sin(\omega t - \phi)] \qquad (2)$$

Here, A' is the DC component of the fluorescence emission, m is the demodulation of the signal $((AC/DC)_{em}/(AC/DC)_{ex}$, or $(AC/DC)_{em}/m_{ex})$, and ϕ is the phase shift. The $E(t)$ and $F(t)$ functions are depicted in Figure 1. Fluorescence lifetimes can be calculated from the phase shift:

$$\tau_p = (1/\omega) \tan \phi \qquad (3)$$

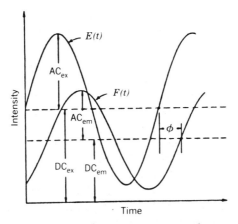

Figure 1. Schematic representation of the excitation $E(t)$ and fluorescence $F(t)$ waveforms. Reproduced from ref. 32, with permission

or from the demodulation

$$\tau_m = 1/\omega \ (1/m^2 - 1)^{1/2} \tag{4}$$

In PRFS, the AC component of the emission function $F(t)$ is multiplied by a square-wave function $P(t)$ of the same modulation frequency ω and is then integrated over time. The function $P(t)$ may alternate between the values 1 and 0, in which case it serves as an "on–off" switch to yield phase-resolved fluorescence intensities (PRFI) of the form

$$F(\phi_D) = A'[m_{ex} m \cos(\phi_D - \phi)] \tag{5}$$

The detector phase angle, ϕ_D, is variable and can be set to any value between $0°$ and $360°$. Other waveforms besides the on–off function can be used for PRFS. For example, the value of $P(t)$ can alternate between 1 and -1. The advantage of this waveform is that all of the AC intensity is integrated, rather than only one-half as in the on–off function.

At a given modulation frequency, ϕ_D can be adjusted to be in phase with the fluorescence of components with a given fluorescence lifetime, thereby selectively enhancing the PRFI of those components relative to other compounds with longer or shorter fluorescence lifetimes. Conversely, the PRFI of a component can be reduced to zero by setting ϕ_D to be $90°$ out of phase with the emission from that component. Alternatively, at a given ϕ_D setting, fluorescence components can be selectively enhanced as a function of fluorescence lifetime through the use of modulation frequency as the selectivity parameter. For example, consider a multicomponent system in which each component contributes equally to the total steady-state intensity at the wavelengths used for the measurements. In terms of equation (5), this means that the factor $A'm_{ex}$ is the same for all of the components. However, the demodulation and phase shift for each component will depend on the fluorescence lifetime of the component and on ω. Let us say, for example, that ϕ_D is adjusted to suppress scattered light. Since scattered light has a lifetime of zero and therefore a phase of zero, ϕ_D must be set to 90° relative to the exciting light to suppress the scattered light. Under these conditions, we can predict the limits that will occur for the PRFI of each component in the system as a function of ω. For components with short lifetimes, m will approach unity and ϕ will approach zero. Since $\cos(90° - \phi)$ will approach zero as lifetime, and therefore ϕ, approach zero, the PRFI will approach zero as τ approaches zero. On the other hand, for components with long lifetimes, ϕ approaches 90° and $\cos(90° - \phi)$ approaches unity, but m approaches zero. Therefore, at a given modulation frequency and with $\phi_D = 90°$, PRFI as a function of fluorescence lifetime will rise from zero to a maximum value and descend back to zero.

Figure 2 shows the theoretical dependence of the $m \cos(90° - \phi)$ term on fluorescence lifetime for several different modulation frequencies. As the modulation frequency is increased, the fluorescence lifetime at which the maximum value for $m \cos(\phi_D - \phi)$ occurs will decrease. In other words, shorter-lifetime components are enhanced at higher modulation frequencies, and longer-lifetime components are enhanced at lower modulation frequencies. Clearly, modulation frequency is a powerful parameter through which

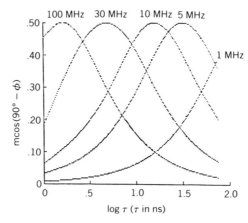

Figure 2. Theoretical dependence of phase-resolved fluorescence intensity on fluorescence lifetime for five modulation frequencies. The detector phase angle of 90° corresponds to the setting required to suppress scattered light intensity. Reproduced from ref. 31, with permission.

fluorescence lifetime "windows" can be created in order to selectively observe fluorescent compounds in a given fluorescence lifetime range. By varying ω, successive windows may be scanned in order to improve the selectivity for multicomponent determinations of fluorescent compounds and to study the dynamic behavior of fluorescent molecules in samples as a function of fluorescence lifetime.

3. INSTRUMENTATION

Numerous instruments have been designed for phase-modulation fluore-scence lifetime determinations, from the first instrument used by Gaviola (9) to the state-of-the art multifrequency instruments. Modern instruments generally resemble modern fluorometers and use similar components, includ-ing arc lamp or laser sources, monochromator or filter wavelength selectors, and PMT detectors. Excitation modulation can be accomplished by an acousto-optical device such as the Debye–Sears ultrasonic modulator, in which the modulation frequencies are limited to a single fundamental frequency and several harmonic overtones. Alternatively, multifrequency instruments have been designed in which an electro-optical modulator such as a Pockels cell is used to achieve a wide range of essentially continuously selectable modulation frequencies (10–13). Performance has been greatly enhanced through the use of cross-correlation (14–16) in which the phot-omultiplier gain is modulated at frequency $\omega + \Delta\omega$ to produce a cross-correlation signal of $\Delta\omega$ which is isolated with a low-pass filter.

The first PRFS experiments were performed by Veselova et al. (17) using an instrument based on a design by Bonch-Bruevich et al. (18). The instrument used (a) an optical diffraction modulator operating at 11.2 MHz, (b) a light dividing system, (c) photomultiplier tube (PMT) detection, and (d) a phase-meter unit which measured the phase difference between the dual inputs from the sample and a scattering solution. Phase-resolved spectra were acquired by using either an optical or electrical phase-shifter to adjust the detector phase to be out of phase with the emission of the component to be suppressed.

Modern instruments for PRFS have been described in which argon ion lasers with external amplitude modulators are used for excitation, and PMT detectors are combined with commercial lock-in amplifiers to accomplish phase-sensitive detection (19, 20). Phase-modulation instruments with con-tinuous xenon arc lamp excitation, Debye–Sears acousto-optic modulation, and cross-correlation electronics have been described for PRFS (3, 5). More recently, multifrequency cross-correlation spectrofluorometers for phase-modulation fluorescence lifetime determinations and PRFS have been devel-oped. For example, the instrument shown in Figure 3 has an electro-optic

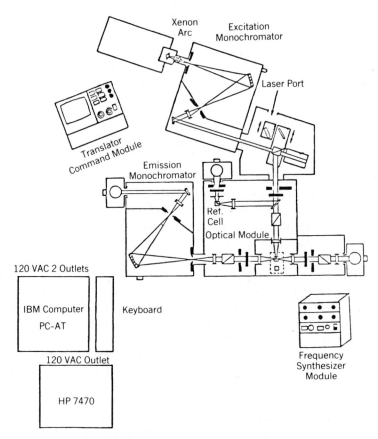

Figure 3. Schematic diagram of the SLM 48000S multifrequency phase-modulation spectro-fluorometer (SLM instruments, Urbana, Illinois.)

Pockels cell modulator which provides a frequency range of 1–250 MHz, adjustable in 1-MHz increments.

4.   APPLICATIONS

4.1.   Phase-Resolved Spectra

If a mixture of two fluorescent components A and B is measured with the detector set 90° out of phase with one of the components, the individual spectrum of the other component will be observed (Figure 4). In other words, if $\phi_D = \phi_A \pm 90°$, the phase-resolved spectrum of B will be obtained, and vice

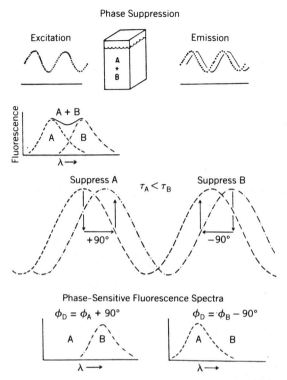

Figure 4. Intuitive description of phase suppression for a fluorophore mixture. Reproduced from ref. 5, with permission.

versa. The spectrum of the observed (nonsuppressed) component will be attenuated by a factor equal to the sine of the difference between the phase of the observed component and the phase of the suppressed component. Therefore, as the fluorescence lifetimes of A and B approach each other, the difference between the phases of A and B approaches zero, the sine of the difference approaches zero, and the intensity of the observed component approaches zero. This is simply another way of saying that phase resolution is poor when the fluorescence lifetime difference between the components is small.

Several studies have demonstrated the resolution of the spectra of the individual components in mixtures of two PACs. The original work by Veselova et al. demonstrated two applications of PRFS (17). First, the spectra of 3-amino-N-methylphthalimide ($\tau = 2.25$ ns) and 3,6-diacetylamino-N-methylphthalimide ($\tau = 10.7$ ns) were resolved from a 1:1 mixtures of the two. Second, the spectra of anthracene in the monomeric ($\tau = 3.59$ ns) and aggrega-

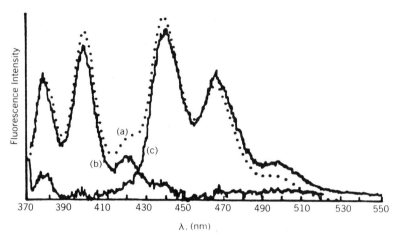

Figure 5. (*a*) The steady-state emission spectrum of a 1:1 mixture of anthracene and perylene in methanol elicited by a 362-nm excitation. (*b*) The PRFS spectrum obtained by nulling the contribution of perylene at 450 nm. (*c*) The PRFS spectrum obtained by nulling the contribution of anthracene at 390 nm. Reproduced from ref. 2, with permission.

ted ($\tau = 7.56$ ns) forms were resolved for a 2×10^{-4} M solution of anthracene that was quick frozen at $-196°C$. Other mixtures of PACs that have been spectrally resolved with PRFS include perylene/anthracene (Figure 5) (2) and dibenzo[*a,h*]anthracene/dibenzo[*c,g*]carbazole (5).

4.2. Quantitative Analysis of Multicomponent Systems

In the absence of synergistic effects, the phase-resolved fluorescence intensity [PRFI, expressed as $F(\phi_D)$ in equation (5)] of a mixture is simply the sum of the PRFIs of the fluorescent components in the mixture. At a given set of wavelength, detector phase angle, and modulation frequency conditions, we obtain

$$\text{PRFI}_{\text{total}} = \sum_{i=1}^{n} \text{PRFI}_i = \sum_{i=1}^{n} I_i C_i \tag{6}$$

where I_i is the molar PRFI of component i, and C_i is the concentration of component i. The value for I_i is found by measuring a standard solution of component i under the same conditions used to measure $\text{PRFI}_{\text{total}}$. For an n-component system, $\text{PRFI}_{\text{total}}$ and the I values are experimentally determined at m sets of experimental conditions to generate m independent linear equations of the form of equation (6). If $n = m$, the concentrations of the components can be found through either an exact solution or a best fit to the equations. If $m > n$, the system is overdetermined and a least-squares best fit solution is required.

A four-component system of anthracence and three chloroanthracene derivatives was analyzed through the use of an overdetermined system in which different combinations of emission and excitation wavelengths and detector phase angle were used at a single modulation frequency (21). The results are summarized in Table 1 in terms of the determination errors. A second four-component system of anthracene derivatives was analyzed by PRFS using two modulation frequencies (22). The fluorescence emission spectra of the four components are shown in Figure 6, and the determination errors are summarized in Table 2.

In the above-mentioned four-component studies, the measurement conditions were tailored to the particular components known to be in the solutions. In order to broaden the applicability of PRFS to include samples in which the identities of the fluorescent components are not necessarily known, it is necessary to use a more generalized data format in which the experimental parameters are systematically varied. One such data format has been described in which PRFI is plotted as a function of synchronously scanned wavelength on one axis and detector phase angle on the other (23). The advantages of synchronous excitation relative to conventional emission or

Table 1. Errors of Determination for a Single Four-Component Mixture Using Eight Equations (Four Φ_D's)[a]

Solution Number	Components[b]				$\%E^c$	$|\%E|^d$
	A	1CA	2CA	9CA		
1	1.276	1.175	1.292	1.880	−0.33	1.62
2	0.638	1.175	1.292	1.880	1.44	2.75
3	1.276	0.588	1.292	1.880	−0.86	2.94
4	1.276	1.175	0.646	1.880	−0.33	0.79
5	1.276	1.175	1.292	0.940	−0.12	0.20
6	1.701	0.157	0.172	2.507	1.48	14.6
7	1.701	0.783	0.862	2.507	0.09	2.47
8	1.701	0.0157	0.0172	2.507	16.6	124

[a] 30 MHz, at $\lambda_{ex} = 360$ nm and $\lambda_{em} = 382$ and 418 nm. From ref. 21, with permission.
[b] Concentration (micromolar) in cuvette. A = anthracene; 1CA = 1-chloroanthracene; 2CA = 2-chloroanthracene; 9CA = 9-chloroanthracene.
[c] Average % error.
[d] Average |% error|.

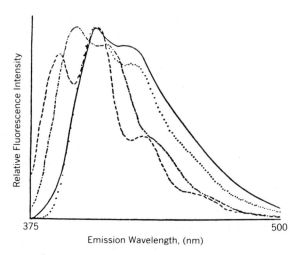

Figure 6. Steady-state emission spectra of 1-chloroanthracene (solid line), 9-phenylanthracene (dot–dash line), 9,10-diphenylanthracene (dotted line), and 9-vinylanthracene (dashed line), excitation at 360 nm. Reproduced from ref. 21, with permission.

Table 2. Error of Determination for the Individual Components in a Single Four-Component Mixture as a Function of Both Modulation Frequency and Emission Wavelength[a]

| Equations Generated | Relative Error (%)[b] | | | | %E[c] | \|%E\|[d] |
	1CA	9PA	DPA	9VA		
8[e]	241	−16100	−1970	366	−4370	4670
8[f]	287	−95.0	−14600	38.0	−3600	3770
12[g]	0.41	−91.0	14.5	416	84.9	130
12[h]	17.8	40.9	−11.9	247	73.4	79.4
16[i]	12.5	36.8	−9.87	59.3	24.7	29.6
24[j]	0.25	1.96	−0.19	−1.42	0.15	0.96

[a] From ref. 22, with permission.
[b] 1CA = 1-chloroanthracene; 9PA = 9-phenylanthracene; DPA = 9,10-diphenylanthracene; 9VA = 9-vinylanthracene.
[c] Average % error.
[d] Average |% error|.
[e] 18 MHz, emission at 385 and 425 nm.
[f] 30 MHz, emission at 385 and 425 nm.
[g] 18 MHz, emission at 385, 395, and 425 nm.
[h] 30 MHz, emission at 385, 395, and 425 nm.
[i] 18 and 30 MHz, emission at 385 and 425 nm.
[j] 18 and 30 MHz, emission at 385, 395, and 425 nm.

excitation scanning include spectral band narrowing and a reduction in the number of spectral bands for each component (24, 25). The use of synchronous excitation–PRFS at one or more modulation frequencies has been shown to increase the determination accuracy for multicomponent systems (23, 26). Determination errors are shown in Table 3 for the synchronous excitation–PRFS and steady-state synchronous excitation analyses of mixtures containing two to six PACs. Accuracy was significantly better for the PRFS determinations for the systems containing four or more components.

A different approach has also been described for PRFS multicomponent determinations in which phase-resolved spectra of the mixtures are fit with

Table 3. Summary of Errors for Phase-Resolved Fluorescence Spectroscopy (PRFS) and Steady-State (SS) Synchronous Excitation Determinations of 9-Phenylanthracene (9PA), 9,10-Diphenylanthracene (DPA), Benzo[k]fluoranthene (BkF), Benzo[a]pyrene (BaP), 1,3,6,8-Tetraphenylpyrene (TPP), and Benzo[ghi]perylene (BgP)[a]

	PRFS		SS[b]	
Mixture	%E[c]	|%E|[d]	%E[c]	|%E|[d]
Two-component (9PA, DPA)	1.2	2.4	−0.2	1.7
Two-component (DPA, BkF)	0.7	3.2	0.7	1.8
Three-component (9PA, DPA, BkF)	−0.9	2.8	−1.6	2.8
Four-component (9PA, DPA, BkF, BaP)	−1.2	4.4	−2.0	4.9
Five-component (9PA, DPA, BkF, BaP, TPP)	0.9	5.8[e]	−1.5	8.0
Six-component (9PA, DPA, BkF, BaP, TPP, BgP)	−3.2	7.5[e]		[f]

[a] From refs. 23 and 26, with permission.
[b] Steady-state determinations with 24 equations generated using 24 wavelengths.
[c] Relative errors (%) averaged for all solutions.
[d] Average relative error magnitudes (%).
[e] PRFS determinations with 96 equations generated using 24 wavelengths and four detector phase angles (180°, 225°, 270°, and 315°).
[f] Error undefined due to a value of zero obtained for the concentration of the component in one or more solutions.

steady-state spectra of the individual components using nonlinear least-squares analysis (25). The phase-resolved spectra are collected at a series of detector phase angles at a single modulation frequency. For example, 15 individual phase-resolved spectra were used for the determination of mixtures of 9,10-diphenylanthracene ($\tau = 5.87$ ns) and 9-methylanthracene ($\tau = 4.47$ ns). This approach differs from the simultaneous-linear-equation approach described above in that (a) many more measurements are required, (b) standards and mixtures are measured under different experimental conditions (steady-state for the former, PRFS for the latter) and (c) fluorescence lifetimes of the components must remain constant over the spectral wavelength range.

4.3. Phase-Resolved Suppression of Scattered Light

Steady-state fluorescence spectra often contain significant intensity contributions from scattered light. As discussed in Section 2, scattered light has a lifetime of zero on the scale of fluorescence experiments and can be suppressed in PRFS by setting the detector phase at 90°. This will suppress all scattered light contributions, including Raman, particulate, and Rayleigh scattered light. Phase resolution has been used to suppress the Raman scattered light signal of water in the fluorescence emission spectra of quinine bisulfate (27) and rhodamine 6G (28).

The determination of PACs can be enhanced through the phase-resolved suppression of scattered light. For example, PRFS has been used to suppress scattered light background in the synchronous-excitation fluorimetric determination of five PAC compounds (29). Scattered light is particularly troublesome in synchronous excitation determinations that require a small scanning

Table 4. Limits of Detection Calculated for Five PACs[a]

PAC	PRFS		Steady-State	
	I[b]	II[c]	I[b]	II[c]
Anthracene	18	27	5.2	93
BkF	2.6	5.7	1.0	9.0
BaP	22	28	3.2	68
Perylene	2.2	4.1	0.2	4.5
BgP	98	150	12	130

[a] Limits of detection (nanomolar) in cuvette. BkF, BaP, and BgP are defined in Table 3.
[b] Excludes uncertainties in sensitivity and y-intercept.
[c] Includes uncertainties in sensitivity and y-intercept.

difference ($\Delta\lambda$) between the emission and excitation wavelengths, since the spectral region traversed by the synchronous scan is so close to the Rayleigh and particulate scattered light region. Using a $\Delta\lambda$ of 3 nm, which is typical for PAC determinations, detection limits were significantly improved for three of the five PACs studied when PRFS was used to suppress the scattered light (Table 4). The steady-state and PRFS synchronous spectra for a mixture of three of the PACs are shown in Figure 7.

Phase-resolved fluorescence spectroscopy can also be used to eliminate scattered light contributions in total luminescence spectra. This has been demonstrated for crude oils (Figure 8) and human serum samples (Figure 9) (30). Suppression of scattered light should improve the analysis of highly scattering samples, first by allowing the direct observation of spectral features that would otherwise be obscured by the scattered light peaks and, second, by enabling the use of higher sample concentrations in order to increase the intensities of minor fluorescent components. Phase-resolved suppression of scattered light can play an important role in the identification and quantitation of PACs in highly scattering samples and in the spectral fingerprinting of such samples.

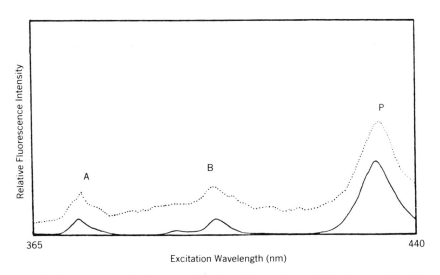

Figure 7. Synchronous excitation spectra ($\Delta\lambda = 3$ nm) for a mixture of (A) anthracene (47.1 nM), (B) benzo[*a*]pyrene (50.2 nM), and (P) perylene (10.4 nM) acquired using steady-state measurements with blank correction (dotted line) and phase resolution with scattered light suppressed (solid line). Reproduced from ref. 29, with permission.

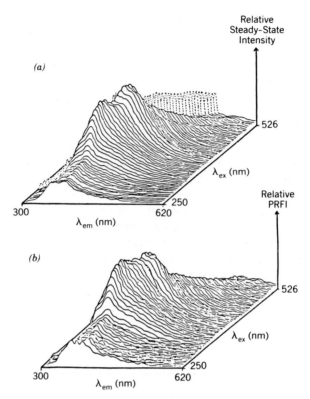

Figure 8. Total luminescence spectra of Mayan crude oil (5000-fold dilution in cyclohexane). (*a*) Steady-state spectrum, with scattered light indicated by dashed lines. (*b*) Phase-resolved spectrum acquired with scattered light suppressed, at 30-MHz modulation frequency. Reproduced from ref. 30, with permission.

4.4. Qualitative Analysis

It was shown in Section 2 that modulation frequency can be used in conjunction with phase-resolved suppression of scattered light to create fluorescence lifetime "windows." This has been demonstrated for a simple two-component system (31) consisting of benzo[k]fluoranthene (BkF, τ = 8 ns) and benzo[b]fluoranthene (BbF, $\tau = 29$ ns). Total luminescence spectra of BkF, BbF, and a mixture of the two were acquired first under steady-state conditions and, second, using PRFS at three modulation frequencies with scattered light suppressed. The individual steady-state spectra of BkF and BbF are shown in Figure 10. The steady-state and PRFS spectra of the mixture are shown as contour plots in Figure 11 and as three-dimensional

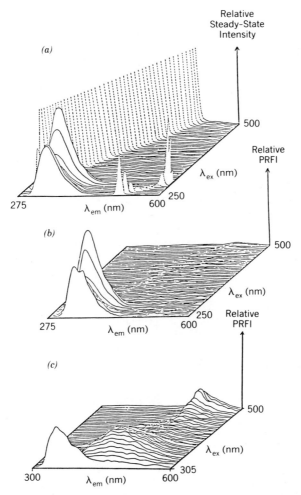

Figure 9. Total luminescence spectra of pooled serum. (*a*) Steady-state spectrum of 20-fold dilution of serum, with first- and second-order scattered light indicated by dashed lines. The first-order scattered light extends off-scale. (*b*) Phase-resolved spectrum of 20-fold dilution of pooled serum with scattered light suppressed. (*c*) Phase-resolved spectrum of five-fold dilution of pooled serum showing visible region only, with scattered light suppressed. Spectra in parts *b* and *c* were acquired at 30-MHz modulation frequency. Reproduced from ref. 30, with permission.

projections in Figure 12. As expected, the longer-lived BbF emission dominates the PRFS spectrum of the mixture at the lower modulation frequency (6 MHz), whereas the shorter-lived BkF emission is the dominant contributor to the spectrum at the higher modulation frequencies (18 and 30 MHz). From this example, it is clear that multifrequency PRFS can be used to selectively

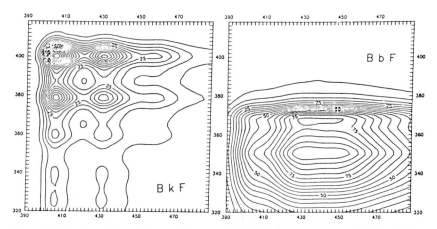

Figure 10. Contour plots of the steady-state total luminescence spectra of BkF and BbF. Excitation wavelength and emission wavelength are plotted on the vertical and horizontal axes, respectively, in nanometers.

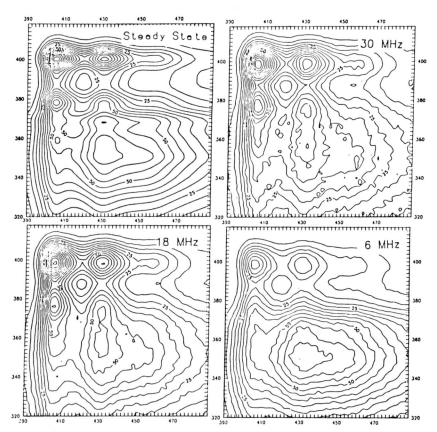

Figure 11. Contour plots of the steady-state and phase-resolved total luminescence spectra of a mixture of BkF and BbF. Axes same as in Figure 10.

216

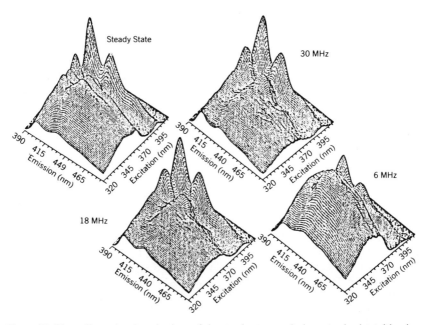

Figure 12. Three-dimensional projections of the steady-state and phase-resolved total lumine-scence spectra of the mixture of BkF and BbF. Reproduced from ref. 31, with permission.

enhance or reduce the intensity contributions of fluorescent compounds as a function of their fluorescence lifetimes.

5. CONCLUSIONS

The use of steady-state fluorescence spectroscopy for the analysis of complex samples containing numerous fluorescent PACs is limited by the ability to resolve the fluorescence contributions of the individual PACs in the spectral domain. The additional resolution that can be achieved through the temporal dimension of fluorescence lifetime greatly increases the power of fluorescence spectroscopy to analyze complex, multicomponent samples. Phase-resolved fluorescence spectroscopy provides a convenient means for the incorporation of fluorescence lifetime information into fluorescence spectra. The availability of commercial phase-modulation spectrofluorometers with phase-resolution capabilities has greatly increased the accessibility of the PRFS technique to the analytical community and will hopefully result in greater exploration of the potential of PRFS for the analysis of PAC-containing samples.

REFERENCES

1. L. B. McGown and F. V. Bright, *Anal. Chem.*, **56**, 1400A (1984).
2. J. R. Mattheis, G. W. Mitchell, and R. D. Spencer, Phase-Resolved Nanosecond Spectrofluorometry: Theory, Instrumentation and New Applications of Multicomponent Analysis by Subnanosecond Fluorescence Lifetimes, In *New Directions in Molecular Luminescence*, D. Eastwood, Ed., ASTM STP 822, American Society for Testing and Materials, Philadelphia, p. 50 (1983).
3. R. D. Fugate, J. D. Bartlett, Jr., and J. R. Mattheis, *Bio. Tech.* **May/June** (1984).
4. K. Berndt, *Opt. Commun.*, **56**, 30 (1985).
5. J. R. Lakowicz and H. Cherek, *J. Biochem. Biophys. Meth.*, **5**, 19 (1981).
6. E. Gratton and D. M. Jameson, *Anal. Chem.*, **57**, 1694 (1985).
7. J. R. Lakowicz, *Principles of Fluorescence Spectroscopy*, Plenum Press, New York (1983).
8. L. B. McGown and F. V. Bright, *Crit. Rev. Anal. Chem.*, **18**(3), p. 245 (1987).
9. E. Gaviola, *Z. Phys.*, **42**, 853 (1927).
10. M. Hauser and G. Heidt, *Rev. Sci. Instrum.* **46**, 470 (1975).
11. E. Gratton and M. Limkeman, *Biophys. J.*, **44**, 315 (1983).
12. E. Gratton, D. M. Jameson, N. Rosato, and G. Weber, *Rev. Sci. Instrum.*, **55**, 486 (1984).
13. J. R. Lakowicz and B. P. Maliwal, *Biophys. Chem.*, **21**, 61 (1985).
14. J. B. Birks and W. A. Little, *Proc. Phys. Soc. London Sec. A*, **66**, 921 (1953).
15. A. Schmillen, *Z. Phys.*, **135**, 294 (1953).
16. R. D. Spencer and G. Weber, *Ann. N.Y. Acad. Sci.*, **159**, 361 (1969).
17. T. V. Veselova, A. S. Cherkasov, and V. I. Shirokov, *Opt. Spectrosc.*, **29**, 617 (1970).
18. A. M. Bonch-Bruevich, V. A. Molchanov, and V. I. Shirokov, *Bull. Acad. Sci. U.S.S.R. Phys. Ser.*, **20**, 541 (1956).
19. J. N. Demas, W. M. Jones, and R. A. Keller, *Anal. Chem.*, **58**, 1717 (1986).
20. A. van Hoek and A. J. W. G. Visser, *Anal. Instrum.*, **14**, 143 (1985).
21. F. V. Bright and L. B. McGown, *Anal. Chem.*, **57**, 55 (1985).
22. F. V. Bright and L. B. McGown, *Anal. Chem.*, **57**, 2877 (1985).
23. K. Nithipatikom and L. B. McGown, *Anal. Chem.*, **58**, 2469 (1986).
24. T. Vo-Dinh, *Anal. Chem.*, **50**, 396 (1978).
25. T. Vo-Dinh, R. B. Gammage, A. R. Hawthorne, and J. H. Thorngate, *Environ. Sci. Technol.*, **12**, 1297 (1978).
26. K. Nithipatikom and L. B. McGown, *Appl. Spectrosc.*, **41**, 395 (1987).
27. S. Keating-Nakamoto, H. Cherek, and J. R. Lakowicz, *Anal. Biochem.* **148**, 349 (1985).
28. J. N. Demas and R. A. Keller, *Anal. Chem.*, **57**, 538 (1985).
29. K. Nithipatikom and L. B. McGown, *Anal. Chem.*, **58**, 3145 (1986).
30. K. Nithipatikom and L. B. McGown, *Appl. Spectrosc.*, **41**, 1080 (1987).
31. D. W. Millican, K. Nithipatikom, and L. B. McGown, *Spectrochim. Acta.*, **43B**, 629 (1988).
32. E. Gratton, D. M. Jameson, and R. D. Hall, *Annu. Rev. Biophys. Bioeng.*, **13**, 105 (1984).

CHAPTER

8

MASS SPECTROMETRY OF POLYCYCLIC AROMATIC COMPOUNDS

RONALD A. HITES

School of Public and Environmental Affairs
and Department of Chemistry
Indiana University
Bloomington, Indiana

1. INTRODUCTION

Mass spectrometry (MS) has been used for the analysis of polycyclic aromatic compounds (PACs) since 1951 (1). This early realization that MS was an ideal tool for the analysis of these compounds stemmed from the clear spectra that they produced. Since 1951, there have been hundreds of papers reporting on the use of MS for the qualitative and quantitative analysis of PACs. Many of these applications have been reviewed elsewhere; excellent reviews have been published by Bartle et al. (2), Lee, et al. (3), Schuetzle and Hampton (4), and Howard and Mills (5). This review will not focus on applications; rather, it will focus on the processes taking place in the ion source of the mass spectrometer. This review will discuss the fragmentation mechanisms of various classes of PACs, will look at the behavior of these compounds under chemical ionization conditions (both positive and negative ion modes), and will explore the application of desorption ionization methods. The formation of ions by collisional processes will also be presented in the context of MS–MS methods.

2. ELECTRON-IMPACT IONIZATION

The electron-impact mass spectra of PACs are usually quite simple. Because of the extensive electron delocalization of these molecules, the molecular ions in these spectra are very intense and fragment ions are small. Because of this lack of fragmentation, it is usually impossible to distinguish among isomers using electron-impact MS alone. Thus, distinguishing among isomers has driven research on the mass spectra of PACs. The paragraphs that follow discuss the electron-impact mass spectra of a series of PACs that are

219

derivatives of phenanthrene and anthracene. The reader should remember that the features observed in the spectra of these relatively small PACs are also observed in the spectra of other, larger PACs. As an intellectual convenience, this discussion will focus on this one series.

The mass spectrum of phenanthrene (see Figure 1A) is almost identical to that of anthracene. Both spectra are dominated by a molecular ion at m/z 178. Losses of one to four hydrogens are observed; the loss of H_2 predominates. Because of the intensity of the molecular ion, the doubly charged molecular ion at m/z 89 is of medium intensity. The only other fragment ions are at m/z 152, due to a loss of C_2H_2 from the molecular ion, and a doubly charged version of that ion at m/z 76. Note that the abundances of the singly and doubly charged $(M - C_2H_2)$ ions are about the same. The relative abundances of the doubly charged molecular ions are a strong function of the size of the aromatic system. The molecule's ability to stabilize a second charge increases as the number of pi electrons increases (which is, of course, related to the number of carbon atoms in the molecule); this effect is shown in Figure 2 (6). For example, the M^{2+} ion of a large polycyclic aromatic hydrocarbon (PAH) such as ovalene ($C_{32}H_{14}$) is 43% relative to M^+.

In general, in the mass spectra of all PAHs, the molecular ion is always the most intense peak. Loss of two hydrogens is frequently the second most intense peak. The doubly charged molecular ion is of medium intensity, and both singly and doubly charged $(M - C_2H_2)$ ions are present. Furthermore, without gas chromatographic retention data or other supplementary data, it is impossible to distinguish among isomeric compounds.

The mass spectra of alkylated PAHs are also dominated by molecular ions and fused-ring tropylium ions. For example, the electron-impact mass spectrum of 1-methylphenanthrene is shown in Figure 1B. The mass spectra of the five different methylphenanthrene isomers and of the three different methylanthracene isomers all resemble one another. Note the intense molecular ion at m/z 192 and the intense $(M - H)^+$ ion at m/z 191. This latter ion may result from the expansion of the ring to incorporate the methyl carbon. The rest of the mass spectrum is similar to that of a PAH with a molecular formula of $C_{15}H_{11}$. For example, the fused-ring tropylium ion at m/z 191 loses two hydrogens to give an ion of medium intensity at m/z 189; the m/z 191 ion can also lose a C_2H_2 group to give the ion at m/z 165. The other ion of significant intensity in the mass spectrum is the doubly charged molecular ion at m/z 96.

The structure of the fused-ring tropylium ion is probably different, depending on where the methyl group was attached. For example, 1-, 2-, 3-, and 4-methylphenanthrene would all give a fused-ring tropylium ion with the seven-membered ring at the end; see structure I in Figure 3. 9-Methylphenanthrene would give a fused-ring tropylium ion in which the seven-membered ring is in the middle; see structure II in Figure 3. A similar argument holds for 1- and 2-

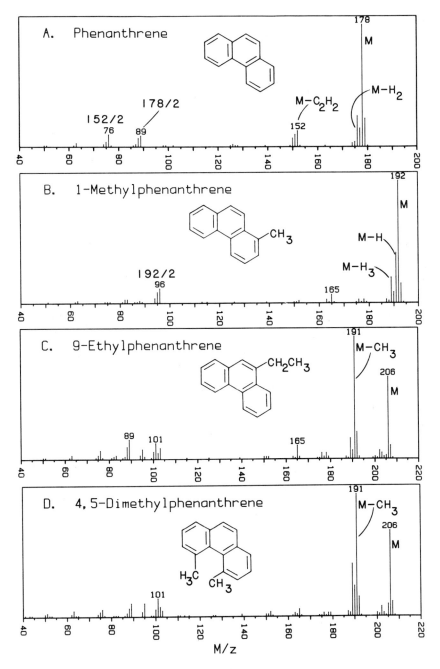

Figure 1. Electron-impact mass spectra of 15 phenanthrene- and anthracene-related compounds. All data are from the Chemical Information System (CIS, Inc., Baltimore, Md.).

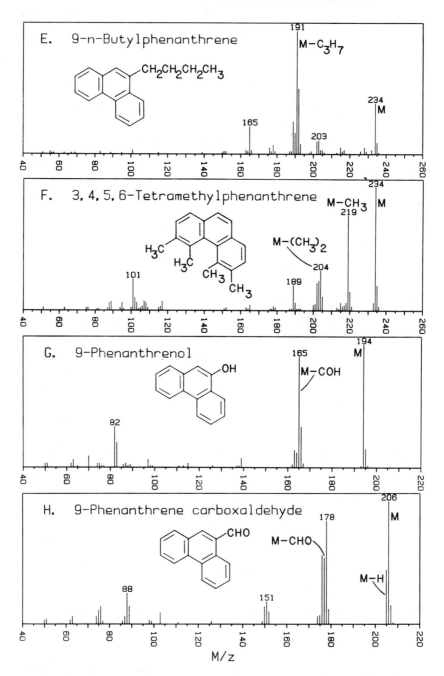

Figure 1 (*Continued*)

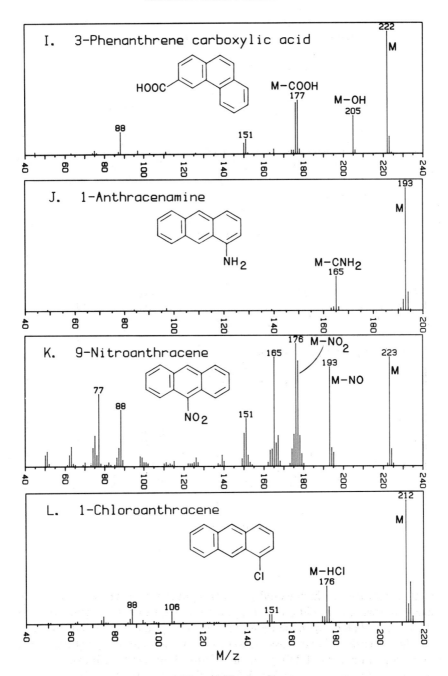

Figure 1 (*Continued*)

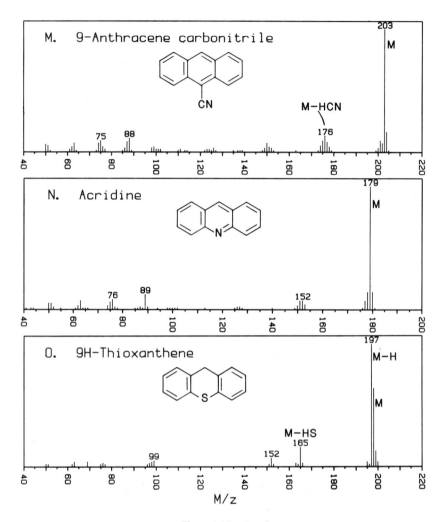

Figure 1 (*Continued*)

methylanthracene, which would give a tropylium ion of type **III**, whereas 9-methylanthracene would give a tropylium ion of type **IV**. It is significant that the spectra of all of these methyl-substituted compounds are identical to one another. Clearly, it is not possible to distinguish among the structures of the four tropylium ions (**I–IV**) given in Figure 3 by electron-impact MS. It remains to be seen whether MS/MS techniques can make this distinction.

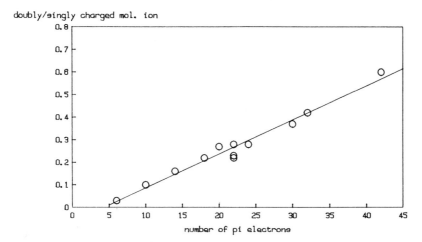

Figure 2. The relative intensity of the doubly to singly charged molecular ions vs. number of pi electrons (in this case, equals number of carbon atoms) in the electron-impact spectra of PAHs. Replotted from ref. 6.

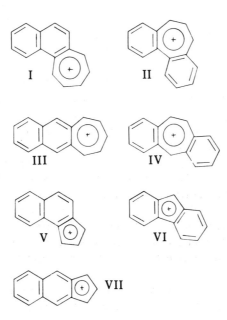

Figure 3. Possible structures of the m/z 191 (**I** to **IV**) and m/z 165 (**V** to **VII**) ions in the electron-impact spectra of phenanthrene- and anthracene-related compounds (see Figure 1).

The ion at m/z 165 occurs in most of the spectra of anthracene- and phenanthrene-based compounds. This ion presumably has one of the three structures given in Figure 3 (see structures V–VII). The exact structure of this ion depends on which of the four fused-ring tropylium ions it came from, although a rearrangement into the most stable form is possible. If this rearrangement takes place, it is quite likely that the structure of m/z 165 is a fluorene ion (see structure VI in Figure 3).

The electron-impact mass spectra of the higher alkyl homologues of PAHs are also dominated by the molecular ion and by fused-ring tropylium ions. It is usually impossible to tell the site(s) of alkyl substitution. In fact, it is frequently difficult to tell anything beyond the total number of alkyl carbons. For example, Figure 1C shows the spectrum of 9-ethylphenanthrene, and Figure 1D shows the spectrum of 4,5-dimethylphenanthrene. Note that these spectra are very similar and, without reference data, it would be impossible to tell whether the alkyl carbons were present as either two methyl groups or one ethyl group. Furthermore, the mass spectra of other ethyl-substituted phenanthrenes and anthracenes are similar to those shown in Figure 1C and 1D, as are the spectra of other dimethyl-substituted phenanthrenes and anthracenes. There are, however, quantitative differences among these spectra. For example, the ratio of the intensity of the molecular ion at m/z 206 to that of the fused-ring tropylium ion at m/z 191 varies from 0.7 for 4,5-dimethylphenanthrene (the most sterically crowded isomer) to 6.3 for 2,7-dimethylphenanthrene (one of the least crowded). It is not clear whether these variations are a result of differences in experimental conditions or reflect real differences in the structures of the molecules.

Alkyl homologues of PAHs with several alkyl carbons also show spectra with the same features. In this case, it is usually possible to distinguish the nature of the alkyl substitution. For example, Figure 1E shows the mass spectrum of 9-butylphenanthrene, and Figure 1F shows that of 3,4,5,6-tetramethylphenanthrene. Clearly, these spectra can be distinguished. On the other hand, it is still impossible to tell where the alkyl substituents are attached on the polycyclic aromatic carbon skeleton. Note that the mass spectrum of the tetramethyl-substituted compound (Figure 1F) simply shows successive losses of methyl groups to give ions at m/z 219, 204, and 189, whereas the butyl-substituted compound (Figure 1E) shows a strong preference for cleavage next to the alpha carbon to give a fused-ring tropylium ion at m/z 191. Also, note the presence of m/z 165 in the spectrum of 9-butylphenanthrene.

The electron-impact mass spectra of hydroxy-substituted PACs are, of course, dominated by the molecular ions. In addition, there is a substantial loss of COH. An example is shown in Figure 1G for 9-phenanthrenol. Note the intense molecular ion at m/z 194 and the intense $(M–COH)^+$ ion at m/z 165. Presumably, this ion has the structure of the fluorene radical cation, as shown in Figure 3, structure **VI**. The spectra of isomeric hydroxy-substituted compounds are almost indistinguishable; thus, it is not possible to fix the position of the hydroxy group or to determine the precise structure of the carbon skeleton.

The electron-impact mass spectra of aldehyde-substituted PACs are also dominated by the molecular ion. These spectra show abundant ions resulting from (a) the loss of a hydrogen, (b) the loss of carbon monoxide, and (c) the loss of the aldehyde group. For example, the mass spectrum of 9-phenanthrene carboxaldehyde (Figure 1*H*) shows a large molecular ion at m/z 206, the loss of a hydrogen to form the ion at m/z 205, the loss of the aldehyde group to form the ion at m/z 177, and a rearrangement ion presumably giving an ion with the structure of phenanthrene itself at m/z 178. As with other such compounds, distinguishing among isomers is impossible. For example, the mass spectrum of 9-anthracene carboxaldehyde is identical to that of its phenanthrene isomer.

Compounds with a carboxylic acid group substituted on the polycyclic aromatic ring system give electron-impact mass spectra dominated by the molecular ion and by ions formed as a result of the loss of hydroxy and carboxy groups. The example spectrum of 3-phenanthrene carboxylic acid (Figure 1*I*) shows these ions at m/z 205, $(M-OH)^+$, and at m/z 177, $(M-COOH)^+$. Distinguishing among isomers is also very difficult in this case, although the database for isomeric compounds is very small.

The mass spectra of polycyclic aromatic amines are dominated by the molecular ion and by a loss of CNH_2. The example for 1-anthracenamine (Figure 1*J*) shows this fragment ion very clearly at m/z 165. Presumably, this ion is formed as a result of excision of the amino group and the associated carbon from the ring system, giving an ion of type **VII** as shown in Figure 3. The mass spectrum of 2-anthracenamine is identical to Figure 1*J*; and thus, isomers cannot be reliably distinguished.

The electron-impact mass spectra of nitro-substituted PACs are more complicated than most of the other compound classes that have been discussed above. An example is 9-nitroanthracene (see Figure 1*K*). There are five very intense ions in its spectrum because of (a) the molecular ion at m/z 223, (b) the $(M-NO)^+$ ion at m/z 193, (c) the $(M-NO_2)^+$ ion at m/z 177, (d) an intense ion at m/z 176, and (e) an intense ion at m/z 165. Presumably, the ion at m/z 165 results from a loss of CO from the ion at m/z 193, which, in turn, resulted from a rearrangement that put an oxygen on the number 9 carbon atom. The m/z 165 ion has the structure of a fluorene radical cation (see structure **VI** in Figure 3). There is some debate about the "M−NO" ion. This ion could also result from the reduction of the nitro group to an amino group $(M - O_2 + H_2)^+$. Under electron-impact conditions, it has been suggested that about 50% of the "M − NO" ion is formed because of such a reduction (7).

The substitution of one or more halogen atoms for one or more hydrogen atoms on the polycyclic aromatic system gives simple spectra, which are dominated by the molecular ion and a loss of HX. For example, the electron-impact mass spectrum of 1-chloroanthracene (Figure 1L) shows a molecular ion at m/z 212, a chlorine-37 isotope peak at m/z 214, and an ion formed as a result of the loss of HCl at m/z 176. The same type of mass spectrum is seen if a nitrile group is substituted for one of the hydrogens. The example given in Figure 1M is the electron-impact mass spectrum of 9-anthracene carbonitrile. It shows an intense molecular ion at m/z 203 and a small loss of HCN to give an ion at m/z 176.

Heterocyclic PACs give electron-impact mass spectra that are very similar to the equivalent hydrocarbon. For example, the mass spectrum of acridine (Figure 1N) looks virtually identical to that of anthracene and phenanthrene with the exception that the molecular ion has been shifted up by 1 dalton. Other compounds that are isomeric with acridine give identical spectra; clearly, heterocyclic PAC isomers cannot be distinguished from one another by electron-impact MS. Incidentally, the MS of aza-arenes has been reviewed by Schmitter and Arpino (8).

The substitution of a sulfur atom into the polycyclic aromatic ring system produces a somewhat different mass spectrum because the $(M - H)^+$ ion is usually more intense than the molecular ion. For example, the mass spectrum of 9H-thioxanthene, shown in Figure 1O, shows an $(M - H)^+$ ion that is almost twice as abundant as the molecular ion; this ion probably has the following structure:

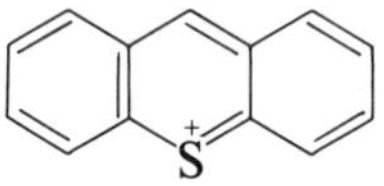

$(m/z = 197)$

The only other fragment ion of significance is an ion at m/z 165, formed as a result of the loss of HS from the molecular ion. Presumably, this ion has the structure of the fluorene radical cation (see VI in Figure 3); this ion has been seen repeatedly in this series of spectra.

3. POSITIVE CHEMICAL IONIZATION

Chemical ionization (CI) MS is a technique based on ion–molecule reactions. The most classic chemical ionization experiment is done in an ion source that has been pressurized to about 1 torr with methane. The initial electron-impact reaction produces CH_4^+ and CH_3^+, which undergo reactions with methane molecules to produce CH_5^+ and $C_2H_5^+$ ions. These ions will, in turn, ionize other molecules by chemical processes. For example, CH_5^+ will donate a proton to an analyte, giving an $(M+H)^+$ ion. Hydrogen abstraction reactions also occur; $C_2H_5^+$ will abstract a proton from an analyte to give an $(M-H)^+$ ion. This CI technique was developed in 1966 by Munson and Field (9). The goal of the technique was to ionize molecules softly, without much transfer of internal energy. The hope was that molecular ion abundances could be enhanced for those compounds which, under 70-eV electron-impact ioniz- ation conditions, did not give molecular ions. This turned out to be the case, and CI is now widely used as a complementary technique to electron-impact ionization.

Given that PAHs produce abundant molecular ions under electron-impact conditions, it seems ironic that chemical ionization has been applied to the analysis of these compounds. Clearly, CI is not used to increase the abundance of the molecular ion but is used for other reasons. Chemical ionization in the positive-ion mode, and in the negative-ion mode as will be discussed later, is useful for distinguishing among isomeric compounds. For example, under the correct chemical ionization conditions, phenanthrene and anthracene can be induced to give different mass spectra. Isomeric com- pounds with heteroatoms also can be made to give different mass spectra by the use of the proper chemical ionization conditions.

Hunt and Sethi (10) were the first to investigate the chemical ionization mass spectra of PAHs. They used deuterated reagent ions formed from D_2O, C_2H_5OD, or ND_3. Seven PAHs, ranging from naphthalene to chrysene and including 0–2 methyl groups, were studied. There was no deuterium exchange for any of these PAHs when D_2O was used as the reagent gas. However, when C_2H_5OD was used, there were varying amounts of deuterium incorporation. Naphthalene and phenanthrene incorporated an average of five deuterium atoms into the $(M+D)^+$ ion, chrysene incorporated an average of one, and anthracene and pyrene incorporated an average of 0.2. The only isomeric pair of molecules studied were phenanthrene and anthracene; these could be distinguished from their positive chemical ionization spectra when C_2H_5OD was used as the reagent gas.

When ND_3 was used as the reagent gas, naphthalene and phenanthrene gave only molecular ions; they failed to give an $(M+D)^+$ ion at all and showed no deuterium exchange. Under the same conditions, anthracene, 9,10- dimethylanthracene, pyrene, and chrysene all showed $(M+D)^+$ ions and varying amounts of deuterium incorporation. This ranged from almost no deuterium exchange for chrysene to an average of 0.7 deuteriums for pyrene. The protonation behavior of these various compounds is obviously related to their proton affinities relative to the proton affinities of C_2H_5OD and ND_3. The isomeric pair, phenanthrene and anthracene, could easily be distinguished by ionization with ND_3. The application of these deuterium exchange techniques for the analysis of PAHs and for distinguishing among isomers has not been pursued by other workers.

Buchanan (11) has reported on another application of positive chemical ionization for the analysis of PAHs; she used NH_3 and ND_3 reagent gases to differentiate primary, secondary, and tertiary amines. A sample was run with NH_3 as the reagent gas and then again with ND_3. The compounds are ionized by a transfer of a hydrogen or deuterium; in addition, hydrogen atoms on the nitrogen atom are exchanged with either hydrogen or deuterium. Clearly, the exchange with hydrogen produces an ion with no shift in mass, but exchange with deuterium does:

$$R_{3-n}NH_n + NH_4^+ \longrightarrow (R_{3-n}NH_{n+1})^+ + NH_3$$
$$R_{3-n}NH_n + ND_4^+ \longrightarrow (R_{3-n}ND_{n+1})^+ + NH_nD_{3-n}$$

Thus, by comparing the mass spectrum obtained with NH_3 versus that with ND_3 and noting the difference in additional mass, one can determine how many deuteriums exchanged onto the nitrogen atom and can thus determine the degree of the amine (n in the above equation). For example, primary amines, such as amino-substituted aromatic compounds, show a difference in mass between ionization with NH_3 and ND_3 of 3 daltons. Secondary amines show a difference of 2 daltons, and tertiary amines, which include both alkyl-substituted anilines and heterocyclics, such as acridine, show a difference of only 1 dalton. This technique has been applied to the analysis of complex mixtures resulting from coal liquefaction processes, and numerous aromatic amines and aza-arenes have been partially identified.

This same experimental strategy has been used for the analysis of oxygen-containing aromatic compounds by using methanol and deuteromethanol (12). For example, an alkylphenol can be ionized with $CH_3OH_2^+$ to give a protonated alkylphenol molecular ion at $(M+1)^+$. If one ionizes with perdeuteromethanol, the reactant ion is then $CD_3OD_2^+$, which ionizes by deuterium transfer and exchanges the active phenolic hydrogen to produce an ion at $(M+3)^+$. A comparison of the two mass spectra shows that there is a difference of 2 daltons, which indicates a phenolic functionality. Ethers, on the other hand, do not have an active hydrogen and are ionized simply by hydrogen or deuterium transfer. In this case, the difference between the spectra from these two ionization techniques is only 1 dalton. This technique has also been used for the analysis of compounds found in coal liquefaction products.

The problem of distinguishing isomeric PAHs with chemical ionization MS has been addressed in most detail by a series of papers from our laboratory; we have used charge exchange chemical ionization MS with argon–methane mixtures as the reagent gas (13–16). This gives different mass spectra of isomeric PAHs; and furthermore, the differences are predictable based on the ionization potential of the molecule, which can be calculated for those compounds for which standards are not available. Under these ionization conditions, the methane forms CH_5^+, which ionizes the analyte by hydrogen transfer, giving an ion at $(M+H)^+$. The Ar^+ ionizes by simple charge exchange, giving an ion at M^+ (17). We have found that the ratio of the $(M+H)^+$ ion to the M^+ ion, as measured by the $(M+1)/M$ ratio, correlates very well with the first ionization potential of the molecule. Since the ionization potentials of PAHs depend on the exact structures of each molecule, this particular ionization technique can produce different spectra for different isomers. Thus, in one GC/MS experiment, components of complex mixtures can be characterized by a gas chromatographic retention index, by a combined electron-impact and methane CI mass spectrum, and by the $(M+1)/M$ ratio. This ionization technique adds to the information content of a mass spectrum and allows the positive confirmation of individual PAH isomers that may or may not be chromatographically resolvable.

A simple example of the application of this technique should make its power clear (14). The gas chromatogram of a mixture of PAHs extracted from an experimental carbon black showed three large peaks (among others). The electron-impact mass spectra of these three peaks showed that all three compounds had molecular weights of 202; presumably their formula was $C_{16}H_{10}$. Peak A gave an $(M+1)/M$ ratio of 1.533, peak B gave an $(M+1)/M$ ratio of 1.428, and peak C gave an $(M+1)/M$ ratio of 0.623. At an elemental composition of $C_{16}H_{10}$, there were four reasonable structures for these compounds: fluoranthene, pyrene, aceanthrylene, and acephenanthrylene. Based on our known correlation of the $(M+1)/M$ ratio with ionization potential, we calculated that the ionization potential of the compound represented by peak A was 8.555 eV, peak B was 8.512 eV, and peak C was 8.025 eV. Calculated ionization potentials of the four hypothetical compounds were 8.466 eV for fluoranthene, 8.029 eV for pyrene, 8.124 eV for aceanthrylene, and 8.500 eV for acephenanthrylene (18). Allowing for about a 1% error, we noted that peak A was either fluoranthene or acephenanthrylene, peak B was either acephenanthrylene or fluoranthene, and peak C was pyrene. We were easily able to obtain standards of fluoranthene and pyrene and to verify that peak A was fluoranthene and that peak C was pyrene; clearly, peak B was acephenanthrylene. We were easily able to synthesize a standard of acephenanthrylene and to verify that this assignment was correct

(19). However, even in the absence of the standard of acephenanthrylene, it was certain that peak B was not aceanthrylene.

Keough (20) has distinguished PAH isomers using positive chemical ionization with dimethylether as the reagent gas. Some compounds (anthracene, tetracene, and benz[a]anthracene) show intense $(M + C_2H_5O)^+$ adduct ions, whereas others (naphthalene, phenanthrene, chrysene, and triphenylene) do not. This behavior seems to be predictable: The compounds that form adduct ions are known to undergo Diels–Alder cycloadditions in solution, and the others are not (20). Keough only investigated seven PAHs; clearly, this technique warrants more work.

4. NEGATIVE ION TECHNIQUES

The formation of negative ions from PACs is a potentially useful analytical technique. Negative ions can be formed from the analyte in many different ways; the two that have been used most often for PAC analysis have been electron-capture ionization and reactions with oxygen. Both of these techniques use chemical ionization sources; that is, ion sources operating at about 1 torr of pressure. In most cases, the objective of these studies has been to differentiate among PAHs. There has been, in addition, some attention given to nitrogen-containing PACs. The following sections will first review the work using electron-capture ionization and will then review the work using oxygen reactions for ionization.

4.1. Ionization by Electron Capture

Oehme (21) studied the electron-capture negative-ion mass spectra of 24 PAHs. About half of these compounds give $(M - H)^-$ as the most intense peak, and the others give M^- as the most intense peak. Some of these data are summarized in Table 1. The different compounds responded differently to the electron-capture process. Oehme reported this difference as relative response factors, that is, the response of the analyte under electron-capture conditions relative to its response under electron-impact conditions. These relative responses varied from 0.01 for benzo[e]pyrene to 170 for dibenzo [def,mno]chrysene. Some isomeric pairs were startlingly different in their response. For example, benzo[e]pyrene responded about 10^4 times less well than benzo[a]pyrene. These relative response factors are summarized in Table 1. The reason for these large variations in response are related to the electron affinity of the molecule, which is, in turn, related to its LUMO. This relationship will be discussed below.

The same general idea was also published by Iida and Daishima (22). These authors studies 21 different PAHs, which were divided into three groups. The first group gave $(M - H)^-$ as the most intense peak in its electron-capture negative-ion mass spectrum. The second group gave M^- as its most intense peak, and the third group gave $(M + H)^-$ as the most intense peak. These data are summarized in Table 1. The reader will note that there are some discrepancies between these data and those reported by Oehme. For example, Oehme observed that M^- was the most intense peak in the mass spectrum of anthracene, whereas Iida and Daishima observed that the $(M + H)^-$ ion was the most intense peak. In general, however, there is reasonable agreement between these sets of data. Iida and Daishima also measured response factors relative to electron-impact conditions, and these data are summarized in Table 1. The relative response factors varied from 0.04 for naphthalene to 7.7 for benzo[j]fluoranthene. This is less of a dynamic range than that observed by Oehme, but, in general, there is a rough correlation between the two sets of values.

Buchanan and Olerich (23) carried out similar experiments using methane to produce electron-capture negative-ion spectra of 35 PAHs. Although these authors did not report relative negative-ion response factors, they did point out that certain compounds ionized and that others did not. For example, they could observe no ionization for pyrene or benzo[e]pyrene, whereas benzo[a]pyrene and fluoranthene gave good mass spectra. Buchanan and Olerich pointed out a relationship between electron affinity and ionization and suggested that PAHs with electron affinities (EAs) greater than 0.5 eV will ionize, while those with EAs less than 0.5 eV will not. These authors also noticed that fluorene gave an ion at $(M + 14)^-$ and not at M^-. Subsequent work by Buchanan et al. (24) has shown that this ion has the structure of 9-fluorenone, and it results from oxidation in the ion source. The data by Buchanan and Olerich are summarized in Table 1. In this case, zero in the relative intensity column means that the compound was not ionized; a blank in both columns means that the compound was not measured. There is a reasonable correlation between these data and those of Oehme and those of Iida and Daishima.

Hilpert et al. (25) studied the response of benzo[a]pyrene under electron-capture negative-ionization conditions. They determined that quantities of benzo[a]pyrene as low as 1 pg can be detected in the selected ion monitoring mode and that the response was linear over three orders of magnitude.

Brotherton and Gulick (26) also studied the methane electron-capture negative-ion mass spectra of 13 PAHs. In many cases, molecular anions (M^-) were observed; in some cases, $(M + H)^-$ and $(M - H)^-$ ions were particularly abundant. Unfortunately, the relative response factors were reported in arbitrary units by these authors, and thus it is not possible to directly compare

Table 1. Electron-Capture Negative-Ion Mass Spectral Data of Oehme (21), Iida and Daishima (22), Buchanan and Olerich (23), Brotherton and Gulick (26), and Low et al. (27)[a]

	MW	LUMO (-eV)	EA (eV)	Oehme	Iida and Daishima	Buchanan and Olerich	Brotherton and Gulick	Low et al.
Dibenzo[def,mno]chrysene	276	1.316	M	170		M		
Indeno[1,2,3-cd]pyrene	276	1.257	M	120		M	M	27
Naphthacene	228	1.182		M	0.11	M	2	
Benzo[j]fluoranthene	252	1.154	M	42	7.7	M		
Benzo[rst]pentaphene	302	1.139		M		M		
Benzo[a]pyrene	252	1.095	0.64	120	0.89	M	380 M	18
Perylene	252	1.088		1.5	0.36		13	
Benzo[ghi]perylene	276	1.068	0.51	64		M	M	0.8
Naphtho[1,2,3,4-def]chrysene	302	1.050				M		
Coronene	300	1.010	0.32			0		
Benzo[ghi]fluoranthene	226	0.995				M		
Benz[e]acephenanthrylene	252	0.980		38	3.4	M	700 M	20
Fluoranthene	202	0.939	0.63	8.9	0.44	M	1300 M	0.84
Acenaphthylene	152	0.923	0.77		0.55	M	M	0.3
Benzo[k]fluoranthene	252	0.923	M	36		M	1500 M	12
Pyrene	202	0.878	0.45 ?	0.03 M+H	0.13	0 M+H	10 M-H	0

Compound	MW	LUMO					
Benzo[e]pyrene	252	0.857	0.35 ?	0.01 M+H	0.07	0 M	6 M−H ?
Anthracene	178	0.841	0.49 M	0.54 M+H	0.1	0 M	31 M−H 0.02
Azulene	128	0.839		M	0.59 M	0	
Benzo[b]triphenylene	278	0.830	0.34			0	
Benz[a]anthracene	228	0.829	0.42 ?	6.5 M+H	0.05	0 M	34 M−H 0
Dibenz[a,h]anthracene	278	0.828	0.37	M+H	0.18		M 16
Dibenz[a,j]anthracene	278	0.820	0.33			0	
Picene	278	0.744	0.29	M−H	0.22	0	
Chrysene	228	0.716	0.26	M−H	0.09	0 M−H	7 M−H 0.17
Benzo[c]phenanthrene	228	0.686	0.18	M−H	0.08	0	
Benzo[a]fluorene	216	0.658	M−H	0.12		M−H	
Benzo[b]fluorene	216	0.570	M−H	0.16			
Triphenylene	228	0.558	?	0.04 M−H	0.15	M−H	4
Phenanthrene	178	0.480	0.03 ?	0.04 M−H	0.13	0 M−H	6 M−H 0
Fluorene	166	0.369	M−H	? M−H	0.16 M+14		M−H 0.02
Acenaphthene	154	0.363					M−H 0.01
Naphthalene	128	0.331	−0.06	M−H	0.04	0	M−H 0
Indene	116	0.118		M−H	0.1		M−H 0

[a] Each entry gives the most intense ion observed and the ratio of negative- to positive-ion response (except for the data of Brotherton and Gulick, for which the response ratios have been normalized such that pyrene = 10). The first three columns give the molecular weight, the energy of the lowest unoccupied molecular orbital (LUMO), and the electron affinity. The LUMOs are from ref. 18, and the EAs are from ref. 23. A blank entry means that value was not measured.

them to other data. Nevertheless, these data are given in Table 1. It is noted that benzo[a]pyrene, benz[e]acephenanthrylene, fluoranthene, and benzo[k]fluoranthene are good responders in this study, whereas naphthacene, benzo[e]pyrene, chrysene, triphenylene, and phenanthrene are poor responders. In general, this agrees with the work reported previously. Brotherton and Gulick also studied the negative-ion response in seven other reagent gases. In general, there was a high correlation between the response factors using all of the reagent gases. Thus, for example, fluoranthene and benzo[a]pyrene were among the highest responding compounds in all of the reagent gases.

Low et al. (27) carried out a detailed study of the electron-capture negative-ion mass spectra of 21 PAHs. They studied the dependency of the spectra on ion source temperature and pressure, and they showed that the formation of M^- and $(M - H)^-$ ions is competitive and depends on the precise ion source conditions. The data from these authors are summarized in Table 1. These authors found that compounds such as phenanthrene, pyrene, and benzo[a]anthracene could not even be detected in the negative-ion mode, whereas compounds such as benzo[k]fluoranthene, benz[e]acephenanthrylene, dibenz[a,h]anthracene, benzo[a]pyrene, and indeno[1,2,3-cd]pyrene gave responses at least 10–20 times greater than those in the positive-ion mode. These authors concluded that the PAHs could be divided into two groups. The first group consisted of those with electron affinities greater than 0.5 eV; these compounds gave molecular anions and little, if any, $(M - H)^-$ ions. Compounds with electron affinities less than 0.5 eV gave $(M - H)^-$ ions and very little, if any, M^- ions. Buchanan and Olerich also noted this threshold (23).

Table 1 includes the data from the studies by Oehme (21), Iida and Daishima (22), Buchanan and Olerich (23), Brotherton and Gulick (26), and Low et al. (27). The compounds have been sequenced based on the energy of their lowest unoccupied molecular orbital (LUMO). These LUMO values are given as negative electron volts, and they have been calculated from molecular orbital theory (18). For a few compounds, the electron affinities are also given. In general, there is a correlation $(r = 0.85)$ between the LUMOs and the electron affinities.

Table 1 shows three categories of compounds. The first category includes all of those compounds with LUMOs as low as 0.9 eV; this also includes compounds with electron affinities above 0.5 eV. In this category, the electron-capture negative-ion spectra are dominated by the molecular anion, and the relative negative-ion sensitivities are usually greater than 1. In some cases, such as benzo[k]fluoranthene, benz[e]acephenanthrylene, and benzo[a]pyrene, these relative responses are substantially greater than 1. This is the group of compounds for which electron-capture negative-ionization MS would be most useful. The second category consists of compounds with LUMOs between 0.9 and 0.8 eV. In this category, molecular anions $(M - H)^-$ and $(M + H)^-$ are predominant, and relative sensitivities are generally less than 1, implying that the electron-capture process is less efficient than the electron-impact process. The third category consists of those compounds with LUMO values less than 0.8 eV. In all of these cases, the electron-capture mass spectrum is dominated by $(M - H)^-$ ions, and the relative response factors are substantially less than 1. Clearly, a simple molecular orbital calculation to determine the LUMO value of the potential analyte could establish into which category the analyte would fall. If the LUMO value was above 0.9 eV, analysis by electron-capture negative-ionization MS would be preferred. If the value was below 0.7 eV, analysis by such a technique would be con-traindicated.

Negative-ion MS is also useful for PACs that have been substituted with electronegative moieties, such as nitro groups. Thus, negative-ion MS has found some application for the analysis of nitrated PACs. In the early 1980s, there was a considerable interest in this class of compounds because of two facts. First, these compounds were found to be associated with the soot produced by diesel engines. Second, nitroaromatics were found to be mutag-enic—in fact, mutagenic without metabolism. This was such an important topic at the time that reviews have been published on the presence and analysis of nitroaromatics in the environment (28).

Ramdahl and Urdal studied the negative-ion mass spectra of 15 different nitro-substituted PACs using methane electron-capture negative-ionization techniques (29). They found that the mass spectra of all of these compounds were dominated by the molecular anion. Most spectra showed fragment ions resulting from losses of 16, 30, and 46 daltons. Some of the spectra are shown in Figure 4. It seems clear that the $(M - 16)^-$ ions are formed as a result of the loss of oxygen from a nitro group, giving a nitroso-substituted anion. The loss of 30 daltons, as discussed before, is due to the loss of NO with a rearrangement to give an oxide-type structure. Reductions in the ion source to give amino-substituted polycyclic species are also possible (7). The loss of 46 daltons (e.g., at m/z 246 in 1,3-dinitropyrene and at m/z 291 in 1,3,6-trinitropyrene) is clearly the loss of a nitro group. It is interesting to note that in the di- and trinitro compounds (see Figure 4, middle and bottom), the loss of two nitro groups to give ions at m/z 200 and m/z 245, respectively, is a relatively unlikely process. The sensitivity for detecting these compounds is excellent under negative-ion conditions. Limits of detection of 2 pg were observed for optimum cases. Unfortunately, differentiating among isomers is not possible; for example, 1-, 2-, and 4-nitropyrene all gave the same mass spectrum. The relative improvement in detection by the use of electron-capture negative-ion MS relative to electron impact is uniformly about a

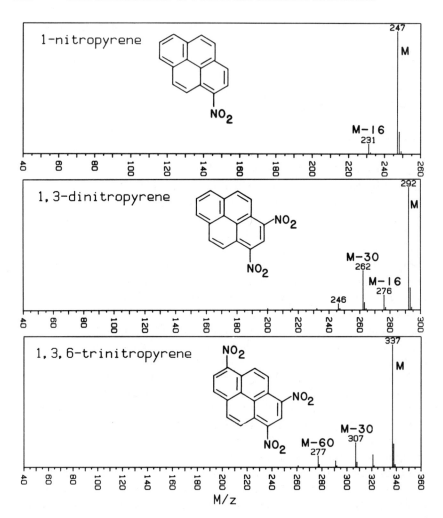

Figure 4. Electron-capture negative-ion mass spectra of three nitropyrenes. Replotted from ref. 29.

factor of 4 for the 15 different nitro compounds studies. Ramdahl et al. (30) identified nitro-PAHs in urban air particulates using this technique.

Newton et al. (31) identified several nitro aromatic compounds in diesel exhaust particulates using methane electron-capture negative-ion MS. Several nitroanthracenes and phenanthrenes with 0–3 alkyl carbons were present. Like Ramdahl and Urdal, these authors found that the spectra were dominated by molecular anions, and they observed no evidence of isomer selectivity.

Oehme (32) studied several classes of polar-substituted PACs by methane electron-capture negative ionization. He found that electron-capture negative ionization was more sensitive than electron impact for compounds such as 9-fluorenone. Again, there was no isomer selectivity with this technique.

4.2. Ionization by Reactions with Oxygen

Hunt et al. (33–35) carried out a series of studies in which negative ions were formed by reacting the analyte with ions such as O_2^-, OH^-, and O^-. These ions were formed in a chemical ionization source by using oxygen as the reagent gas. Unfortunately, hot metal filaments tend to be oxidized in such an environment. This caused rather severe operational difficulties with the mass spectrometer. For this reason, Hunt et al. (33) developed a Townsend discharge ion source to produce the same reactions. These reactions with oxygen are true chemical ionization reactions; that is, the ionziation is caused by an ion–molecule reaction in the ion source. This is different than the electron-capture reactions described in the previous section. Thus, it is important for our nomenclature to distinguish between CI negative-ion MS (the subject of this section) and electron-capture negative-ion MS (the subject of the previous section).

In their first study, Hunt et al. (33) demonstrated that the Townsend discharge ion source, operating with oxygen as the reagent gas, gave mass spectra of acenaphthalene, benzofuran, and anthracene that show essentially two ions. Ions were observed at masses corresponding to the molecular anion, usually about 40% of the total sample ionization, and at $(M + 15)^-$, usually about 60% of the total ionization. The $(M + 15)^-$ ion was shown to be $(M + O_2 - OH)^-$. That is, one of the hydrogens had been displaced by an oxygen, presumably giving an oxide-type structure.

Further work by Hunt et al. (34) demonstrated that the Townsend source, when operating with a 9:1 mixture of oxygen and hydrogen, gave interpretable mass spectra of PAHs. The reactions taking place in the ion source are initially reactions of oxygen with an electron to give O^+, O_2^+, O^-, and O_2^-. The O^- ion goes on to react with hydrogen to produce water. PAHs react with O_2^+ by charge-exchange reactions. In the negative-ion mode, PAHs undergo electron-capture reactions to form molecular anions, which, in turn, react with O_2 (uncharged) to give the $(M + 15)^-$ ion and an OH radical. PAHs can also undergo reactions with O_2^- to give the $(M + 15)^-$ ion and an OH radical. The negative-ion mass spectrum of pyrene (although it was called coronene) obtained under these conditions was presented. The spectrum showed an intense ion at m/z 217, which is due to the $(M + O_2 - OH)^-$ ion. No molecular anion at m/z 202 was observed. This is different than the previous report from this laboratory in which, for example, anthracene gave an intense molecular anion. The difference in the two studies is due to the addition of hydrogen to the reagent gas mixture. This presumably eliminates the reactions that tend to form the molecular anion when the Townsend discharge source is operated with pure oxygen. The mass spectrum of dibenzothiophene was also presented under these ionization conditions. It showed an ion at m/z 216, which is due to $(M + O_2)^-$ and a small molecular anion. The $(M + O_2)^-$ ion is probably due to the addition of two oxygens to the sulfur atom to give a sulphone-type ion.

Using this oxygen–hydrogen reagent gas mixture operating in a Townsend discharge source, Hunt and Sethi (35) showed that isomers could be distinguished. The spectra of benzo[ghi]perylene and indeno[1,2,3-cd]pyrene, both of elemental composition $C_{22}H_{12}$, were presented. Benzo[ghi]perylene gave both the $(M + O_2 - OH)^-$ ion at m/z 291 and the molecular anion at m/z 276; these ions were of about equal abundance. Indeno[1,2,3-cd]pyrene gave the same pair of ions, but the molecular anion at m/z 276 was 10 times more abundant than the $(M + O_2 - OH)^-$ ion at m/z 291. Clearly, more data are needed to establish whether this ionization technique can distinguish many isomers and to determine how this behavior relates to the electronic properties of the molecule.

As described above, Hunt and Sethi (10) also investigated the use of deuterated reagent gases for the analysis of PAHs. The only reagent gas investigated in the negative-ion mode was deuteroammonia, ND_3. In all cases, there was extensive deuterium incorporation into the $(M - H)^-$ ion. This was not an isomer specific process; for example, anthracene and phenanthrene gave almost identical spectra. The average degree of deuterium incorporation was: benzo[a]pyrene, 3.6; naphthalene, 4.1; pyrene, 4.3; chrysene, 4.8; anthracene, 5.3; and phenanthrene, 5.3. It seems unlikely that this technique will find general utility; it only serves to complicate the spectrum.

The obvious experimental difficulties of using oxygen as a reagent gas have caused Oehme (36) to investigate an alternate gas to obtain similar ionization reactions; he used a mixture of methane and nitrous oxide (N_2O). The initial electron-impact-induced ion–molecule reactions give OH^-, CH_3, and N_2. The hydroxide ion can react with PAHs in a number of ways. It can charge exchange, producing molecular anions; it can abstract a proton, giving $(M - H)^-$ ions; or it can add to the analyte, giving $(M + OH)^-$ ions. Other ions due to the addition of NO or N_2O to the $(M - H)^-$ ion are also observed; these occur at $(M + 29)^-$ or $(M + 43)^-$.

There is considerable difference in the mass spectra of some isomeric compounds (see Figure 5). The $C_{20}H_{12}$ isomers (benzofluoranthenes and benzopyrenes) sometimes give distinguishable spectra. For example, benzo[e]-pyrene gives an $(M - H)^-/M^-$ ratio of 1.35, whereas benzo[a]pyrene

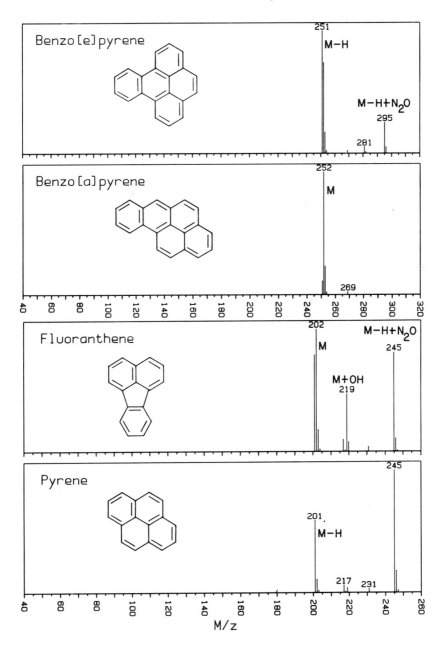

Figure 5. Negative-ion mass spectra obtained with N_2O as the reagent gas of two isomeric pairs of PAHs. Replotted from ref. 21.

gives a ratio of 0.11. In addition, benzo[e]pyrene gives an ion of medium intensity at $(M - H + N_2O)^-$, but benzo[a]pyrene does not. Fluoranthene and pyrene give different spectra. The $(M - H)^-/M^-$ ratio for fluoranthene is 0.79, whereas for pyrene it is 5.40. Both of these compounds have intense ions at $(M - H + N_2O)^-$, but their spectra differ dramatically in the $(M + OH)^-$ ion abundance at m/z 219. This last difference parallels the findings of Hunt et al. (34).

The response factors, relative to electron impact for these various compounds, were also measured. Almost all compounds showed at least a factor-of-10 improvement in sensitivity over electron impact by the use of the methane–nitrous-oxide reagent gas (36). This is compared to the use of methane only, in which only about half of the compounds showed improved sensitivity over electron impact. There is a correlation between the relative sensitivities with methane only and with methane–nitrous-oxide mixture ($r = 0.79$). Clearly, there is a substantial advantage in the use of the methane–nitrous-oxide mixture, both in terms of absolute sensitivity and in terms of isomer differentiation.

Oehme also applied this technique to the analysis of polar- PAHs (32,37). He studied 35 compounds, including nitrogen heterocycles and keto-polycyclic aromatics. In most cases, the methane–nitrous-oxide mixture gave higher responses than did the methane electron-capture system. However, most of the 33 compounds did not respond as well under the negative-ion conditions as they did under electron-impact conditions. Only seven compounds responded better under the negative-ion conditions; these were acenaphtho[1,2,6]pyridine, benz[a]acridine, 9H-fluorenone, 9,10-anthracenedione, 11H-benzo[a]fluorene-11-one, 11H-benzo[b]fluorene-11-one, and 7H-benzo[c]fluorene-7-one. Clearly, the nitrous-oxide–methane mixture would be the method of choice for the analysis of keto-polycyclic aromatics. In some cases, there was also significant isomer specificity; for example, quinoline and isoquinoline gave $(M - H)^-/(M + OH)^-$ ratios of 1.9 and 2.7, respectively. Acridine could be distinguished from its isomer benzo[b]quinoline by the presence of an intense M^- ion in the spectrum of the former but not in the latter (37). In most other cases, however, there was no isomer specificity.

5. DESORPTION IONIZATION TECHNIQUES

So far in this review, we have been discussing techniques that require the analyte molecule to be vaporized. Typically, this is done by introducing the analyte into a gas chromatograph, the effluent of which is fed into a mass spectrometer's ion source. For most PACs of molecular weights below about 300, this is a powerful technique and it is the method of choice. There are, however, larger PACs that are not volatile enough to pass through a gas chromatograph. For these compounds, techniques that bypass the vaporization step will be important. These techniques are called *desorption ionization techniques* because the sample is desorbed from the solid phase directly into the ion phase.

Desorption ionization techniques are categorized by the nature of the primary desorption particle and the physcial state of the analyte. If the primary particle is an ion and the analyte is in the solid state, it is usually known as *secondary ion mass spectrometry* (SIMS). If the primary particle is a neutral atom and the analyte is in the liquid phase, it is usually known as *fast atom bombardment* (FAB). If the primary particle is a photon, it is known as *laser desorption mass spectrometry* (LDMS). In all cases, the incident particles must be of high energy. For example, the atoms in fast atom bombardment usually have 10–20 keV of energy.

The earliest work on the SIMS of PAHs was reported by Cooks and his students (38,39). The compounds of interest were simply deposited on a silver foil which was then exposed to a 5-keV beam of Ar^+ ions. Both anthracene (38) and coronene (39) gave clear molecular ions; in addition, anthracene gave a weak $(M + Ag)^+$ ion.

Later, Ross and Colton (40) developed another technique for obtaining the secondary ion mass spectra of these compounds. In this case, PAHs were loaded onto carbon from a chloroform solution, and the carbon was burnished onto a silver foil which was then bombarded with 4.4-keV Ar^+ ions. Weak secondary ion signals were observed for M^+ of naphthalene, acenaphthylene, acenaphthene, phenanthrene, fluoranthene, and pyrene. A strong signal was observed for anthracene. Signals at $(M + Ag)^+$ were also observed for all of these compounds, with the exception of anthracene. Ross and Colton claimed that the carbon matrix offered several advantages for the SIMS analysis of these compounds. These include: steady emission of molecular ions over a long time period; a detection limit of about 2 ng; and the analysis of nonpolar compounds. These authors suggest that this SIMS technique may be useful for the analysis of "real-world carbons," but they did not provide such data.

Ross and Colton (41) also suggested that liquid metal substrates would be useful for the SIMS analysis of PAHs. They pointed out that a liquid metal substrate would provide an inert, smooth, conducting surface. These authors used a liquid gallium alloy as a substrate, and they demonstrated that this surface was an effective substrate for the analysis of phenanthrene, anthracene, fluoranthene, and pyrene. None of these compounds gave molecular ion signals on silver or indium or gave FAB signals from glycerol, but all of these compounds gave good signals when coated on the liquid gallium alloy. The samples were coated on the alloy by depositing 1 or 2 μl of a chloroform solution of the PAHs onto the gallium. Generally, microgram quantities of PAHs were used, and molecular ion signals were present for at least 30 min. It seems that the liquid metal substrate provides a mobile surface for the sample molecules and allows the bombarded area to be replenished during the course of the experiment. This apparently results in intense and long-lived molecular ion emission.

Colton et al. (42) also investigated ammonium chloride burnished onto silver foil as a substrate, but they found that this substrate gave much less signal than carbon for phenanthrene. However, just the opposite was true for 9-aminophenanthrene. Clearly, the chemical properties of ammonium chloride improved the secondary ion emission of molecules containing basic functional groups. These authors also obtained a good signal for 9-amino-phenanthrene from the liquid gallium substrate that they had previously used for the hydrocarbons. They pointed out that the liquid metal substrate requires a dynamic ion beam and a sample concentration of 10–40 μg for optimum performance. This is unlikely to be a routine technique for trace polycyclic aromatic analysis.

Groenewold et al. (43) studied the use of antimony trichloride ($SbCl_3$) as an SIMS substrate. Molten $SbCl_3$ solutions are known to ionize PAHs by a reversible one-electron oxidation, giving a PAH molecular cation that is stable and that dissolved in the melt. Based on this idea, Groenewold et al. simply mixed about 4 mol% pyrene with $SbCl_3$ in a glove box. About 10 mg of the mixture was placed on the probe tip and was heated on a hot plate until the mixture melted; it gave a bright green solution. The probe tip was cooled, was removed from the glove box, and was inserted into the mass spectrometer. It was bomarded with a 5-keV Ar^+ beam and was heated in the ion source. The results were quite dramatic. The $SbCl_3$ background at m/z 191–195 was about half of the pyrene M^+ signal at m/z 202 when the probe was at about 21°C (see Figure 6, top). When the probe was heated to 60°C, the pyrene signal was about 1000 times greater than the $SbCl_3$ background (see Figure 6, bottom). $SbCl_3$ clearly enhances the secondary ion emission of pyrene, perhaps by the formation of a charge-transfer complex. This is a very interesting idea, and it certainly warrants further work with other PAHs.

Fast atom bombardment (FAB) uses an energetic atom beam to bombard a liquid matrix, usually a glycerol solution of the analyte. With one exception, FAB has not been used for the analysis of PAHs primarily because PAHs are not soluble in glycerol. The exception is a study by Dube (44), who used a hydrocarbon oil as the liquid matrix. He obtained spectra of eight PAHs using a beam of 5-keV krypton atoms. The spectra were identical with the electron-impact spectra of these compounds, showing intense M^+ ions but no $(M + H)^+$ ions. Dube concluded that the neutral, gas-phase PAH molecules above the surface layer were being ionized by the fast atom beam. This is not a

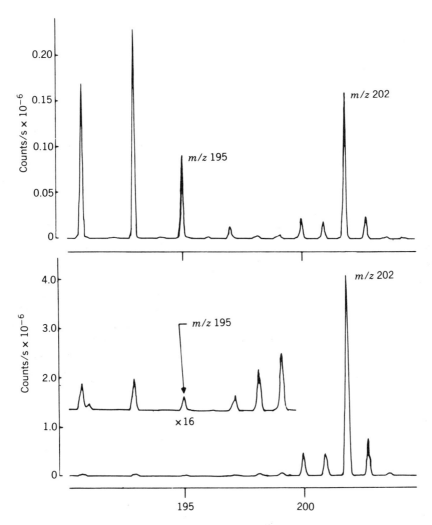

Figure 6. SIMS spectra of a 4 ml % solution of pyrene in SbCl₃ at 21°C (top) and at 44°C (bottom). Reproduced, with permission, from ref. 43.

true FAB effect, and it suggests that very large PAH, for which FAB would be most useful, would not be ionized. Furthermore, FAB seems to work best for those species that are relatively polar, if not already ionic. Thus, it seems unlikely that FAB per se will be useful for the analysis of PAHs.

If a functional group is added to the polycyclic aromatic skeleton such that the compound becomes soluble in glycerol, FAB becomes a very useful technique. Grigsby et al. (45) used FAB for the analysis of nitrogen-containing compounds in fossil fuels. He carried out a model study on 20 nitrogenous bases. These compounds all gave M^+, $(M + H)^+$, and $(M - H)^+$ ions from a glycerol matrix. There were even indications of isomer selectivity. For example, quinoline gave an $(M + H)/M$ ratio of 0.69, whereas isoquinoline gave an $(M + H)/M$ ratio of 1.2; acridine gave an $(M + H)/M$ ratio of 6.7, whereas phenanthradine (an isomer of acridine) gave an $(M + H)/M$ ratio of 2.5. The base fraction of an anthracene oil was also studied. Comparison of these data with high-resolution data obtained under electron-impact conditions showed good agreement.

A laser can also be used for desorption ionization. One such instrument uses a frequency quadrupled Nd-YAG laser (with a wavelength of 265 nm) that is focused through a microscope to a spot size of about 5 μm. A shot from this laser vaporizes and ionizes a portion of the sample. The resulting ions are accelerated into a time-of-flight mass spectrometer.

One application of this laser desorption technique has been the direct analysis of soot from an experimental oil shale retort (46). The sample was mounted for laser analysis by simply touching an electron microscope grid to the powder and shaking off the excess material. The resulting mass spectrum shows positive ions at m/z 252, 276, and 300, which correspond to compounds such as benzo[a]pyrene, benzo[ghi]perylene, and coronene (or their isomers).

This instrument has been used for a detailed study of PAHs by Van Vaeck et al. (47). After optimization of the laser energy at low power, these authors found very different results between the positive- and negative-ion modes. In the positive-ion mode, the spectra resembled those obtained by electron-impact ionization. M^+ was the most intense ion in the spectra of the 21 compounds they studied; significant ions were also present at $(M - 2H)^+$ (see Figure 7, top). Doubly charged ions were not usually abundant nor were $(M + H)^+$ ions. In the negative-ion mode, a few compounds showed spectra resembling those of Iida and Daishima (22); that is, $(M + H)^-$ and/or $(M - H)^-$ ions were prevalent (see Figure 7, bottom). In general, however, in the negative-ion mode, most compounds simply disintegrated to give C_n^- and C_nH^- ions. Obviously, these carbon cluster ions do not contain much structural information. Later, Hercules et al. (48) verified these results.

Hercules and co-workers (49) also measured the negative-ion mass spectra of nitro-aromatics with the laser microprobe mass analyzer. They observed variable amounts of NO_2^-, C_3N^-, and $(M - NO)^-$ ions. M^- ions were not present, but $(M + O - H)^-$ ions were usually abundant. When PAHs were mixed with 1,3,5-trinitrobenzene before laser ionization, intense $(M + O - H)^-$ ions were produced; for example, coronene (MW = 300) gave an ion at m/z 315 and not at m/z 300. The authors termed this effect *chemical ionization in the solid state.*

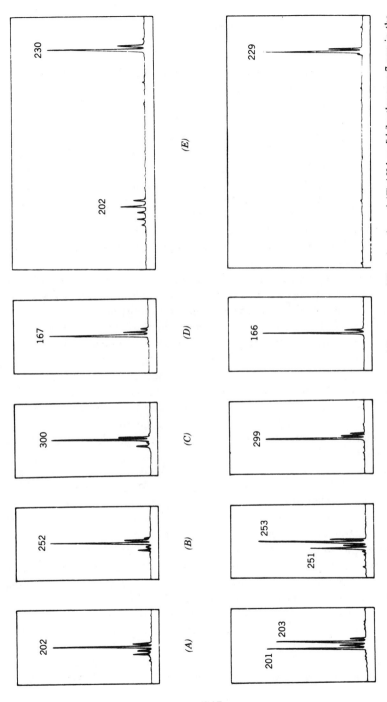

Figure 7. Laser desorption spectra of (A) pyrene, (B) benzo[a]pyrene, (C) coronene, (D) carbazole and (E) 1H-benz[de]anthracene-7-one in the positive-ion mode (top) and negative-ion mode (bottom). Reproduced, with permission, from ref. 47.

247

6. MULTIPHOTON IONIZATION

Ions can be formed by interactions with photons of the correct energy. Lasers are capable of supplying a sufficiently high flux of photons to ionize organic compounds with sufficient efficiency for analytical applications. For most organic compounds, the energy required to remove an electron from the molecule is in the range of 7–11 eV. Light of this energy has wavelengths of 110–180 nm, which is in the vacuum ultraviolet range and is thus not conveniently available. On the other hand, light at wavelenghts between 250 and 500 nm is available from tunable dye lasers. Photons of these wavelengths have energies between 5 and 2.5 eV, which is considerably less than the ionization potential of the molecules of interest. Therefore, in order to ionize these molecules, the simultaneous absorption of two or three photons of this energy range is required—thus the term *multiphoton ionization* (50).

Multiphoton ionization is most efficient when a real electronic excited state can be reached by one of the photons. Therefore, when the wavelength matches the energy required for an electronic transition in the molecule, the process is called *resonance enhanced multiphoton ionization*. For example, the first excited single state of naphthalene lies at an energy of 4.15 eV relative to the ground state. This corresponds to a wavelength of 299 nm. Thus, a laser operated at this wavelength can reach the first excited state with one photon. Another photon of this energy will ionize the molecule since the ionization potential of naphthalene is 8.12 eV. Generally, fragmentation is not observed, but it can be induced by increasing the power density of the laser. Multiphoton ionization is frequently far more efficient than electron-impact or chemical ionization. In addition, PAHs are particularly suitable for multiphoton ionization because they are such excellent chromophores in the region of 250–350 nm.

Rhodes et al. (51) demonstrated rather remarkable sensitivities using a gas chromatograph for an inlet system, an excimer laser producing photons at 4.98 eV, and a time-of-flight mass spectrometer for mass resolution. Using selected ion monitoring at m/z 128, they were able to demonstrate a lower limit of detection of 200 fg of naphthalene (see Figure 8). They were also able to demonstrate isomer specificity. For example, triphenylene was not ionized, but chrysene was, when using 4.00-eV photons. The ionization potential of chrysene is 7.8 eV, and thus two 4.0-eV photons exceed its ionization potential. Conversely, the ionization potential of triphenylene is 8.1 eV, and two 4.0-eV photons are not sufficient to reach this ionization potential. Clearly, this is significant selectivity that should prove useful for many analytical applications.

The combination of multiphoton ionization with Fourier transform MS has been demonstrated by Sack et al. (52) and by Carlin and Freiser (53); both

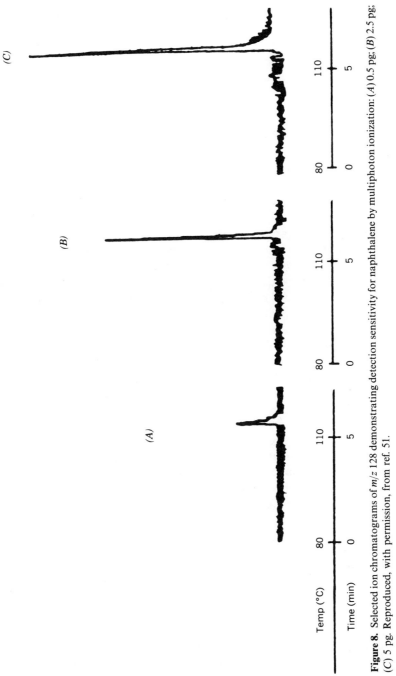

Figure 8. Selected ion chromatograms of m/z 128 demonstrating detection sensitivity for naphthalene by multiphoton ionization: (A) 0.5 pg; (B) 2.5 pg; (C) 5 pg. Reproduced, with permission, from ref. 51.

249

groups used a laser operating at 266 nm (4.66 eV) and mass resolutions in excess of 20,000. Detection limits for several PAHs were in the low picogram range, under both high- and low-resolution conditions. Thus, one might have a technique that is both sensitive *and* high in resolution. Laser desorption has also been combined with multiphoton ionization for the analysis of PAHs (54).

7.   TANDEM MASS SPECTROMETRY

Tandem MS is a simple idea that requires complicated instrumentation. Basically, the idea is to operate two mass spectrometers in series. The ions leaving the first mass spectrometer either fragment on their own or are induced to fragment by colliding them with a neutral gas (or occasionally with photons). The resulting fragment ions are mass separated by the second mass spectrometer. There are three commonly used modes of operation. The first mode is called a *daughter ion scan.* In this mode, the first mass spectrometer is fixed on an ion of interest, and the second mass spectrometer scans the masses of the fragment ions that result from the dissociation of this ion. The second mode of operation is a parent ion scan. In this mode, the second mass spectrometer is held at a constant mass, and the first mass spectrometer is scanned. In this way, all of those ions that dissociate a fragment ion of a given mass are recorded. In the third mode of operation, called a *neutral loss scan,* both of the mass spectrometers are scanned with a fixed mass offset. For example, if this offset were 15 daltons, only ions that lose 15 daltons (for example, a methyl group) between the two mass spectrometers would be detected.

A discussion of the instrumentation used for tandem MS is beyond the scope of this review, and the reader is referred to a new textbook on the subject (55). Suffice it to say that the mass spectrometers can be either magnetic sector instruments, operating at thousands of volts of ion beam energy, or quadrupole instruments. In the latter case, the region between the two mass spectrometers is usually a quadrupole lens, and thus the instrument is called a *triple quadrupole.* In the quadrupole instrument, the ion beam energy is a few tens of volts. The energy of the ion beam is important when considering the energy of the collisions that take place between the two mass spectrometers.

Tandem MS has been used for the analysis of PACs, and one of these studies has been selected as an example to make the concept clear. Wood et al. (56) identified a series of long-chain alkyl naphthalenes in a coal liquid. These liquid samples were put into the mass spectrometer by a direct insertion probe and were volatilized by heating to 200°C. A small but significant ion was observed at m/z 464. Of course, since there was no preliminary separation of the sample, it was impossible to associate any of the fragment ions with this potential molecular ion. Therefore, tandem MS was used to select for m/z 464, to collide this ion with argon at 2 millitorr pressure and at a collision energy of 20 eV, and to mass resolve the resulting ions. The spectrum of the resulting collisionally induced fragment ions is shown in Figure 9. There is an intense peak at m/z 141 and a weak ion at m/z 183, which is three methylene units higher than m/z 141. Wood et al. (56) formed the hypothesis that m/z 141 was the naphthalenic analogue of the tropylium ion that is a very stable fragment in the electron-impact spectra of alkyl naphthalenes. The difference between m/z 141 and m/z 464 is 323 daltons, which corresponds to 23 additional alkyl carbons. Thus, these authors concluded that a C_{24} alkyl naphthalene was present in this coal liquid. Clearly, it is not possible to say anything about the arrangement of the carbons in this alkyl chain; and thus, distinguishing among isomers by this technique is usually very difficult. There is more discussion on this below.

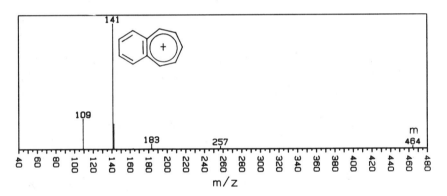

Figure 9. Daughter-ion spectrum of m/z 464 for the molecular ion of a C_{24}-alkylnaphthalene (electron-impact ionization at 20 eV). Replotted from ref. 56.

Shushan and co-workers (57, 58) studied the unimolecular (no collision gas) and collision-induced fragmentation of 19 PAHs in order to distinguish isomers. Their work was done under electron-impact ionization conditions, and they used a double-focusing magnetic sector instrument for mass resolution. They were successful in some cases in distinguishing among isomers; Figure 10 shows their data for four $C_{18}H_{12}$ isomers. Note that the results obtained without collisional activation (dark bars) parallel those with collisional activation (crosshatched bars). With both experimental strategies, it was possible to distinguish among the four isomers in question. For example, triphenylene gave about four times the loss of hydrogen (relative to

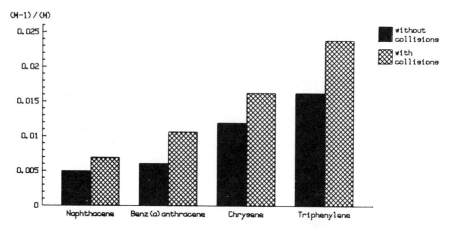

Figure 10. Relative MS/MS abundances of M^- and $(M-H)^-$ ions from the molecular ions of four PAH isomers of composition $C_{18}H_{12}$. Data are with (crosshatched bars) and without (dark bars) collisional activation. Replotted from ref. 58.

M^+) as did naphthacene. Unfortunately, the authors observed variations in the data with experimental conditions, and they commented that "this extreme sensitivity to operating conditions detracted severely from the usefulness of this technique as a routine analytical procedure." Furthermore, there were certain cases (e.g., $C_{22}H_{14}$ isomers) for which the data did not allow the isomers to be distinguished from one another. These authors attempted to correlate trends observed in fragmentation intensities with various molecular parameters but were generally unsuccessful.

Doretti et al. (59) took a similar approach to a more narrowly defined problem, namely, measuring anthracene and phenanthrene in diesel particulates without any previous extraction. They selected m/z 178, collided these ions with N_2 at 8 keV, and measured the ratio of the resulting fragments at m/z 89 (due to $C_{14}H_{10}^{2+}$) and m/z 88 (due to $C_{14}H_8^{2+}$). This ratio was a linear function of the anthracene/phenanthrene ratio. Furthermore, they were able to make an absolute determination of these compounds by including an internal standard, 2,6-diisopropylphenol, with the sample. A detection limit of 1 pg was reported. These authors also measured 3-methylcholanthrene in diesel particulates by a similar approach (60).

Zakett et al. (61) studied nine PAHs by negative-ion charge-inversion MS/MS techniques. The material was evaporated into the ion source from a direct insertion probe. A mixture of isobutane and N_2O was used as the chemical ionization reagent gas. The ions were analyzed by an instrument in which the magnetic sector preceded the electric sector. The collision energy was 7000 eV, the collision gas was about 1 millitorr of nitrogen, and molecular anions were selected by the first mass spectrometer. Both positive and negative ions were then mass resolved by the second mass spectrometer (in this case, the electric sector). No ions were observed when negative ions were collected by the second mass spectrometer (see Figure 11, top). On the other hand, abundant and useful fragment ions were observed when the second mass spectrometer was operated in the positive-ion mode (see Figure 11, bottom). It seems clear that two electrons have been stripped from the molecular anion and that further fragmentation has taken place. This process is called *charge inversion*, and the resulting charge-inversion spectra are more detailed and structurally informative than those obtained by the usual process without charge inversion. Interestingly enough, the charge-inversion spectra of M^- and $(M - H)^-$ ions were similar to each other and resembled the MS/MS spectra of the corresponding $(M)^+$ and $(M + H)^+$ ions. It seems that the spectra are insensitive to the mass and charge of the precursor ion. This is an analytically useful finding since one can use various ionization techniques to produce the precursor ions and still obtain the same analytical (and presumably compound-specific) data.

These authors also used triple quadrupole MS/MS instrumentation to analyze for aza and amino PACs in coal-derived liquids (62). The samples were introduced without separation by the direct insertion probe and were ionized by chemical ionization with isobutane; the collision gas was argon at 2.2 millitorr pressure. In some cases, it was possible to distinguish positional isomers (62). For example, the spectra of 2-, 4-, and 7-methylquinoline are shown in Figure 12; these are the daughter-ion spectra of the protonated molecular ion (m/z 144). Note that the spectra are all quite different, although the same mass numbers tend to appear in each. The ion at m/z 103, for example, is the result of the loss of acetonitrile, and this happens most efficiently when the methyl group is next to the ring nitrogen. Thus, m/z 103 does not appear when the methyl group is on the opposite ring, as in 7-methylquinoline. This ion is smaller when the methyl group is on the same ring, but not adjacent to the nitrogen, as in 4-methylquinoline.

Wood and co-workers have continued to apply tandem MS to the analysis of coal liquids (63, 64). In 1984 they analyzed the basic nitrogen extract of a coal liquid by a variety of techniques, one of which included derivatization with a trifluoroacetyl group (63), and in 1985 they used tandem MS to study the formation of PAHs, especially alkyldibenzothiophenes, in coal liquids (64). Recently, Wood commented that "tandem mass spectrometry has developed into an important analytical tool for the analysis of complex coal-related materials. This is not due to just a single advantage MS/MS brings to the problem, but rather results from the multitudinous facets inherent in the technique" (65).

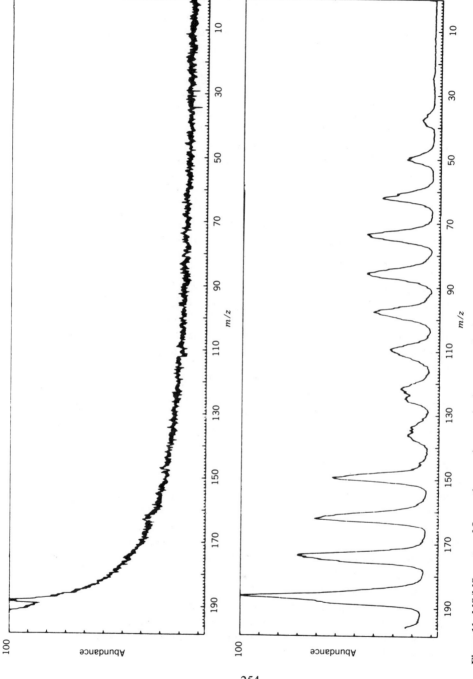

Figure 11. MS/MS spectra of fluoranthene under negative-ion to negative-ion conditions (top) and negative-ion to positive-ion conditions (bottom) using chemical ionization with isobutane. Reproduced, with permission, from ref. 61.

254

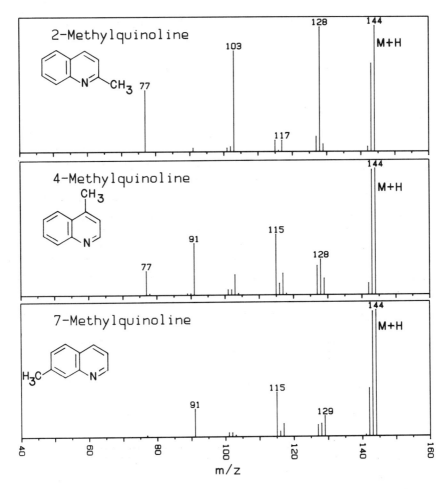

Figure 12. Daughter-ion spectra of the protonated molecular ions of 2-, 4-, and 7-methylquino-line obtained on a triple quadrupole instrument. Replotted from ref. 62. These spectra have been normalized such that the second largest peak is 99% of m/z 144.

Tandem MS has also been applied to the analysis of nitrated PAHs in diesel engine emissions. The strategy here has been to take advantage of a constant neutral loss that is seen in the CI spectra of these compounds. Under these conditions, nitropolycyclics become protonated on one of the nitro oxygens. This newly formed hydroxy moiety is lost to give a fragment ion at $(M + H - OH)^+$. Thus, if the two mass spectrometers are scanned with an offset of 17 daltons, only those ions that lose 17 daltons in a collision will make it through both mass filters. This turns out to be a reasonably specific analysis for nitropolycyclics, given some preliminary clean-up.

Figure 13 shows an example of this type of analysis from Schuetzle et al. (66). The top figure shows the constant neutral loss MS/MS spectrum of 1-nitropyrene recorded with an offset of 17 daltons. An ion appears at m/z 231 as a result of the loss of OH from protonated 1-nitropyrene (which has a molecular weight of 247). Note the simplicity of this spectrum. The bottom panel of Figure 13 shows the constant neutral loss MS/MS spectrum of 20 mg of an unfractionated diesel exhaust particulate extract. The ion at m/z 231 indicates the presence of 1-nitropyrene (or an isomer), and the ion at m/z 207 indicates the possible presence of nitrophenanthrene (or an isomer). Clearly, with this technique it is not possible to distinguish isomeric compounds. For example, one does not know where the nitro group occurs on pyrene or, in fact, whether the carbon skeleton is pyrene or fluoranthene. However, this technique is very rapid because no preliminary separation is required. This technique has also been used by Henderson et al. (67–69) for the detailed analysis of nitro-aromatics in the exhaust of diesel engines operated under

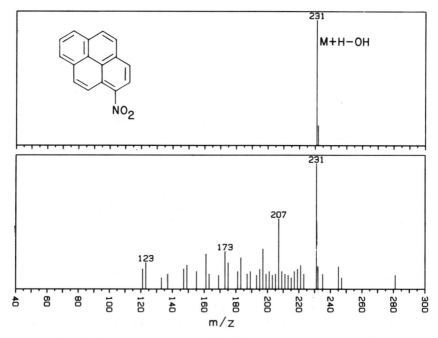

Figure 13. Neutral loss spectrum of 1-nitropyrene (top) and unfractionated diesel exhaust extract (bottom). Replotted from ref. 66. The neutral offset was 17 daltons, and the ion source was operated under methane CI conditions.

various conditions and by Hunt and Shabanowitz (70) for the analysis of aromatic organosulfur compounds in crude petroleum distillates. In the latter case, the neutral loss offset was 33 daltons, which corresponds to the loss of HS from $(M + H)^+$.

8. LIQUID AND SUPERCRITICAL FLUID MASS SPECTROMETRY

As PACs get larger and larger, it becomes progressively more and more difficult to induce them to pass through a gas chromatograph in the vapor phase. For this reason there have been some attempts, generally not very successful ones, to use LC/MS techniques for the analysis of these compounds.

Apffel et al. (71) demonstrated that a direct liquid injection LC/MS interface, which used a jet of helium gas as an aid for the nebulization of the vaporizing LC eluent, gave acceptable results for PAHs of molecular weight up to that of chrysene. These results were not particularly impressive in that these compounds could be easily analyzed by GC/MS techniques.

Krost (72) demonstrated the use of the moving-belt LC/MS interface. PAHs of molecular weight up to that of benzo[a]pyrene could be analyzed by this technique. Unfortunately, the lower limit of detectability, even in the selected ion monitoring mode, was about 10 ng. This is inferior to sensitivities that can be achieved by GC/MS for the same compounds.

Supercritical fluid chromatography has also been interfaced with MS; the laboratory of Smith has been particularly active in this area. Smith et al. (73) demonstrated an interface for capillary column supercritical fluid GC/MS. They were able to separate PAHs of molecular weight up to that of coronene. Because the supercritical fluid was n-pentane, the mass spectra showed considerable proton transfer to the molecules, giving (a) large $(M + H)^+$ ions and (b) large adduct ions at $(M + C_3H_7)^+$, $(M + C_4H_7)^+$, and $(M + C_5H_{11})^+$. Subsequent work with supercritical CO_2 gave spectra that showed only $(M + H)^+$ ions (74).

Later, Smith et al. (75) were able to achieve higher chromatographic resolution, in part, by using pressure programming (at rates of up to 50 bars/min) of the supercritical fluid mobile phase. They were able to analyze a coal-tar extract using these techniques. Unfortunately, all PAHs analyzed were smaller than the benzopyrenes. Similar work was reported by Wright et al. (76). These authors analyzed the PAH fraction of several marine diesel fuel samples. CO_2 was used as the supercritical fluid mobile phase. The chromatographic effluent was routed into an ion source operating with isobutane as the CI reagent gas. Excellent resolution was obtained for compounds of molecular weight up to about 220.

It is clear that no one has demonstrated an application of LC/MS or SFC/MS to PACs that could not more easily be analyzed by GC/MS. It remains to be seen how valuable these techniques will be for very large PACs.

9. SUMMARY

Despite efforts in other areas, electron-impact MS remains the most powerful tool for the analysis of PACs. When combined with capillary column GC, electron-impact MS provides for the complete qualitative and quantitative analysis of extremely complex mixtures of PACs. The major problem from which this technique suffers is its inability to distinguish isomeric compounds. For example, it cannot distinguish benzo[a]pyrene from benzo[e]pyrene, and it cannot determine where functional groups are attached on a given carbon skeleton. Thus, considerable additional work has been directed at identifying, in a predictable and reliable manner, PAC isomers. This problem has been approached by a variety of techniques. These include charge-exchange chemical-ionization and negative-ionization techniques using both electron-capture reactions and reactions with oxygen ions. It now seems clear that if the LUMO of a molecule is more negative than -0.9 eV, then the molecule will respond, frequently with isomer specificity, in the negative-ionization mode. Molecules that have electronegative functional groups, such as nitro groups, also respond well in the negative-ionization mode.

Desorption techniques have been applied to PACs but with results that are not very analytically useful. For example, large PACs (molecular weights above 300–350 daltons) have not been analyzed by these desorption techniques. Multiphoton ionization can occasionally be used to distinguish isomers, but its real power seems to be in its extreme ultimate sensitivity. MS/MS techniques have been shown to be useful, particularly for problems in which preliminary compound separation (e.g., by GC/MS) is not possible or desirable. It is sometimes possible, using MS/MS techniques, to distinguish among isomers.

Clearly, MS will continue to play *the* most powerful role in the analysis of PACs. The challenge is in extending mass spectral techniques to very large polycyclic aromatic systems. The problems of isomer differentiation and of sample volatility multiply almost exponentially as one increases the molecular weight. Mass spectral techniques are on the horizon that may solve these problems. In the meanwhile, the techniques described in this chapter are better on a nanogram per nanogram of sample basis than any others. Use them.

ACKNOWLEDGMENTS

Mark Hermanson and Debbie Weidner provided valuable assistance in the preparation of this manuscript. Ken Busch and Laurel Standley commented on a draft copy. I am grateful to all of them.

REFERENCES

1. M. J. O'Neal, Jr., and T. P. Wier, Jr., *Anal. Chem.*, **23**, 830 (1951).
2. K. D. Bartle, M. L. Lee, and S. A. Wise, *Chem. Soc. Rev.*, **10**, 113 (1981).
3. M. L. Lee, M. V. Novotny, and K. D. Bartle, *Analytical Chemistry of Polycyclic Aromatic Compounds*, Academic Press, New York, p. 242 (1981).
4. D. Schuetzle and C. V. Hampton, in *Mass Spectrometry in Environmental Sciences*, F. W. Karasek, O. Hutzinger, and S. Safe, Ed., Plenum Press, New York, p. 159 (1984).
5. A. G. Howard and G. A. Mills, *Trace Analysis*, Academic Press, New York, p. 213 (1984).
6. E. J. Gallegos, *J. Phys. Chem.* **72**, 3452 (1968).
7. M. A. Quilliam, F. Messier, P. A. D'Agostino, B. E. McCarry, and M. S. Lant, *Spectrosc. Int. J.*, **3**, 33 (1984).
8. J. M. Schmitter and P. J. Arpino, *Mass Spectrom. Rev.*, **4**, 87 (1985).
9. M. S. B. Munson and F. H. Field, *J. Am. Chem. Soc.*, **88**, 2621 (1966).
10. D. F. Hunt and S. K. Sethi, *J. Am. Chem. Soc.*, **102**, 6953 (1980).
11. M. V. Buchanan, *Anal. Chem.*, **54**, 570 (1982).
12. M. V. Buchanan, *Anal. Chem.*, **56**, 546 (1984).
13. M. L. Lee and R. A. Hites, *J. Am. Chem. Soc.*, **99**, 2008 (1977).
14. W. J. Simonsick, Jr., and R. A. Hites, *Anal. Chem.*, **56**, 2749 (1984).
15. W. J. Simonsick, Jr., and R. A. Hites, *Anal. Chem.*, **58**, 2114 (1986).
16. W. J. Simonsick, Jr., and R. A. Hites, *Anal. Chem.*, **58**, 2121 (1986).
17. D. F. Hunt and P. J. Gale, *Anal. Chem.*, **56**, 1111 (1984).
18. R. A. Hites and W. J. Simonsick, Jr., *Calculated Molecular Properties of Polycyclic Aromatic Hydrocarbons*, Elsevier, Amsterdam, p. 39 (1987).
19. S. Krishnan and R. A. Hites, *Anal. Chem.*, **53**, 342 (1981).
20. T. Keough, *Anal. Chem.*, **54**, 2540 (1982).
21. M. Oehme, *Anal. Chem.*, **55**, 2290 (1983).
22. Y. Iida and S. Daishima, *Chem. Lett.*, 273–276 (1983).
23. M. V. Buchanan and G. Olerich, *Org. Mass Spectrom.*, **19**, 486 (1984).
24. M. V. Buchanan, I. B. Rubin, M. B. Wise, and G. L. Glish, *Biomed. Environ. Mass Spectrom.*, **14**, 395 (1987).
25. L. R. Hilpert, G. D. Byrd, and C. R. Vogt, *Anal Chem.*, **56**, 1842 (1984).
26. S. A. Brotherton and W. M. Gulick, Jr., *Anal. Chim. Acta*, **186**, 101 (1986).
27. G. K.-C. Low, G. E. Batley, R. O. Lidgard, and A. M. Duffield, *Biomed. Environ. Mass Spectrom.*, **13**, 95 (1986).
28. D. Schuetzle and T. E. Jensen, in *Nitrated Polycyclic Aromatic Hydrocarbons*, C. M. White, Ed., Alfred Huethig Verlag, Heidelberg, p. 121 (1985).

29. T. Ramdahl and K. Urdal, *Anal. Chem.*, **54,** 2256 (1982).
30. T. Ramdahl, G. Becher, and A. Bjorseth, *Environ. Sci. Technol.*, **16,** 861 (1982).
31. D. L. Newton, M. D. Erickson, K. B. Tomer, E. D. Pellizzari, P. Gentry, and R. B. Zweidinger, *Environ. Sci. Technol.*, **16,** 206 (1982).
32. M. Oehme, *Chemosphere*, **14,** 1285 (1985).
33. D. F. Hunt, C. N. McEwen, and M. T. Harvey, *Anal. Chem.*, **47,** 1730 (1975).
34. D. F. Hunt, G. C. Stafford, Jr., F. W. Crow, and J. W. Russell, *Anal. Chem.*, **48,** 2098 (1976).
35. D. F. Hunt and S. K. Sethi, in *High Performance Mass Spectrometry: Chemical Applications*, M. L. Gross, Ed., American Chemical Society, Washington, D.C., p. 150 (1978).
36. M. Oehme, *Int. J. Mass Spectrom. Ion Phys.*, **48,** 287 (1983).
37. M. Oehme, D. Stockl, and H. Knoppel, *Anal. Chem.*, **58,** 554 (1986).
38. H. Grade and R. G. Cooks, *J. Am. Chem. Soc.*, **100,** 5615 (1978).
39. R. J. Day, S. E. Unger, and R. G. Cooks, *Anal. Chem.*, **52,** 557A (1980).
40. M. M. Ross and R. J. Colton, *Anal. Chem.*, **55,** 150 (1983).
41. M. M. Ross and R. J. Colton, *Anal. Chem.*, **55,** 1170 (1983).
42. R. J. Colton, J. E. Campana, D. A. Kidwell, M. M. Ross, and J. R. Wyatt, *Appl. Surface Sci.*, **21,** 168 (1985).
43. G. S. Groenewold, P. J. Todd, ant M. V. Buchanan, *Anal. Chem.*, **56,** 2251 (1984).
44. G. Dube, *Org. Mass Spectrom.*, **19,** 242 (1984).
45. R. D. Grigsby, S. E. Scheppele, Q. G. Grindstaff, G. P. Sturm. Jr., L. C. E. Taylor, H. Tudge, C. Wakefield, and S. Evans, *Anal. Chem.*, **54,** 1108 (1982).
46. T. Mauney and F. Adams, *Sci. Total Environ.*, **36,** 215 (1984).
47. L. VanVaeck, J. Claereboudt, J. DeWaele, E. Esmans, and R. Gijbels, *Anal. Chem.*, **57,** 2944 (1985).
48. K. Balasanmugam, S. K. Viswanadham, and D. M. Hercules, *Anal. Chem.*, **58,** 1102 (1986).
49. K. Balasanmugam, S. K. Viswanadham, and D. M. Hercules, *Anal. Chem.*, **55,** 2424 (1983).
50. D. F. Hunt, *Int. J. Mass Spectrom. Ion Phys.*, **45,** 111 (1982).
51. G. Rhodes, R. B. Opsal, J. T. Meek, and J. P. Reilly, *Anal. Chem.*, **55,** 280 (1983).
52. T. M. Sack, D. A. McCrery, and M. L. Gross, *Anal. Chem.*, **57,** 1290 (1985).
53. T. J. Carlin and B. S. Freiser, *Anal. Chem.*, **55,** 955 (1983).
54. R. Tembreull and D. M. Lubman, *Anal. Chem.*, **58,** 1299 (1986).
55. K. L. Busch, G. L. Glish, and S. A. McLuckey, *Mass Spectrometry/Mass Spectrometry*, VCH, New York (1988).
56. K. V. Wood, R. G. Cooks, Z. Mudamburi, and P. H. Given, *Org. Geochem.*, **7,** 169 (1984).
57. B. Shushan, S. H. Safe, and R. K. Boyd, *Anal. Chem.*, **51,** 156 (1979).
58. B. Shushan and R. K. Boyd, *Org. Mass. Spectrom.*, **15,** 445 (1980).
59. L. Doretti, A. M. Maccioni, and P. Traldi, *Biomed. Environ. Mass Spectrom.*, **13,** 381 (1986).
60. A. M. Maccioni, P. Traldi, and L. Doretti, *Talanta*, **34,** 483 (1987).
61. D. Zakett, J. D. Ciupek, and R. G. Cooks, *Anal. Chem.*, **53,** 723 (1981).

62. J. D. Ciupek, D. Zakett, R. G. Cooks, and K. V. Wood, *Anal. Chem.*, **54**, 2215 (1982).
63. K. V. Wood, C. E. Schmidt, R. G. Cooks, and B. D. Batts, *Anal. Chem.*, **56**, 1335 (1984).
64. K. V. Wood and R. G. Cooks, *Am. Chem. Soc. Div. Fuel Chem.*, **30**, 68 (1985).
65. K. V. Wood, in *Coal Science and Chemistry*, A. Volborth, Ed., Elsevier, Amsterdam, p. 207 (1987).
66. D. Schuetzle, T. L. Riley, T. J. Prater, T. M. Harvey, and D. F. Hunt, *Anal. Chem.*, **54**, 265 (1982).
67. T. R. Henderson, R. E. Royer, C. R. Clark, T. M. Harvey, and D. F. Hunt, *J. Appl. Toxicol*, **2**, 231 (1982).
68. T. R. Henderson, J. D. Sun, R. E. Royer, C. R. Clark, A. P. Li, T. M. Harvey, D. H. Hunt, J. E. Fulford, A. M. Lovette, and W. R. Davidson, *Environ. Sci. Technol.*, **17**, 443 (1983).
69. T. R. Henderson, J. D. Sun, A. P. Li, R. L. Hanson, W. E. Bechtold, T. M. Harvey, J. Shabanowitz, and D. F. Hunt, *Environ. Sci. Technol.* **18**, 428 (1984).
70. D. F. Hunt and J. Shabanowitz, *Anal. Chem.*, **54**, 574 (1982).
71. J. A. Apffel, U. A. T. Brinkman, R. W. Frei, and E. A. I. M. Evers, *Anal. Chem.*, **55**, 2280 (1983).
72. K. J. Krost, *Anal. Chem.*, **57**, 763 (1985).
73. R. D. Smith, J. C. Fjeldsted and M. L. Lee, *J. Chromatog.*, **247**, 231 (1982).
74. C. R. Yonker, B. W. Wright, H. R. Udseth, and R. D. Smith, *Ber. Bunsenges. Phys. Chem.*, **88**, 908 (1984).
75. R. D. Smith, H. T. Kalinoski, H. R. Udseth, and B. W. Wright, *Anal. Chem.*, **56**, 2476 (1984).
76. B. W. Wright, H. R. Udseth, R. D. Smith, and R. N. Hazlett, *J. Chromatogr.*, **314**, 253 (1984).

CHAPTER

9

LASER MULTIPHOTON IONIZATION SPECTROSCOPY OF POLYATOMIC MOLECULES

KONSTADINOS SIOMOS

Institute of Materials Structure and Laser Physics
Technical University of Crete
Crete, Greece

1. INTRODUCTION

1.1. Scope of the Chapter

Laser multiphoton ionization (MPI) studies of molecular species in the gas, liquid, or solid phase have been carried out under various experimental conditions. The vast majority of these investigations have been conducted in the gas phase (owing to the simplicity and clarity of the experiments undertaken), at either very low ($\sim 10^{-6}$ torr) or moderate (on the order of a few torr) gas pressures, employing a variety of techniques. Unfortunately, there have only been a very few MPI experiments in liquid environments and even fewer at high gas pressures. As the density of the medium increases, so also do the difficulties related both to the detection techniques and procedures and to the theoretical interpretation of the MPI data obtained.

Dense media are the foremost natural environments; thus the effect of the nature and density of the medium on the photophysical and photochemical properties of the molecular species embedded in them is important to both basic and applied research areas of chemistry, physics, biophysics, biology (Christophorou and Siomos, 1984; also see references quoted therein), and medicine (Siomos, 1988). We do not attempt to review the voluminous literature on MPI processes. However, we do provide ample material on multiphoton processes pertinent to our theme. The aspects of multiphoton phenomena have been reviewed by many authors (e.g., Bakos, 1974; Delone, 1965; Letokhov and Chebotayev, 1977; Eberly and Lambropoulos, 1978; Hurst et al., 1979; Georges and Lambropoulos, 1980; Levenson, 1982; Wright, 1982; Parker, 1983; Lin et al., 1984; Christophorou and Siomos, 1984; Smith and Leuchs, 1988).

In this chapter the basic aspects of multiphoton ionization of polyatomic molecules are discussed in relation to the density and nature of the surrounding medium in which the processes occur, and the principal experimental techniques and procedures used are described with special emphasis on MPI techniques suitable for high-density media (liquids). The importance of these processes and techniques in the development of analytical instrumentation with high detection sensitivity, spectral selectivity, and time-resolving capability will be elucidated.

1.2.   Photoionization and Medium-Dependent Processes

Classical one-photon absorption and fluorescence spectroscopy has contributed throughout the years immeasurably toward our understanding of the molecular structure and properties.

Detection techniques based on such one-photon processes have been used extensively for chemical analysis, despite (a) their limitations due mainly to low spectral selectivity and detection sensitivity and (b) unfavorable selection rules.

Excited states of molecules, in general, can be populated by absorbing photons in the ultraviolet (UV) spectral region. However, when electromagnetic radiation of sufficiently high energy, $\hbar\omega$, [e.g., in the vacuum–ultraviolet spectral (VUV) region] interacts with free molecules (AX), it can raise a bound electron into the ionization continuum (see Figure 1a):

$$\dot{A}X + \hbar\omega \rightarrow (AX)^+ + e^- \qquad \text{(direct ionization)} \qquad (1)$$

If I_j is the binding energy of the ejected electron initially in the j orbital and any electronic excitation of the resultant positive ion is neglected, the photoelectron kinetic energy, KE, for process (1) will be

$$KE = \hbar\omega - I_j - E^*_{\text{vib, rot}} \qquad (2)$$

where $E^*_{\text{vib, rot}}$ is the energy that resides as vibrational–rotational excitation energy in $(AX)^+$.

The minimum photon energy $(\hbar\omega)_{\text{min}}$ required to eject the least-bound electron to infinity at rest from AX in its ground electronic vibrational and rotational states, with the resultant positive ion $(AX)^+$ in its lowest electronic, vibrational, and rotational states, is defined as the threshold energy **I** (usually called the ionization potential).

Although the quantity **I** can be precisely determined for atoms, it is often difficult to measure it accurately for molecules because of the possible involvement of vibrational and rotational states in both AX and $(AX)^+$ and

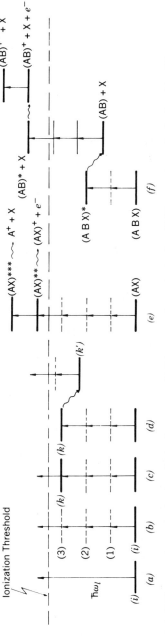

Figure 1. Different pathways in the photoionization process. (a) Direct one-photon ionization. (b) Direct n-photon ionization. (c) m-Photon resonant, to an intermediate state (k), n-photon ionization. (d) n-photon ionization via a lower lying state (k'). (e) n-Photon ionization via resonant (superexcited) autoionizing states. (f) Neutral fragment multiphoton ionization. The solid and dashed lines represent the real and virtual states, respectively.

265

also because of other processes such as predissociation, which becomes rather important in polyatomic systems, especially close to **I**.

It is well known (Berkowitz, 1979; Eland, 1984) that when a photon of energy $\hbar\omega > $ **I** is absorbed by a molecule, besides leading to direct ionization, a process which has been considered so far, it can also lead to indirect ionization. The latter process proceeds via superexcitation to an excited state (AX)** whose energy lies above **I**. Such states are embedded in the ionization continuum and can autoionize (see Figure 1e) by radiationless decay to a degenerate state of the continuum; they can also undergo fast "dissociation" but not, as a rule, radiative deexcitation since this is, by comparison, a much slower process. We may thus write

$$AX + \hbar\omega \rightarrow (AX)** \quad \begin{matrix} \nearrow & AX^+ + e^- & \text{(indirect ionization)} \\ \searrow & A + X & \end{matrix} \qquad (3)$$

Indirect ionization proceeds via a resonance process and usually shows up as distinct structure superimposed on the direct-ionization continuum spectrum.

Although indirect ionization in gases has been known—and extensively studied—for many years, it is only in the last few years that its occurrence in the photoionization of molecules in the liquid phase has been demonstrated by Siomos and co-workers (Siomos and Christophorou, 1980; Siomos et al., 1981a; Kourouklis et al., 1981; see also Section 4.2).

1.2.1. Effects of the Medium on the Spectroscopy of Molecules

The effects of the medium on the spectroscopic characteristics of molecules depend on the nature of the molecular excited states and the medium itself (for detailed discussion see Christophorou and Siomos, 1984). While in low-density environments the spectral energy resolution is limited mainly because of Doppler-broadening effects, in liquids and high-density media the observed broadening is larger. In these media, sinks for rotational–vibrational energy alter both the rates of the nonradiative molecular deactivation processes and the number of nonradiative decay channels and perturb the potentials in which the electrons of a molecule are excited. Furthermore, dispersive forces in the liquid account for the observed red shifts of the absorption spectra of the molecules dissolved in liquids as compared with their gaseous-phase spectra.

With the exception of plasmon states in solids, the change in the oscillator strength distribution in going from a dilute to a dense gas to the liquid is rather small. Low-lying valence states of molecules in dense gases and liquids retain, in general, their identity (e.g., see Robin, 1974; Kourouklis et al., 1982b), as can be seen from the spectra in Figure 2.

In Figure 2a the $^1A_{1g} \rightarrow \, ^1B_{2u}$ transitions of benzene vapor are seen to be virtually identical with and without a high-pressure helium background gas. Similar spectra in liquid perfluoro-*n*-hexane (Kourouklis et al., 1982b) show little change from the low-pressure data. In contrast to the rather moderate effect of the medium on these states, the effect of the medium on "quasi-charged-separated" states, such as Rydberg states, is dramatic. This is clearly illustrated in Figure 2b, where the well-defined 5p–6s Rydberg transition in methyl iodide in low-pressure gases is obscured in the dense gas. The fate of Rydberg transitions in liquid and solid rare gases has been the subject of

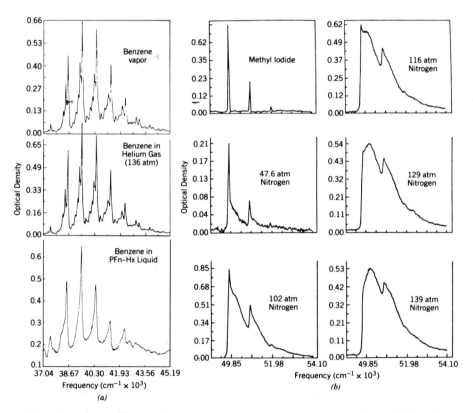

Figure 2. (a) $^1A_{1g} \rightarrow \, ^1B_{2u}$ transitions of benzene vapor before (upper) and after (middle) application of 136 atm of helium gas and in liquid perfluoro-*n*-hexane (lower). (b) Progressive asymmetric broadening of the $5p \rightarrow 6s$ (B system) Rydberg transition in methyl iodide as a function of the background nitrogen-gas pressure [reproduced, with permission from Robin (1974)]. The data for benzene in perfluoro-*n*-hexane are from Kourouklis et al. (1982b).

studies by Rice and Jortner (1966) and Raz and Jortner (1969) who expressed the view that Wannier-like states can, under certain conditions, be observed in condensed media. In organic liquids, however, Wannier-like states were thought to be excessively broadened and thus difficult to observe. In recent studies, however, Siomos and Faidas (1988) have shown that the observed peak in the multiphoton excitation spectrum of dilute solutions of benzene around 380 nm in a number of organic liquids exhibits strong Rydberg-like character. These observations lead to the conclusion that processes (e.g., ionization, electron attachment, electron detachment) responsible for the creation of charge-separated species such as electrons, positive and negative ions, and electron–ion pairs, as well as the physical quantities that characterize these species (e.g., ionization threshold energies, I, and electron affinities, EA), are most affected by the density and nature of the medium in which the elementary processes occur.

1.2.2.   The Electron and Ion State in the Condensed Phase

In vacuum and sufficiently dilute gases, electrons and ions are free. Their energies can be well in excess of thermal. They can be accelerated under the influence of an external electric field and detected practically with unity detection efficiency. In the liquids, electrons and ions are not free (for detailed discussion see Christophorou and Siomos, 1984). Electrons in liquids exist either in a "quasi-free" (e_f) or in a "localized" (e_l) state depending on the nature of the liquid environment and are characterized by their ground-state energies, V_0. Positive V_0 values are interpreted to mean that the electron in the liquid is in a bound "quasi-localized" state, whereas negative V_0 values indicate that the electron in the liquid is "delocalized" and can thus exist as a "quasi-free" particle. The ground-state energy of an excess electron in a liquid is determined by the balance of short-range repulsive and long-range attractive interactions, the latter arising mainly from the polarization of the medium due to the presence of the excess electron.

In the condensed phase, electrons, owing to their characteristic properties, affect considerably the energetics and dynamics of the ionization processes (see Section 1.2.3), and their detection requires the development of high-sensitivity devices, suitable for condensed-phase studies (see Section 3.2.4).

1.2.3.   Effect of the Medium on the Ionization Processes

In view of the preceding discussion, the effect of the medium on the photoionization process and its energetics is expected to be profound. Measurements of the absolute cross section for process (1) do not exist, and only limited data exist on the relative ionization cross section for this process in dense gases and liquids.

There has been, however, an effort to measure the ionization threshold, I_L, for molecules dissolved in liquid media and to compare it with its corresponding value, I_G, in the gas phase. As early as 1931, Reichardt and Bonhoeffer attributed the onset of a UV absorption band in a solution of Hg atoms in water around 240 nm (~ 5.17 eV), in methyl alcohol around 249 nm (~ 4.99 eV), and in n-hexane around 251 nm (~ 4.94 eV) to the photo-ionization of the Hg atoms around these energies. Since $I_G(\text{Hg}) = 10.434$ eV, a lowering of the I_G of Hg by more than 5 eV in water is indicated.

The difference between I_G and I_L could be understood qualitatively using energy balance reasoning and the so-called Born–Haber cycle (Figure 3) (Born, 1919; Haber, 1919). To evaporate a molecule M_L from the liquid to the gas phase M_G, work equal to the vaporization energy Q is necessary. The creation of a positive-ion–electron pair (M_G^+, e_G^-) in the gas phase requires energy equal to the ionization energy I_G of the molecule M_G.

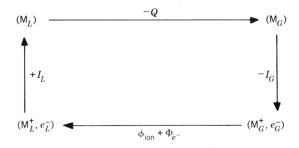

Figure 3. Born–Haber cycle.

The introduction of this gaseous-ion–electron pair into the liquid releases energy equal to $\Phi_{\text{ion}} + \Phi_{e^-}$, while the combination of the M_L^+, e_L^- pair in the liquid frees up energy equal to I_L. Under the assumption that throughout the steps of this cycle (Figure 3) no energy has been lost, we have

$$Q + I_G = \Phi_{\text{ion}} + \Phi_{e^-} + I_L \tag{4}$$

or

$$\Delta I \equiv I_G - I_L = \Phi_{\text{ion}} + \Phi_{e^-} - Q \tag{5}$$

Since $\Phi_{\text{ion}} + \Phi_{e^-} \gg Q$, we have

$$I_L = I_G - \Phi_{\text{ion}} - \Phi_{e^-} \tag{6}$$

It should be noted that $\Phi_{ion} > \Phi_{e^-}$. From equation (6) it can be seen that the value of the ionization threshold I_L of a molecule embedded in a liquid is lower than its gaseous phase value I_G by an amount equal to $S = |\Phi_{ion} + \Phi_{e^-}|$. The major obstacle to an accurate determination of I_L lies in the estimation of S.

In an effort to understand the photoconductance and semiconductance in organic crystals, Lyons (1957) postulated that the difference between the minimum energy I_S needed to ionize a ground-state molecule in a crystal (solid) and I_G arises from the polarization of the region of the crystal that surrounds the positive and negative charges. He assumed that the polarization caused by the electron is the same as that caused by the positive ion, and he expressed I_S as

$$I_S = I_G + 2P^+ \tag{7}$$

Raz and Jortner (1969), suggested that the energy of the "quasi-free" electron in liquefied and solid rare gases could be obtained from the electronic spectra of large-radius Wannier-type impurity states embedded in solids and liquids. Under the assumption that in such systems the conduction band is free-electron-like, the higher (principal quantum number $n > 2$) exciton and impurity energy levels E_n are given by the hydrogenic equation

$$E_n = I_{S,L} - \frac{Rm^*}{\varepsilon^2 n^2} \tag{8}$$

where R is the Rydberg constant, m^* is the effective mass, and ε is the dielectric constant of the medium. These exciton and impurity states correspond to a correlated (bound) electron–hole pair. If the width of the hole state can be neglected, the energy $I_{S,L}$ in equation (8) corresponds to the ionization threshold in the dense medium and should be given by

$$I_{S,L} = I_G + P^+ + V_0 \tag{9}$$

where P^+ is the medium polarization energy resulting from the positive hole. P^+ is always negative and $V_0 \ll P^+$. Equations (8) and (9), derived from semiempirical solid-state theories, contain practically the same basic information as equation (6), and the calculation of I_L from equation (9) depends essentially on the knowledge of P^+ since I_G is reasonably well known for many molecules.

Based on the work of Mott and Littleton (1938), Fowler (1966) derived the expression

$$P^+ = -\frac{e^2}{2s}\left(1-\frac{1}{\varepsilon}\right) \qquad (10)$$

for solid insulators, where s is approximately the lattice spacing. Raz and Jortner (1969, 1970) assumed equation (10) to be applicable in the case of liquefied rare gases and took P^+ (solid) $\approx P^+$ (liquid) for such systems. If in equation (10) s is replaced by the positive-ion radius R_{ion}, we have

$$P^+ = -\frac{e^2}{2R_{ion}}\left(1-\frac{1}{\varepsilon}\right) \qquad (11)$$

which is identical to the expression

$$W_H = -\frac{e^2 z^2}{2R_{ion}}\left(1-\frac{1}{\varepsilon}\right) \qquad (12)$$

derived years earlier by Born (1920) to explain qualitatively the heat of hydration (H) of ions in liquids from the electrostatics point of view. In equation (12), W_H is the work done when a positively or a negatively charged metallic sphere of charge ze and radius R_{ion} is transferred from vacuum into a dielectric medium of dielectric constant ε. Although equation (9) can be used to describe qualitatively the relation between $\mathbf{I}_L$ and $\mathbf{I}_G$, its use to assess the photoionization energetics in the liquid phase quantitatively is hindered by the parameters P^+ and V_0; especially P^+ lacks both exact theoretical representation and direct accurate experimental determination (see Kourouklis et al., 1982a).

2. MULTIPHOTON IONIZATION

2.1. Multiphoton Absorption

Theoretical work on multiphoton absorption was reported as early as 1929 (Goeppert–Mayer, 1929, 1931). However, the first experimental observation both in gases and liquids of multiphoton processes in the visible spectral region has only become possible with the development of intense laser sources (Kaiser and Garrett, 1961; Franken et al., 1961; Abella, 1962; Peticolas et al., 1963; Haseyama and Yoshimura, 1965).

The term *multiphoton absorption*, in contrast to *single-photon absorption* (Figure 1a), is used to refer to processes in which a molecular (atomic) system absorbs more than one photon at a time from the electromagnetic radiation field (Figure 1b). Multiphoton transitions become the predominant processes in any material and at any frequency if the power of the electromagnetic radiation is sufficiently high.

Multiphoton absorption allows highly excited molecular states to be studied (Figure 1c,d,e) using convenient visible radiation. Two-photon transitions, for example, are at an energy twice that of the individual photons absorbed, and thus molecular states, as high in energy as 10 eV, can be reached by means of frequency tunable dye lasers. In this case the excited states have the same parity as the initial ground state. Such transitions, however, are forbidden for single-photon processes, and thus multiphoton absorption may be the only convenient technique for probing molecules in such states.

A number of sophisticated techniques based on the detection of fluorescence, resulting from the decay of the excited state, ions, and electrons, have been deployed to probe multiphoton processes (see Wright, 1982; also see references quoted therein) and contributed to the exploration of new one-photon parity forbidden states in the UV and VUV spectral regions. This, in return, resulted in the development of new analytical techniques, based on multiphoton processes (see Section 4).

2.2. Multiphoton Ionization

2.2.1. Nonresonant (Coherent) Ionization

As we mentioned in Section 1, when a molecule interacts with electromagnetic radiation it can be ionized [process (1)] when the photon energy $\hbar\omega \geq I_G$ (Figure 1a). However, a molecule, in the presence of an intense electromagnetic radiation field, can also ionize even if the photon energy $\hbar\omega < I_{G,L}$ (Figure 1b), by absorbing simultaneously (coherent excitation) several photons (e.g., n photons) from the radiation field. The absorption of n photons from the radiation field in "one elementary quantum act" (coherent excitation) occurs via virtual states (Goeppert-Mayer, 1929, 1931).

Virtual states, as opposed to the long-lived ($\tau \sim 10^{-8}$ s) real intermediate states (Figure 1c, d), have very short lifetimes, typically $\tau \sim 10^{-15}$ s. Therefore the absorption of n photons via virtual states must occur in times shorter than 10^{-15} s. This can only be possible at high photon fluxes, typically on the order of 10^{29} photons cm^{-2} s^{-1}, for $n \gg 2$. Such photon fluxes can only be obtained with lasers. MPI was first predicted by Bunkin and Prokhorov (1964), and the first MPI experimental results were reported by Voronov and Delone (1965), Voronov et al. (1967), Agostini et al. (1968, 1970), Delone and Delone (1969), Petty et al. (1975), and Johnson et al. (1975).

2.2.2. Frequency Dependence of the MPI Cross Section

MPI via virtual states can be theoretically described using nth-order time-dependent perturbation theory. Thus the probability, W, per unit time that a molecular system ionizes absorbing n photons can be written as

$$W^{(n)} = \sigma^{(n)} I_l{}^n \tag{13}$$

where $\sigma^{(n)}$ is the generalized cross section in units $cm^{2n}s^{n-1}$, and I_l the photon flux in units photons $cm^{-2} s^{-1}$. Calculation of $\sigma^{(n)}$ requires summation over all molecular states, including continuum states. For example, for $n = 4$ (see Figure 1b), $\sigma^{(n)}$ can be expressed as

$$\sigma^{(4)} \propto \left| \sum_1 \sum_2 \sum_3 \frac{\langle f|V|3\rangle\langle 3|V|2\rangle\langle 2|V|1\rangle\langle 1|V|i\rangle}{(E_3 - E_i - 3\hbar\omega_1)(E_2 - E_i - 2\hbar\omega_1)(E_1 - E_i - \hbar\omega_1)} \right|^2 \tag{14}$$

where i and f are the initial bound state and final ionization continuum state, respectively. The indices 1, 2, 3 denote the virtual intermediate states in Figure 1b. ω_1 is the laser frequency, and V is the Hamiltonian of the molecule–field interaction. Measurements of absolute cross sections are extremely difficult to carry out. Experimentally, however, the order of the multiphoton ionization process n_{exp}

$$n_{exp} = \frac{\partial \log w}{\partial \log I_l} \tag{15}$$

is more easily measured, and a plot of $\log W$ (or, in practical terms, the logarithm of the photocurrent I_{pc}) vs. $\log I_1$ should give for low photon fluxes and well-characterized systems a straight line of slope n (see Figures 4, and 25). At high intensities, deviation from the expected n-order process is observed (see Figures 4 and 26). This results either from volume saturation effects due to solely geometrical effects in the region of a tightly focused laser beam (Figure 4) or from resonant saturation effects (saturation of the quantum mechanical transition).

In the case of volume saturation, all molecules in the region of a tightly focused laser beam are ionized, while an increasing number of molecules at the edges of the spatial profile of the laser beam contributes to the ionization signal. For a Gaussian beam the volume effect produces an $I_1^{3/2}$ power dependence on the ionization signal (see Figure 4) (Arutyunyan et al., 1970; Speiser and Kimel, 1970; Speiser and Jortner, 1976; Siomos and Christophorou, 1980; Siomos et al., 1983a; Christophorou and Siomos, 1984; Lin et al., 1984; Faidas and Siomos, 1987).

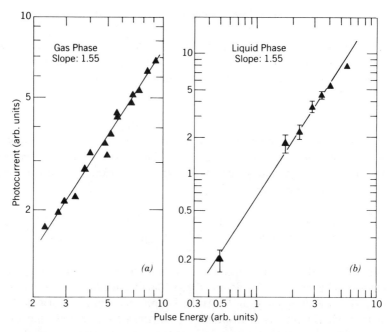

Figure 4. Dependence of the two-photon ionization signal on the incident laser pulse energy. (a) Analine vapor (from Brophy and Rettner, 1979). (b) Pyrene in solution (from Siomos and Christophorou, 1980). The solid lines through the experimental points were obtained by means of a least-squares fitting procedure. The 3/2 (slope values) intensity dependence is due solely to geometrical effects in the region tightly focused laser beam (see text).

2.2.3. Multiple Resonant (Stepwise) Ionization

When the energy of m photons $(m < n)$, interacting with a molecular system, approaches the energy of a real intermediate state, k (Figure 1c), the cross section [equation (14)] for n-photon ionization greatly increases, provided that the m-photon resonant transition is allowed.

It is important to note, however, that expression (14) approaches infinity when the laser photon energy is tuned to a resonant transition, e.g., $\hbar\omega_1 = E_1 - E_i$. The resulting singularity in equation (14) can be avoided by introducing intensity-dependent corrections, related to both laser and transition bandwidths.

For $m = 1$, $n = 2$, one encounters the special case of a two-step photoionization process. In polyatomic molecules, various intra- and intermolecular processes can contribute to the quick loss of the phase memory of the intermediate excited states, so that a two-step photoionization process can be considered as incoherent and thus could be described by rate equations

(Letokhov and Chebotayev, 1977; Antonov and Letokhov, 1981; Parker, 1983; Nikogosyan and Letokhov, 1983). In the liquid phase, where collisional relaxation is very fast, the rate-equation approach can provide useful information about the mechanism of multiphoton ionization processes in dense media (Nikogosyan and Letokhov, 1983).

In stepwise ionization spectroscopy, as the laser excitation wavelength, λ_{exc}, is tuned, the resulting ionization signal I_{pc} shows distinct structure as a result of the excitation of intermediate resonances. The MPI spectrum (photo-current, I_{pc}, vs. laser excitation wavelength, λ_{exc}) measured in this way serves as a fingerprint of the molecules under study and provides a new way in selectively identifying and studying complex molecular systems. Furthermore, resonantly enhanced MPI spectroscopy makes it possible to examine the polarization characteristics of the intermediate resonances observed using both linearly and circularly polarized light. In such cases, the MPI spectra obtained exhibit different intensity distributions, and the ratio Ω of the photoionization signals ($\Omega = I_{pc}^{circular}/I_{pc}^{linear}$) reflect the symmetry properties of the intermediate resonant excited states involved in the MPI process, assuming bound to continuum transitions (Manson and McClain, 1970; McClain, 1974; Parker et al., 1978; Krogh-Jespersen et al., 1979; Robin and Kuebler, 1978; Lin et al., 1984; Faidas and Siomos, 1987, 1988). For example, for all non-totally symmetric two-photon transitions, as well as for all non-Q-branched transitions irrespective of their symmetry, $\Omega = 3/2$ (see Faidas et al., 1985; Faidas and Siomos, 1988; also see references quoted therein).

Furthermore, in the case of a two-step ionization process, two different excitation wavelengths, originating from two different laser-light sources, can be used to excite and ionize a molecular system. Appropriate synchronization of the two lasers can provide additional information about the kinetics of the intermediate excited states and their role in the photophysics of the ionization process itself.

Although resonant transitions can enhance the photoionization signal and provide useful information about high excited molecular states, they can, however, complicate the description of the ionization process.

At resonant conditions, the nth-order dependence of the measured I_{pc} vs. I_1 as given in equation (15) does not hold any more as the intensity of the laser radiation increases. If I_{pc} depends on I_1 as $I_{pc} \propto I_1^S$, then for m-photon resonant n-photon ionization processes we have $m \leqslant S \leqslant n$ and under certain conditions, we have $S < n$ (Ito and Mikami, 1980; Levenson, 1982; Parker, 1983; Siomos et al., 1983b; Faidas and Siomos, 1987; Siomos, 1988). In the case of benzene vapor, for example, it was shown, on the basis of energy reasoning, that benzene can absorb as many as nine (3.17 eV) photons within a 6-ns laser pulse (Zandee and Bernstein, 1979a). This energy exceeds the ionization threshold energy of benzene in the gas phase, I_G, by a factor of three

$[I_G(\text{benzene}) = 9.2 \text{ eV}]$. Similar behavior has also been observed by Siomos and co-workers for benzene dissolved in hydrocarbon liquids (Faidas and Siomos, 1987). These authors observed that the number of photons absorbed increased from three to nine as the laser-light intensity increased (see also Figure 26 in Section 4.2).

It is well known that vibrational dissociation processes, which lead to molecular fragmentation, take place at times on the order of 10^{-12} s. However, since the time required for a molecule to undergo a Franck–Condon transition is $\sim 10^{-18}$ s and the lifetime of virtual intermediate states is $\sim 10^{-15}$ s, there is no time for the molecule under the influence of an intense electromagnetic radiation field to do anything else than either relax to the ground state or absorb another photon(s) from the electromagnetic radiation field and ionize. If, however, the MPI proceeds via resonant intermediate states, the long lifetimes ($\tau \sim 10^{-8}$ s) of real intermediate resonances provide enough time for dissociation processes, which under concomitant multiphoton absorption can lead to extensive molecular fragmentation (Figure 1f) and frequently to atomization of a polyatomic molecule (Zandee et al., 1978; Zandee and Bernstein, 1979a; Rockwood et al., 1979; Robin, 1980; Boesl et al., 1980, 1981; Letokhov, 1983; Lubman, 1987; Antonov and Letokhov, 1981).

Although molecular dissociation processes in gases have been observed and extensively studied for many years, it is only lately that its occurrence in the MPI of polyatomic molecules in the liquid phase has been demonstrated experimentally by Siomos and co-workers (Siomos et al., 1981a; Siomos and Christophorou, 1982a, b).

3. EXPERIMENTAL TECHNIQUES AND PROCEDURES

3.1. The Laser as Spectroscopical Light Source

3.1.1. General Considerations

The development of the laser has opened new avenues in the exploration and understanding of the properties of matter. New measuring techniques based on these powerful light sources have revolutionized the field of spectroscopy during the last two decades. In particular, the development of convenient frequency tunable lasers offers the analytical scientist new and unique capabilities for the use of lasers as an analytical tool.

Despite these promising developments, however, more work needs to be done in order to overcome the existing inertia for adopting these new technologies in the important field of analytical chemical methodology. This inertia could probably be understood from the point of view that there are well-established conventional analytical techniques. With these techniques, one performs the necessary analyses with habitual simplicity and perhaps at a lower cost, often despite the well-recognized need for higher sensitivity and spectral selectivity.

The major properties that put the laser in a unique position over any other light source can be summarized as follows:

(i) *Spectral Power Density.* The spectral power density obtainable from the majority of the laser devices available today may exceed that of incoherent light sources by several orders of magnitudes.

(ii) *Monochromacy and Spectral Resolution.* The narrow emission linewidth of a laser allows spectral resolution which may outdo that of the largest spectrometers by orders of magnitude ($\Delta\lambda_1 \leqslant 0.0001$ nm, see Table 1). Thus the spectral resolution is no longer limited by the instrumental bandwidth but rather by the spectral linewidth of the molecular transition.

(iii) *Coherence.* The small divergence of a laser beam allows the focusing of the laser beam to a waist, limited only by diffraction effects, so that photon densities on the order of $\sim 10^{10}$ W cm^{-2} (or $\sim 10^{29}$ photons cm^{-2} s^{-1}) are easily obtainable.

Furthermore, the unique capability of the laser to deliver bursts of very short light pulses in the subpicosecond time domain ($\Delta t_1 \sim 10^{-15}$ s) introduces a new dimension in the spectroscopical procedures. Thus, for the first time, one can study directly the time evolution of the photophysical and photochemical processes of atoms and molecules and understand their initial interaction mechanisms and phenomena.

3.1.2. The Frequency Tunable Dye Laser

There are today a number of frequency tunable lasers available (e.g., semiconductor lasers, spin-flip Raman lasers, optical parametric oscillators, color center lasers, dye lasers, etc.) (Demtroeder, 1981). These coherent sources emit light from the UV spectral region to the far infrared. The most important frequency tunable laser today is the dye laser. The availability today of reliable commercial dye lasers has allowed laser spectroscopy to become practical.

In a dye laser system, dye molecules (e.g., rhodamine 6G) (see Figure 5) in solution (methanol is the most prominent solvent) are excited by the intense light of an excitation source (e.g., a flash lamp or a laser) (see Table 1) from the lowest vibrational level (v_0') of the electronic ground state S_0 to the higher vibrational level (e.g., v_3'') of the first singlet state S_1 or higher electronic states (e.g., S_2) as shown in Figure 5. From the level v_3'' the molecules relax in 10^{-12}–10^{-10} s, nonradiatively, to the lowest vibrational level v_0'' of S_1, which is the upper level of the laser transition.

Table 1. Dye Laser Characteristics

Pumping Source	Tuning Range (nm)	Bandwidth (nm)	Pulse Duration	Resonator Configuration
Nitrogen laser	360–900	10^{-2}–10^{-4}	~5 ns	(a) Grating, θ, L_1 Telescope, L_2, L_3, Excitation beam, Dye cell, Polarized, Output mirror
Excimer or Nd:YAG laser	320–900	10^{-2}–10^{-3}	~10 ns	(b) Grating, θ, L_1 Telescope, L_2, Excitation beam, Dye cell, Laser oscillator; L_3, L_4, L_5, L_6, L_7, Two-stage amplifier
Flash lamp	370–800	~0.1–0.01	2–20 μs	(c) Output mirror, Fabry Perot interferometer, Flash lamp, Dye cell, Tuning mirror
Argon-krypton laser	400–1000	10–100 MHz	CW	(d) Excitation beam, Tuning mirror, Tuning prism, Tuning prisms, Dye cell, Output mirror

278

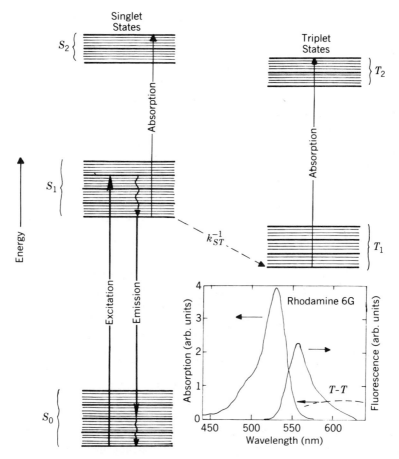

Figure 5. Schematic representation of the energy level diagram of an organic dye molecule in solution. S_0, S_1, and T_1 denote different electronic states. The vibrational sublevels are represented by the heavy horizontal lines. Transitions associate with the lasing process are shown by the heavy vertical and wavy lines.

Given a suitable resonator (see resonator configurations in Table 1) and assuming that no other mechanisms (e.g., radiationless transitions to the triplet state and consecutive absorption processes in the triplet manifold; see Figure 5) cause any depletion of the S_1 state, laser emission occurs from S_1 (v_0'') to e.g., $S_0(v_2')$ and thus laser oscillation can occur over a broad spectral range, as shown in Figure 5. The molecules return finally to the ground state $[S_0(v_0')]$ by a second nonradiative transition from $S_0(v_2') \rightarrow S_0(v_0')$.

Although only the two electronic states S_0 and S_1 appear to be involved in the lasing process, the dye laser scheme, as described above is a four-level one.

By inserting appropriately chosen frequency-selective elements into the dye laser resonator (e.g., prisms, gratings, Lyot filters, interference filters, etalons, etc.) or a combination of them (Walther and Hall, 1970; Siomos, 1972; Haensch, 1972; Wallenstein and Haensch, 1974, 1975), laser radiation with narrow emission line width (Figure 6), tunable over the broad-band gain curve of the dye solution used, can be obtained as shown in Figure 7. It is not the place here to discuss the physics and technology of dye lasers. For the interested reader, there are two excellent books devoted to the dye lasers, namely, Shaefer (1973) and Walther (1976). Nevertheless, some of the most important properties, as well as excitation and tuning techniques, of dye laser systems are presented in Table 1.

3.2. Ionization Detection Techniques and Procedures

The study of MPI processes requires, depending on the state (gas, liquid, or solid) of the molecules under study and the information needed to be obtained, different experimental techniques and detection methods. Charged particles such as electrons and positive and negative ions can generally be easily detected because of their high collection efficiency. Modern charge-sensitive devices can be used practically with unity detection efficiency. These devices, however, must operate in well-controlled vacuum environments, imposing restrictions on the sample vapor pressure. At sample vapor pressures of several millitorr, ionization chambers (cells) can be employed successfully, with a detection efficiency of a few thousand electron–ion pairs, although proportional counters could detect single thermal electrons.

However, as the sample density increases, the detection sensitivity decreases. At high pressures and in the liquid phase, the detection efficiency depends not only on the charge–particle detector arrangement but also on the characteristic properties of the electrons and ions in dense media (e.g., electron and ion mobility, electron scattering processes). Thus MPI spectra in the liquid phase were only obtainable after the development by Siomos and co-workers (Siomos and Christophorou, 1980) of a highly sensitive MPI conductivity technique, suitable for liquid-phase studies.

3.2.1. The Photoionization Cell Detector: Measurements of Total Ionization Currents

The most well known photoionization detector is the proportional counter invented by Rutherford and Geiger (1908). It consists of a thin wire (collector) and a cylindrical electrode surrounding it. A cross-sectional view of such a detector is seen in Figure 8a. The collector is maintained at a voltage typically of $+1000$ V. The cell is filled with ~ 100 torr P-10 gas (a gas mixture of 90%

(a)

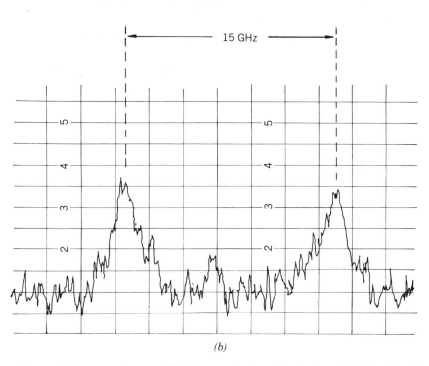

(b)

Figure 6. Spectrum of an unstabilized narrow-band tunable dye laser. (*a*) Fabry Perot interferometer (FPI) spectrum (free spectral range: 15 GHz). (*b*) Densitometer tracing of the FPI spectrum. Laser light bandwidth: $\Delta\lambda_1 = 0.0034$ nm. Reproduced, with permission, from Siomos (1972).

281

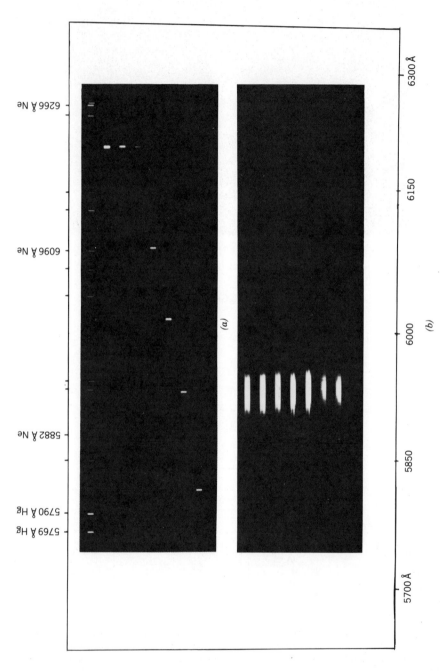

Figure 7. Multiple exposure spectrographic plate. (a) Tunable narrow-band emission. (b) Wideband emission of a rhodamine 6G dye laser. Reproduced, with permission, from Siomos (1972).

282

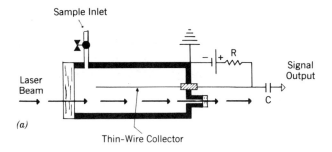

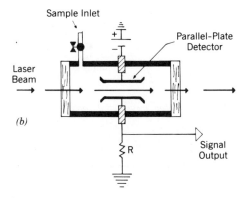

Figure 8. Cross-sectional view of photoionization-detector cells. (*a*) Proportional counter-type detector. (*b*) Parallel-plate-type detector.

Ar and 10% CH_4) and with several torr pressure of the molecular sample under study. The electric field between the electrodes drives the free electrons, created by the ionization of the sample molecules, to the collector. Under the conditions mentioned above, and with a collector wire with a diameter of 0.05 mm, charge amplification by collisions can occur. The electron detection sensitivity may then increase by a factor of 10^4.

If amplification is not needed, a parallel-plate device (Figure 8*b*) provides the simplest geometry for an ionization detector. This photoionization detector operates at a sample pressure of a few torr. A voltage, typically on the order of 150 V, is applied across the electrode plates. With the appropriate choice of detection electronics and noise-rejection shielding, a detection sensitivity of $\leqslant 10^3$ electron–ion pairs can be achieved using these devices.

In a parallel-plate photoionization detector the output-voltage signal as a function of time, following the sudden creation of electron–ion pairs, can be derived in terms of the drift velocities of the charged species as they separate and finally are collected under the influence of an external applied electric field (see also Figure 13). Figure 9 shows schematically a typical experimental setup for laser MPI studies. It consists of a frequency tunable dye laser, the polarization optics, the photoionization cell, (containing the parallel-plate detector and the sample under study), and the detection electronics. The induced ionization current, resulting from the MPI of the sample molecules, is measured as a function of λ_{exc}.

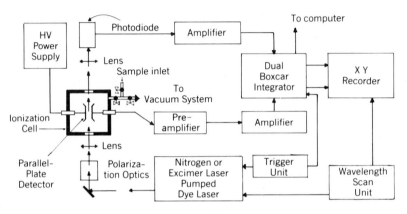

Figure 9. Layout and principle of a laser multiphoton ionization technique.

At very low pressures ($< 10^{-5}$ torr), electrons and ions (positive or negative) can be accelerated under the influence of an external electric field and easily detected, practically with unity-detection efficiency, using other types of charge-sensitive devices. Such devices are: electron multipliers, channeltrons, and multichannel plates. The latter are frequently used in ionization mass spectrometers (see Section 3.2.2) because of their fast response time (< 1 ns).

3.2.2. The Multiphoton Ionization Mass Spectrometer: Selection and Identification of Ionic Species

The photoionization detectors described above, despite their attractive features (namely, of being simple and very sensitive), are limited because they cannot provide specific information about the ionic species produced in the MPI process. In order to identify these species, mass spectrometers are employed as the photoionization detector. The unique combination of laser MPI (as the ion source) and a mass spectrometer (see Figure 10) allows the photoionization current to be measured both as a function of λ_{exc} and the mass of the ions or fragments produced, following multiphoton ionization of the molecules under study.

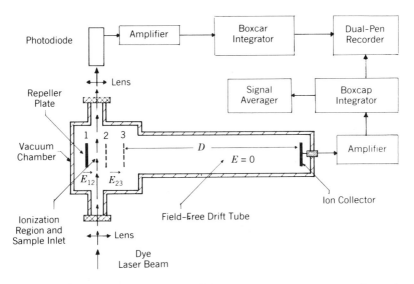

Figure 10. Cross-sectional view of a time-of-flight laser mass spectrometer and associated detection electronics for multiphoton ionization experiments.

The ions in a laser MPI mass spectrometer are produced during the short time duration of the laser pulse (typically 6 ns) and in a small volume ($\sim 10^{-5}$ cm^3) defined by the dimensions of a well-focused laser beam. It seems, therefore, that mass analysis, employing time-of-flight (TOF) mass spectrometers, is a natural choice. A TOF mass spectrometer, in its simplest form, consists of an ion source and an ion collector situated on both sides of an evacuated tube (Wiley and McLaren, 1955) (Figure 10). The ions formed in the ionization region are accelerated toward the collector by either one or a series of external electric fields (E_{12}, E_{23}) applied between the repeller plate 1 and the grid electrodes 2 and 3 (Figure 10). The accelerating electric fields can be either continuous or pulsed. In either case the velocity of the ions in the field-free drift region D (Figure 10) is a function of the ratio of their charge, q, to their mass, m. Hence, when the ions reach the collector, they have been separated into bunches that correspond to q/m. In the case of singly charged ions, the lightest group will arrive at the collector first, followed by the heavier ones. Thus each laser ionizing pulse results in a mass spectrum that can, in principle, be displayed on an oscilloscope or a recorder. By tuning the λ_{exc} of a dye laser and recording both the photoionization current and the mass spectrum, a two-dimensional photoionization mass spectrum is obtained (Zandee and Bernstein, 1979b) which provides, in addition to the spectral selectivity, identification and relative abundances of the fragment ions produced, following MPI (Figure 11).

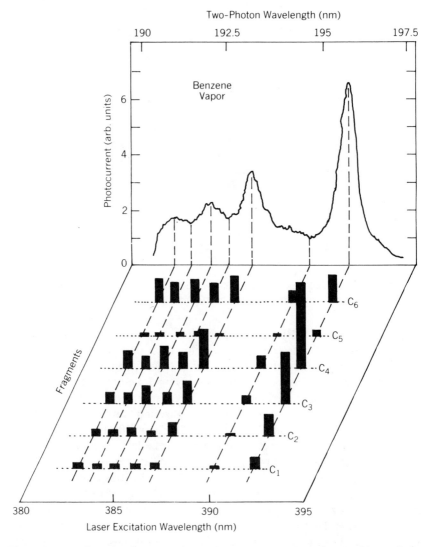

Figure 11. Two-dimensional mass spectrum of benzene vapor. Reproduced, with permission, from Zandee and Berstein (1979a, b).

By adding an electrostatic electron-energy analyzer to an MPI mass spectrometer, Compton and co-workers (Miller and Compton, 1981) developed a more versatile instrument suitable for measuring photoelectron-energy distribution spectra. With this instrument important information can be obtained pertaining to the photofragmentation mechanisms involved following MPI.

3.2.3. Multiphoton Ionization Gas Chromatograph–Mass Spectrometer: Analysis of Complex Molecular Systems

The analysis of complex mixtures of polyatomic molecules (e.g., environmental samples) requires devices with high sensitivity, increased selectivity, and identification power. Combining the unique capabilities of an MPI–mass spectrometer and a gas or liquid chromatograph, Reilly and co-workers (Rhodes et al., 1983) and Nielsen (1983) developed an MPI–gas (liquid) chromatograph–mass spectrometer suitable for such analyses.

In an MPI–gas chromatograph–mass spectrometer, the gas (liquid) chromatograph is connected into the sample inlet (see Figure 10) and thus into the ionization region of the MPI–mass spectrometer. That way the molecular components, which have been selectively separated by the gas chromatograph, are further selectively photoionized and detected in the mass spectrometer.

3.2.4. Liquid-Phase Multiphoton Ionization Spectrometer: In Situ Detection of Organic Compounds

The techniques and devices described above have been developed for gaseous-phase studies, where they have been applied successfully. An understanding, however, of the interaction of chemical compounds in dense media (e.g., quasi-liquid and liquid environments), that is, the way of participating in various processes and mechanisms of nature, requires the development of sensitive and selective techniques suitable for *in situ* liquid-phase studies.

3.2.4.1. Basic Knowledge and Detection Techniques. In the liquid phase it has been found that the ionization threshold, I_L, of polyatomic molecules is ~ 3 eV lower (Siomos and Christophorou, 1980, 1982a, b; Siomos et al., 1981a; Christophorou and Siomos, 1984) than its corresponding ionization threshold, I_G, in the gas phase. This understanding, combined with the availability today of frequency tunable dye lasers as well as laser-associated technologies, led to the development by Siomos and co-workers of a novel MPI technique suitable for liquid-phase studies (Siomos and Christophorou, 1980).

This technique proved most useful not only for accurate measurements of I_L but also for the study of molecular states embedded in the ionization continuum—and thus subjected to autoionization—as well as for the study of photophysical processes that compete with ionization (dissociation). The use of frequency tunable dye lasers and appropriate advanced electronics increased substantially the detection sensitivity, as well as spectral resolution and polarization efficiency. Unlike the situation in low-pressure gases, where the detection sensitivity for free charges (electrons, ions) approaches single-particle detection efficiencies, the detection efficiency of charged species in the liquid phase is limited.

The MPI technique developed by Siomos and co-workers (Siomos and Christophorou, 1980) at the Oak Ridge National Laboratory is shown schematically in Figure 12. It consists of: a frequency tunable dye laser; the polarization optics; the photoionization cell containing the ionization detector and the solution under study; the vacuum system; and the detection electronics.

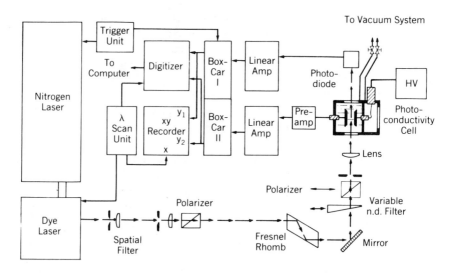

Figure 12. Layout of the liquid-phase laser multiphoton ionization technique. Reproduced, with permission, from Siomos and Christophorou (1980).

3.2.4.2. The Photoionization Detector: Measurement of Electron and Ion Currents in the Liquid Phase.

The photoionization detector is a parallel-plate device immersed in the liquid or solution under investigation, and the induced photocurrent resulting from the moving charges (electrons and ions), produced via photoionization, is measured as a function of the excitation-photon energy.

The two parallel plates (see Figure 13a) form a capacitor (of capacitance C_0) that is coupled to a high-impedance ($R \sim 10^{11}$ Ω) preamplifier. The two plates were 2-cm-diameter disks separated by a distance d, typically of the order of 0.1 cm, and were maintained at a potential difference V of 2 kV, which corresponds to an applied electric field E of 20 kV cm^{-1}.

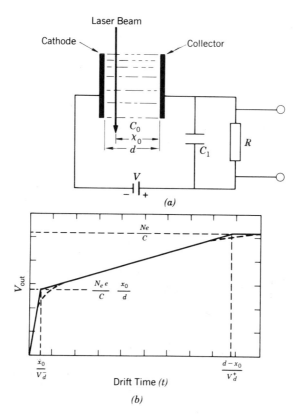

Figure 13. (*a*) Schematic of the charge-sensitive parallel-plate conductivity detector. (*b*) Voltage–pulse-time characteristics resulting from the combined motion of electrons and positive ions (see text).

Let us assume that in such a system, N_e electrons are created via photoionization at each laser pulse [the time duration of the laser pulse is very much shorter than the maximum drift time t_d $(=d/w)$] at a distance x_0 (see Figure 13*a*) from the collector and that no electron attachment occurs. These will begin to move toward the collector under the influence of E with an average drift velocity w, inducing a current $i = N_e e w/d$ in the external circuit. If the time constant of the external circuit is large compared with t_d and if C is the total capacitance of the system, then the rate of the change on the anode potential is

$$dV'/dt = -\frac{N_e e w}{Cd}, \qquad 0 \le t \le t_d \tag{16}$$

whence

$$V'(t) = -\frac{N_e ewt}{Cd + \text{const}}$$

$$= V - \frac{N_e ewt}{Cd} \tag{17}$$

where V is the potential at the anode at $t = 0$ and where the effects of diffusion have been neglected. The maximum amplitude of the (negative) output voltage pulse $V_{out} = V - V'(t_d)$ induced on the collector by the passage of the electrons from the plane of formation* (a distance x_0 from the collector) to the collector is

$$V_{out} = \frac{N_e e x_0}{Cd} \tag{18}$$

If V_{out} is a result of the combined motion of N_e electrons and N_+ positive ions, moving with drift velocities w and w_+ ($w \gg w_+$), respectively, we have

$$V_{out}(t) = \frac{N_e wt}{Cd} + \frac{N_+ ew_+ t}{Cd} \tag{19}$$

Equation (19) shows that V_{out} will first rise linearly with a slope $N_e ew/Cd$ to a level given by equation (18) (see also Figure 13b) because of the motion of the electrons; then it will continue to rise an additional amount $N_+ e(d - x_0)/Cd$ much more slowly because of the motion of the positive ions with a slope $N_+ ew_+/Cd$, reaching finally the maximum output level Ne/C (Figure 13b), where $N(= N_+ = N_e)$ is the number of charges of one kind (positive or negative) created by photoionization. These features are indeed borne out in practice, as can be seen from Figure 14 for a $\sim 10^{-3}$ M solution of benzene in liquid $n-$Pt and TMSi.

The frequency tunable dye laser pumped by a nitrogen or excimer laser was operated with a repetition frequency of ~ 20 Hz. The bandwidth of the spectral distribution of the dye laser was ~ 0.02 nm and was tuned over the spectral region from 360 to 900 nm to within ± 0.1 nm, which is sufficient for MPI measurements in the liquid phase. Great care has been exercised concerning the spatial characteristics of the laser-beam profile (see Figure 15). The laser beam had ~ 1-mm diameter after spatial filtering and was focused onto the photoconductivity cell with various focal-length lenses to a waist of

---

* This condition is satisfied since the interaction volume defined by the well-collimated laser beam is very small, and thus the interaction distance dx is much smaller than d.

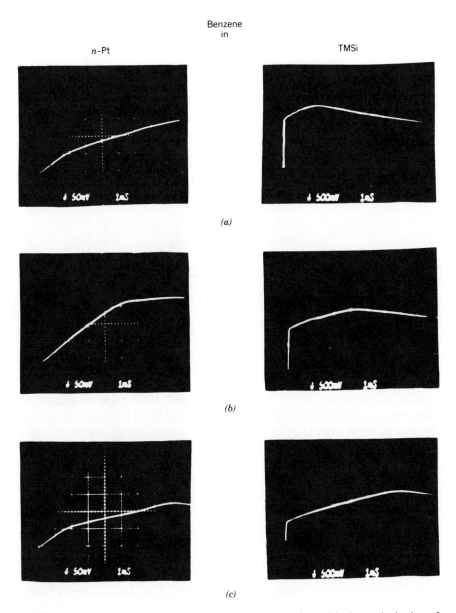

Figure 14. Oscilloscope voltage–pulse traces resulting from the multiphoton ionization of benzene in *n*-pentane and tetramethylsilane for different laser beam positions between the electrode plates (see Figure 13*a*). (*a*) Laser beam close to the cathode. (*b*) Laser beam at the middle of the detector plates. (*c*) Laser beam close to the anode. The electric field across the electrodes was $10 \, \text{kV cm}^{-1}$.

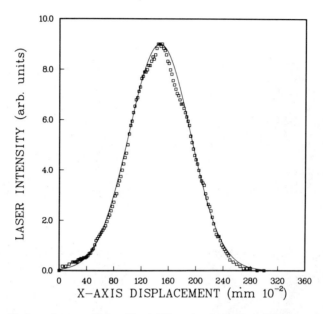

Figure 15. Laser-beam spatial profile at 406 nm (see text). From Siomos et al. (1983).

~ 40-μm diameter. This focusing condition was necessary at extreme low-intensity signals ($\sim 10^{-14}$ A). Under better signal/noise ratio conditions, as, for example, in the case of high-mobility liquids (TMSi), the waist of the focal point was larger. The laser beam on exiting the photoconductivity cell was monitored by a photodiode of well-known spectral responsivity (Figure 12) and was sampled by a boxcar integrator. The preamplifier integrated output voltage after linear amplification and shaping were averaged using a second boxcar integrator. The laser intensity and the photoionization signals from the respective integrators after digitalization were fed into a computer for analysis. The individual MPI spectra (I_{pc} vs. λ_{exc}) corresponding to each laser-dye solution were corrected for variations of the laser intensity by dividing them by the appropriate power of the laser intensity, I_1, before they were combined in their respective regions of overlap. The power, S, to which I_1 was raised was determined exactly by conducting I_{pc} vs. I_1 measurements (see Figures 4, 25, and 26).

Utilizing this technique, Siomos and co-workers (Siomos, 1988) have studied a number of aromatic molecules in various hydrocarbon liquids and in tetramethylsilane (TMSi) and obtained the first *high-resolution MPI* spectra in liquid media.

4. MULTIPHOTON IONIZATION OF POLYATOMIC MOLECULES: EXPERIMENTS AND RESULTS

4.1.   Gaseous-Phase Experiments

4.1.1.   Non-Mass-Selective Measurements

4.1.1.1. Nonresonant Multiphoton and Multiple Resonant Ionization Measurements. Whereas the first experiments on the multiphoton ionization of atoms were conducted in 1965 (Delone, 1965), the first MPI studies of molecules did not begin until 1975, with the pioneering work by Petty and co-workers (Petty et al., 1975) on iodine molecules and by Johnson (Johnson, 1975; 1976a, b) on benzene vapor.

The three-photon ionization spectrum (total ionization current vs. λ_{exc}) of iodine molecules (Petty et al., 1975) has been measured over the spectral region from 360 to 380 nm using a frequency tunable dye laser and a photoionization cell detector (see Figure 8a) in a typical experimental setup as shown in Figure 9. The observed resonances over the spectral range ~ 370–380 nm corresponding to two-photon energy from 54,000 to 48,780 cm^{-1} (no structure was observed above 380 nm) were assigned to two-photon allowed transitions to an intermediate molecular state of g-symmetry.

The first MPI spectrum of benzene (total ion current, I_{pc}, vs. λ_{exc}) was measured over the laser excitation wavelength from ~ 370 to 410 nm (see Figure 16a) in a photoionization cell detector similar to the one shown in Figure 8a filled with ~ 50 torr of benzene vapor. The experimental setup used was similar to that in Figure 9. The observed spectrum was ascribed to a two-photon resonant three-photon ionization process. Detailed analysis (Johnson, 1976a) of the observed resonances (see Figure 16a) indicated that the peak at $\lambda_{exc} = 391$ nm could probably be assigned to a two-photon allowed state with either a $^1E_{1g}$ or $^1E_{2g}$ symmetry. Although on the basis of the experimental results by Johnson (Johnson, 1976a) it is not possible to decide between the two states, multiphoton excitation (MPE) and MPI studies of benzene in dilute solutions by Siomos and co-workers (Faidas and Siomos, 1987) provided the first direct experimental evidence of the Rydberg-like nature of the two-photon allowed peak at $\lambda_{exc} \sim 391$ nm (in the liquid phase this transition is observed at $\lambda_{exc} \sim 386$ nm (see also Section 4.2.2.2.1 and Figure 21). These findings strongly support the assignment of this transition to a $(3S)^1E_{1g}$ state of benzene, predicted to exist in this spectral region (Krogh-Jespersen et al., 1979; Scott and Albrecht, 1981).

In a similar way, the MPI spectrum of benzene was studied by Neusser and Schlag (1984) over the spectral regions from 36,500 to 39,000 cm^{-1} ($512 \leq \lambda_{exc} \leq 548$ nm) and was compared with the two-photon excitation

(TPE) spectrum (Wunsch et al., 1977) of benzene over the same spectral region. Even though both spectra have been measured over the same energy region, only the blue side of the MPI spectrum ($\sim 38,500$–$39,000$ cm^{-1}) resembles the TPE spectrum. Below $38,500$ cm^{-1}, both spectra are substantially different. From the comparison of the two spectra, the authors concluded that MPI spectroscopy is a more sensitive technique for investigating highly excited states close to the ionization continuum (which cannot fluoresce appreciably) and hence is more suitable for the detection of Rydberg states.

The MPI spectrum of *trans*-1,3-butadiene (Johnson, 1976b) has been measured over the excitation laser wavelength region from ~ 370 to 410 nm corresponding to a two-photon energy from $54,000$ to $48,780$ cm^{-1} (three-photon ionization region) and from ~ 425 to 470 nm (four-photon ionization region) corresponding to the 142–155-nm region in the VUV absorption spectrum (see Figure 16b). The resonances observed to the shorter wavelengths above the three-photon ionization threshold ($\lambda_{exc} < 405$ nm) are ascribed to vibronically induced, two-photon allowed transitions with the

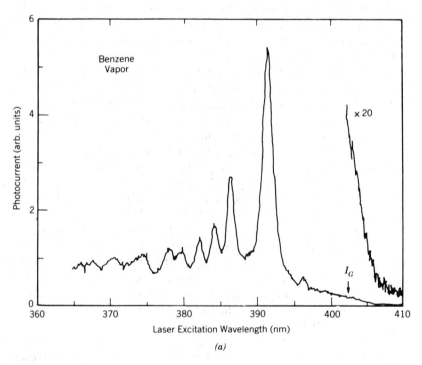

Figure 16. The multiphoton ionization spectra of (*a*) benzene vapor and (*b*) trans-1,3-butadiene vapor. Reproduced, with permission from Johnson (1976a, b).

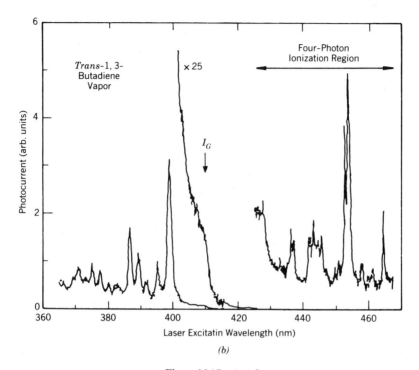

Figure 16 (*Continued*)

strongest peak at $\lambda_{\text{exc}} = 399$ nm. This peak corresponds to a two-photon energy of 50,113 cm^{-1} and has been assigned to the 0–0 transition of *trans*-1,3-butadiene. The 0–0 transition of *trans*-1,3-butadiene was theoretically predicted to be at 50,144 cm^{-1} (McDiarmid, 1975).

Using resonance-MPI techniques, new electronic states have been observed in iodine (Dalby et al., 1977) and ammonia (Nieman and Colson, 1978); and molecules of significance to biochemistry and synthetic chemistry, such as pyrrole, N-methyl pyrrole, and furan, have been studied by Compton and co-workers (Cooper et al., 1980) over the laser wavelength region from 365 to 650 nm. A strong two-photon allowed transition at 41,193 cm^{-1} was observed in the MPI spectrum of N-methyl pyrrole and was attributed to an $^1A_2$ state. Accurate ionization threshold values have been derived for pyrrole $(I_G = 8.207 \pm 0.003$ eV) and for N-methyl pyrrole $(I_G = 7.94 \pm 0.02$ eV), proving the ability of MPI spectroscopy to determine accurate ionization threshold values.

Miller and Compton (1981), using an electrostatic electron-energy analyzer, have studied the photoelectron spectrum of NO in a two-photon resonant four-photon ionization process, whereas Glownia et al. (1982) have recorded the first photoelectron-energy distribution spectra of ammonia following four-photon ionization. In the ammonia experiments, electrons with nearly zero kinetic energy were observed for practically all resonant-ionization transitions, with exception of the C state from which direct ionization was observed.

4.1.1.2. Two-Step Photoionization Measurements. The early experiments on two-step photoionization were carried out by Letokhov and co-workers (Andreev et al., 1977) on formaldehyde molecules. The formaldehyde molecules, in an ionization cell, were first excited to the $^1A_2$ electronic state by a nitrogen laser ($\lambda_{exc} = 337$ nm) and were consequently ionized using an H_2 laser ($\lambda_{exc} = 160$ and/or 116 nm) synchronized within 2 ns with the nitrogen laser. By delaying the second laser pulse in a well-defined manner with respect to the first one, it was possible to investigate the rate of depopulation of the resonant $^1A_2$ state of formaldehyde. Williamson and Compton (1979), using two dye lasers synchronously pumped by the same nitrogen laser, studied higher excited states in the iodine spectrum by first populating monophotonically the $B^3\Pi$ state of iodine vapor and consequently tuning the frequency of the second dye laser over high-order resonances. Parker and El-Sayed (1979) studied the MPI spectrum of 1,4-diazabicyclo(2,2,2)octane (DABCO) over the laser excitation wavelength region from 400 to 470 nm using two-step excitation. The lower excited state of DABCO was first populated by two-photon absorption and was then ionized by a second frequency tunable dye laser. The observed dependences in the fluorescence and photocurrent intensities for resonant (one laser) and two-step MPI indicated that significant population losses occurred at the third photon absorbed in a two-photon resonant four-photon ionization process as a result of molecular dissociation at energies corresponding to three-photon absorption. This competitive channel may be avoided in a two-step MPI process using two different lasers for excitation and ionization, thus enhancing the overall ionization efficiency.

Molecular intersystem crossing, as well as other nonradiative molecular transitions, could also be studied using two-step MPI techniques by appropriately timing the excitation and ionization laser pulses. In this way, Duncan et al. (1981) studied the intersystem crossing in benzene by exciting a supersonic beam with a frequency tunable dye laser and ionizing the excited benzene molecules with an excimer laser ($\lambda_{exc} = 248$ nm). The quantum yield for the intersystem crossing was found to be on the order of 80% and in agreement with other measurements.

4.1.2.   Mass-Selective Multiphoton Ionization Measurements

4.1.2.1.   Multiphoton Ionization Mass Spectrometry. As discussed earlier (see Section 2.2.3 and Figure 1*f*), there is often strong molecular fragmentation following MPI processes. Letokhov and co-workers (Antonov et al., 1978), in their pioneering work, studied the fragmentation of polyatomic molecules (e.g., benzaldehyde and benzophenone) using a nitrogen-laser pumped dye laser as the excitation light source and an H_2 laser as the ionizing radiation source in a two-step multiphoton ionization process. Boesl et al. (1978), using a commercial quadrupole mass spectrometer, observed the mass spectrum of benzene in a one-photon resonant two-photon ionization process ($\lambda_{exc} = 259$ nm). At low laser light intensities the benzene mass spectrum showed only the single benzene parent-ion peak ($C_6H_6^+$); at higher intensities ($> 10^7$ W cm^{-2}), however, smaller molecules (e.g., $C_2H_x^+$, $C_3H_x^+$, and $C_4H_x^+$) were observed. Zandee and Bernstein (1979a, b) conducted the first real MPI ($n > 2$) experiments of molecules with mass analysis for identification and determination of the relative abundance of the fragment ions. In their pioneering studies, the wavelength of a frequency tunable dye laser was tuned over the respective wavelength region of interest while concurrently the mass spectrum of the positive ions produced via MPI were recorded. Figure 11 shows the so-called two-dimensional spectrum of benzene, obtained at relative moderate laser light intensities (weak focusing conditions). Each bar represents the sum of the contributions of all positively charged fragment ions with a given number of carbon atoms ($C_i \equiv \Sigma_k C_i H_k$) under high laser light intensities ($> 10^9$ W cm^{-2}; strong focusing conditions). Under strong focusing conditions the overall MPI efficiency increased by a factor of ~ 50, thus allowing for unit mass resolution. At these higher laser light intensities (see also Antonov et al., 1980; Robin, 1980; Reilly and Kompa, 1980; and Antonov and Letokhov, 1981), extensive fragmentation of the benzene molecules was observed with the C^+ peak being the most abundant fragment (22% of the total ion yield). It is interesting to note that the C^+ peak is not observed in the electron-impact mass spectrum of benzene. On the basis of appearance potentials argumentation (the appearance potential for C^+ in benzene was estimated to be ~ 26 eV), the formation of the C^+ peak indicated that under these experimental conditions at least nine photons of 3.17 eV energy were absorbed per benzene molecule during the 6-ns duration of the laser pulse. The mechanism responsible for this extensive fragmentation has not yet been fully understood. Further experiments are therefore required in order to verify the observed results (see also Robin, 1980).

In similar experiments (Zandee and Bernstein, 1979a) the *trans*-1,3-butadiene mass spectrum was recorded at $\lambda_{exc} = 387$ nm. It was estimated that under the authors' experimental conditions $\sim 10^{-4}$ molecules, within the focal region of the laser beam, were ionized at each laser pulse. Furthermore, an analysis of the resonance MPI mass-fragmentation pattern, as in the case of benzene, indicated that although only three photons of 3.2 eV energy should have been sufficient to photoionize trans-1,3-butadiene ($I_G = 9.06$ eV) in the gas phase, at least six photons have been absorbed per molecule and laser pulse.

A comparison between the results of benzene and trans-1,3-butadiene shows that at comparable laser light intensities benzene undergoes more extensive ionic fragmentation and that the number of the photons absorbed per molecule is larger for the molecule with the higher density of states.

The introduction of MPI mass spectrometry, as discussed above, triggered a number of successful attempts to utilize this technique for performing chemical analysis. In their work, Lubman and Kronick (1982), using an Nd:YAG laser pumped dye laser and a 1.5-m time-of-flight mass spectrometer, studied the fragmentation patterns of a number of aromatic molecules. Although they were unable to distinguish between isomeric compounds in effusive molecular beams. They could, however, examine successfully and analyze, on the basis of MPI fragmentation patterns, a mixture of aniline, naphthalene, 2-methyl naphthalene, and 1,4-dimethyl naphthalene in a cooled molecular beam.

Using a supersonic beam technique in conjunction with an MPI mass spectrometer, Lubman et al. (1985) succeeded in discriminating between the two isomers azulene and naphthalene. The ratio between the azulene signal and napthalene background was $> 500:1$.

In further interesting studies, Tembreull et al. (1985) and Lubman (1987) succeeded in discriminating among four isomers of dichlorotoluene. The resonant two-photon ionization spectra of the four isomers studied are shown in Figure 17. These spectra were obtained by expanding several parts per million of each compound in a 1-atm back pressure of Ar into a vacuum of $\sim 10^{-6}$ torr. The dichlorotoluene molecules were then excited over the spectral region from 273–282 nm by means of a frequency doubled dye laser. The energy of the laser pulse was ~ 0.1–0.3 mJ, and the laser light spectral bandwidth was ~ 0.5 cm^{-1} in the UV spectral range. Similar experiments were also carried out for the isomers of cresol (Tembreull and Lubman, 1984). It was found that a discrimination of $1:300$–$1:500$ was possible between any of two isomers with a detection limit at < 20 parts per billion (ppb) in 1 atm of nitrogen.

An interesting application of MPI mass spectrometry is the molecular isotopic analysis in molecular supersonic beams. In molecular isotope analysis, differences in vibrational frequencies from the ground state to an excited state, due to isotopic mass shifts, are measured. The detection of ^{13}C in aniline in a supersonic beam was first demonstrated by Leutwyler and Even

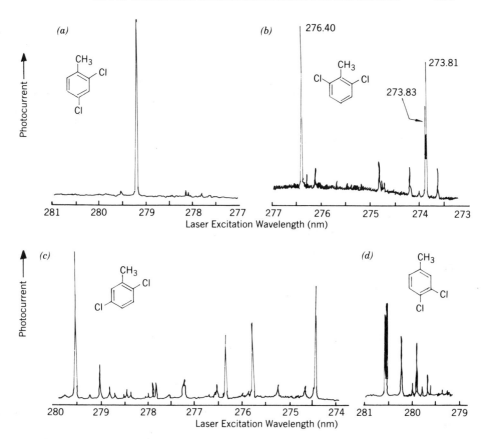

Figure 17. The multiphoton ionization spectra of four isomers of dichlorotoluene as a function of the laser excitation wavelength obtained using a multiphoton ionization time-of-flight mass spectrometer. Only the isotopic molecular ions were monitored. (*a*) 2,4-Dichlorotoluene. (*b*) 2,6-Dichlorotoluene. (*c*) 2,5-Dichlorotoluene. (*d*) 3,4-Dichlorotoluene. Reproduced, with permission, from Tembruell et al. (1985).

(1981). The observed isotopic frequency shifts of $\sim 4 \text{ cm}^{-1}$ were well within the spectral resolution, Δv_1, of commercially available pulsed dye laser systems ($0.5 \leq \Delta v_1 \leq 1.5 \text{ cm}^{-1}$). Lubman et al. (1985) have studied the MPI mass spectra of the isotopes of dichlorotoluene in which the natural abundance of Cl isotopes $^{35}Cl_2$: ^{35}Cl: $^{37}Cl_2$ is 9:6:1. The results showed that the relative ratios of the isotopic peaks were enhanced for the particular isotope whose transition energy corresponded to the selectively chosen λ_{exc}. The observed enhancement for each isotopic combination differed by a factor of ~ 16. An interesting finding of this work is that isotopic selectivity for large molecules in an MPI mass spectrometer is possible even at relatively low laser light spectral resolution.

4.1.2.2. Multiphoton Ionization Gas (Liquid) Chromatography–Mass Spectrometry.

Klimcak and Wessel (1980) demonstrated (see Table 2) for the first time the usefulness of laser MPI spectroscopy as a detector of gas chromatographic (MPI/GC) effluents. In this way, optical selectivity was also obtained for molecular isomers, without the need of supersonic molecular beams by initially separating the isomers (e.g., phenanthrene–anthracene) in a gas chromatograph and subsequently detecting them in a photoionization cell, by tuning the laser ionizing radiation either to $\lambda_{exc} = 310$ nm (anthracene) or $\lambda_{exc} = 285$ nm (phenanthrene). The minimum detectable amount of anthracene with a signal/noise ratio of 2 was estimated to be 10 pg, while for phenanthrene it was 50 pg (Table 2).

Excimer laser ($\lambda_{exc} = 308$ nm)-induced MPI has been utilized for ion generation in a capillary gas chromatograph–mass spectrometer (MPI/GC–MS) to selectively separate and detect polyatomic hydrocarbon (e.g., chrysene and triphenylene) by Rhodes et al. (1983). Selective ionization

Table 2. Detection Limits of Polycyclic Aromatic Hydrocarbons (PAHs) Using Various Detection Techniques

PAH	Detection Limits (pg)[a]			
	MPI/GC[b]	MPI/GC–MS[c]	MPI/GC–FTMS[d]	LD/MPI–MS[e]
Naphthalene	0.30	0.20	7.7	
Biphenyl		5.00	78.9	
Acenaphthalene	10.00	0.55	34.3	
Fluorene		0.23	76.2	
Anthracene	10.00	0.92		
Phenanthrene	50.00	0.30		
Pyrene	6.00	1.10		
1,2-Benzanthracene	4.00			
PTH-alanine				~5
PTH-proline				~5
PTH-valine				~5

[a] Signal/noise ratio = 2.
[b] From Klimcak and Wessel (1980).
[c] From Nielsen (1983).
[d] From Sack et al. (1985). FTMS; Fourier transform mass spectrometer.
[e] From Engelke et al. (1987).

has been demonstrated based upon the small differences in the ionization threshold values between chrysene ($I_G = 7.8$ eV) and triphenylene ($I_G = 8.1$ eV) in a TOF mass spectrometer. Furthermore, Nielsen (1983), in a series of interesting experiments, using MPI/GC–MS, succeeded in selectively identifying and detecting polycyclic aromatic hydrocarbons (PAHS) in complex mixtures with improved detection sensitivity. Using MPI/GC–MS, a number of PAHs were studied and identified (see Table 2).

4.1.2.3. Laser Desorption–Multiphoton Ionization Mass Spectrometry. The study and detection of biological molecules requires special care since such molecules are nonvolatile and thermally unstable. A promising technique for studying such biological compounds, providing synchronously improved selectivity and high detection sensitivity, is laser desorption. MPI mass spectrometry (LD–MPI/MS). In this technique, a pulsed infrared laser (e.g., CO_2) is used to induce fast heating on a surface that desorbs molecules that have been previously absorbed in this surface before they have the time to decompose. The desorbed molecules are then swept by means of a pulsed supersonic beam into the ionization region of a mass spectrometer. Using this technique a number of biological molecules (e.g., catecholamines, indoleamines, amino acids, small peptides, and porphyrins) have been investigated (Lubman, 1987; Engelke et al., 1987; Tembreull and Lubman, 1987). Engelke et al. (1987) were the first to quantitatively analyze molecules absorbed on surfaces by LD–MPI/MS. In their studies they were able to detect a number of PTH-amino acids (e.g., PTH-alanine, PTH-proline, PTH-valine) with a sensitivity on the order of a few picomoles (see Table 2).

4.2. Liquid-Phase Experiments

4.2.1. Multiphoton Ionization at Fixed Excitation Photon Energy

Although laser-induced MPI of molecules in the gas phase, as we have seen in previous sections, has been the subject of numerous experimental and theoretical investigations, relatively little attention has been given to similar processes in the liquid phase (Christophorou and Siomos, 1984).

Whereas the first experiments on the photoionization of atoms (Hg) and small molecules (HI, NaI, NaBr, KCl) in liquid water and n-hexane were conducted in the years around 1930 (Warburg and Rump, 1928; Franck and Scheibe, 1928; Reichardt and Bohnhoeffer, 1931), the systematic study of the photoionization of polyatomic molecules in liquid media did not begin until around 1960 with the pioneering work by Pilloff and Albrecht (1966, 1968) and Yamamoto et al. (1966).

Gibbons et al. (1965), in their studies on aromatic molecules (naphthalene, phenanthrene, perylene, etc.) in 3-methylpentane (3-MP) glass at 77 K, found that the photoionization signal depended on the square of the excitation light intensity ($\lambda_{exc} = 253.7$ nm). This finding led to the proposition that the photoionization mechanism involves a stepwise excitation with the lowest triplet state as the intermediate resonant step, namely,

$$AX + \hbar\omega \rightarrow (AX)^* \rightarrow {}^3(AX)^* \qquad \text{(first step)}$$

and

$${}^3(AX)^* + \hbar\omega \rightarrow {}^3(AX)^{**} \rightarrow (AX)^+ + e^-$$

where ${}^3(AX)^{**}$ is a superexcited state that autoionizes (see Figure 1e). Similar studies by Pilloff and Albrecht (1966) on N,N,N',N'-tetramethyl-p-phenylenediamine (TMPD) in liquid 3-MP at room temperature showed the multiphotonic nature of the photoionization process in this system. Further studies by a number of other authors (e.g., Nakato et al., 1967; Gary et al., 1968; Pilloff and Albrecht, 1968; Babenko et al., 1969; Hauser and Jarnagin, 1970) supported the two-photon ionization mechanism of TMPD in liquids. Takeda et al. (1971), however, suggested that for TMPD in solution, at room temperature, besides the above two-photon process via the triplet state, a one-photon process from the ground state of TMPD also occurs and leads to ionization. Later work by Kellmann and Tfibel (1980), as well as earlier studies (e.g., Richards et al., 1970; Tamir and Ottolenghi, 1970; Grellmann and Watkins, 1971; Taniguchi et al., 1972; Alchalal et al., 1972), questioned the *sole* role of the triplet state in the two-photon (stepwise) ionization of aromatic molecules in solution and argued that both singlet and triplet states can serve as intermediate resonances. Richards et al. (1970), in particular, suggested that the two-photon mechanism for pyrene in solution of $\lambda_{exc} = 347.1$ nm could be a "direct two-photon absorption from the ground state to a higher excited state which ionizes". This mechanism, of course, is nothing else but a one-photon resonant two-photon ionization process via the first singlet state of pyrene, as verified by Piciulo and Thomas (1978) and Hall and Kenney–Wallace (1978) in a double-pulse excitation experiment ($\lambda_{exc} = 347.1$ nm).

In photoionization studies of liquid benzene at $\lambda_{exc} = 355$ nm, Scott et al. (1979) showed that the photoionization signal (I_{pc}) depended on the excitation laser light intensity I_1 in the form of $AI_1^2 + BI_1^3$ and suggested that in liquid benzene, photoionization occurs via a three-photon process involving a two-photon intermediate excited state (or states) of ~ 6.98 eV, which does not ionize to a "significant extent" and which relaxes to a "relatively long-lived state" which then ionizes by absorbing a third photon. In a subsequent study of multiphoton ionization of liquid benzene at $360 \leq \lambda_{exc} \leq 440$ nm, Scott et al. (1982) proposed that simultaneous two-photon absorption indirectly populates the lowest singlet *excimer state* of benzene, from which a third photon excites a superexcited state that preionizes. At $\lambda_{exc} = 337$ nm (7.3 eV) they observed, along with the three-photon process, a competing two-photon ionization process as well. They argued, however, that such a process is "less efficient" than the three-photon ionization route they proposed. Miyasaka et al. (1985), conducting transient absorption measurements in liquid benzene following two-photon excitation at $\lambda_{exc} = 355$ nm, found that the transient absorption observed originated from a transient excimeric species whose formation time was ~ 57 ps. They identified, interestingly enough, that the precursor of these excimeric species is the geminate electron–ion pair, whose formation time is ≤ 10–20 ps. These findings, together with similar studies by Hamanoue et al. (1981), strongly suggest that two-photon ionization is possible at $\lambda_{exc} = 355$ nm (6.98 eV) and that ionization precedes excimer formation (see also Siomos et al., 1981b; Faidas and Siomos, 1987).

4.2.2. Multiphoton Ionization with Variable Excitation Photon Energy

4.2.2.1. Experimental Results for Various Molecules. Siomos and co-workers (Siomos and Christophorou, 1980; Siomos et al., 1981b) reported the first systematic MPI spectra of polyatomic molecules in dilute solutions, using a sophisticated MPI technique (see Section 3.2.4) suitable for liquid-phase studies. The first MPI spectra of pyrene in *n*-pentane (*n*-Pt) were studied by monitoring the photocurrent (I_{pc}) of a $\sim 10^{-6}$ M solution as a function of the laser excitation wavelength, λ_{exc}, in the spectral region 360–530 nm, corresponding to two-photon transitions in the spectral range 180–265 nm (55,555–37,736 cm^{-1}). The MPI spectrum of pyrene in *n*-Pt is shown in Figure 18. It comprises nine separate corrected spectra combined by normalization in the respective regions of overlap. Measurements of I_{pc} vs. I_1 at several laser wavelengths around 400 nm (the intensity dependence of the photoionization signal could not be measured for $\lambda_{exc} > 450$ nm because of the weak signals near the ionization threshold) indicated a direct two-photon ionization (TPI) process, to be involved in the photoionization of pyrene in dilute solutions (concentration $\leq 10^{-6}$ M). The TPI signal vs. λ_{exc} in Figure 18 increases as λ_{exc} decreases up to $\lambda_{exc} \sim 400$ nm and then rises very steeply. Such a sharp rise is expected in a TPI process, as discussed previously (see Section 2.2), in the case of one-photon resonant two-photon ionization process. The first two resonances in the TPI spectrum at $\lambda_{exc} \sim 375$ nm (6.60 eV) and $\lambda_{exc} \sim 370$ nm (6.69 eV) are a result of two-photon ionization via the lowest vibrational levels of the first singlet state (dotted line in Figure 18; Kourouklis et al., 1982b). On the long-wavelength side the photocurrent approaches "zero level" at $\lambda_{exc} \sim 516$ nm, which corresponds to a two-photon energy of 4.80 ± 0.02 eV. The authors associated this value with the I_L of

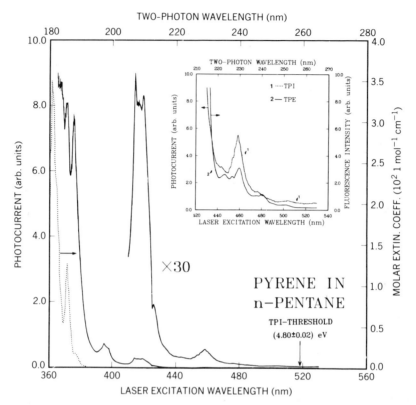

Figure 18. The multiphoton ionization spectrum of pyrene in *n*-pentane. *Inset*: The two-photon ionization spectrum (. . . .) and the two-photon excitation (——) spectra of pyrene in solution. Both spectra were normalized at $\lambda_{exc} = 480$ nm. The dotted line represents the one-photon absorption spectrum of pyrene in *n*-pentane.

pyrene in *n*-Pt, which is 2.61 eV lower than the corresponding I_G value ($= 7.41$ eV; Boschi and Schmidt, 1972). The structure in the TPI spectrum (Figure 18) was attributed to *autoionization* and was ascribed (Siomos and Christophorou, 1980; Siomos et al., 1981a) predominantly to two-photon excitation of one-photon forbidden transitions, since pyrene molecules belong to the D_{2h} symmetry group. This interpretation is consistent with the excellent agreement between the peaks observed in the two-photon excitation (TPE) and the TPI spectra of pyrene (see also Salvi et al., 1983) in solution (see inset in Figure 18) over the spectral region $420 \leq \lambda_{exc} \leq 530$ nm.

For pyrene the ground π-electron state is the totally symmetric even (g) parity $^1A_{1g}$ state of pyrene, and hence higher π-electronic states of odd (u) parity ($^1B_{3u}$ or $^1B_{2u}$) can be reached from $^1A_{1g}$ via allowed one-photon transitions. In addition to these purely electronic states, there are even-parity "vibronics" that can be built upon peaks as a result of transitions to purely electronic odd-parity states. In pyrene, one-photon transitions to $^1A_{1g}$ or $^1B_{1g}$ states are forbidden (Robin, 1974). These states can, however, be reached via two-photon transitions from the $^1A_{1g}$ ground state of pyrene. Similar results were obtained by Siomos and co-workers for TMPD and fluoranthene (Siomos et al., 1981a; Kourouklis et al., 1981; Siomos and Christophorou, 1982b).

The TPI and the TPE spectra of fluoranthene measured with linearly polarized light for λ_{exc} from 360 to 560 nm (curves 1 and 3) and from 440 to 860 nm (curve 2), respectively, are shown in Figure 19. As in the case of pyrene, the MPI spectrum of fluoranthene (curve 1 in Figure 19) shows distinct structure. It differs, however, from that of pyrene (see Figure 18) in a well-defined way: Instead of showing the expected sharp increase of the pyrene spectrum with increasing excitation energy, the photocurrent in Figure 19 declines sharply when the ionization proceeds via the one-photon resonant first singlet-state manifold ($\lambda_{exc} \leq 415$ nm) while showing simultaneously

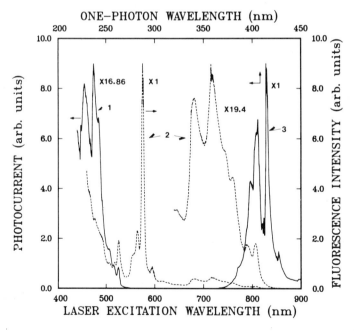

Figure 19. The mutiphoton ionization (curves 1 and 3) and two-photon excitation (curve 2) spectra of fluoranthene in *n*-pentane. The one-photon resonant two-photon ionization spectrum (curve 3) is plotted at the corresponding two-photon wavelength to allow for comparison with the two-photon excitation spectrum.

distinct peaks (curve 3 in Figure 19). The same structure is revealed by the TPE spectrum (curve 2 in Figure 19) (Faidas et al., 1985).

The structure observed in the TPI spectrum of fluoranthene has been attributed, as in the case of pyrene, to autoionization of discrete states embedded in the ionization continuum reached by two-photon absorption. The correspondence between the maxima in the TPE (curve 2) and the TPI (curve 1) spectra in Figure 19, on the one side, and between the one-photon absorption (solid line) and TPE (dashed line) spectra in Figure 20 on the other, is consistent with this.

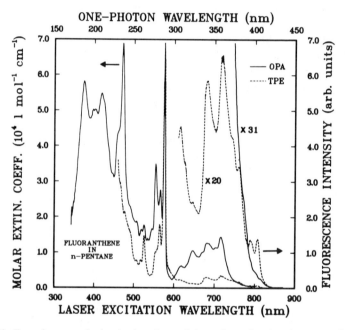

Figure 20. Two-photon excitation (- - -) and one-photon absorption (——) spectra of fluoranthene in *n*-pentane.

The fluoranthene molecule has a C_{2v} symmetry and does not possess a center of inversion; all excited states might therefore be one- and two-photon allowed, although the spectral distributions for one- and two-photon excitation might be different, owing to vibronic coupling (Siomos et al., 1981a; Siomos and Christophorou, 1982a). It is seen from Figure 19 that at $\lambda_{exc} \sim 430$ nm ($\sim 23,760$ cm^{-1}), corresponding to the onset of the first absorption band of fluoranthene, the TPI signal increases sharply as expected for one-photon resonant two-photon transitions. The TPI signal, however, unexpectedly decreases for $\lambda_{exc} < 414$ nm. It is known that the long ($\sim 10^{-8}$ s) lifetimes of intermediate resonant states, as opposed to the short (10^{-15} s) lifetimes of virtual states, not only contribute to the probability of upward transitions but also allow for other processes which, under favorable time and energy conditions, compete effectively with MPI. Since the decline in the TPI signal sets in when TPI occurs via the one-photon resonant first singlet-state manifold (S_1) and since autoionization is usually faster than dissociation, Siomos and co-workers attributed this decrease in the TPI signal to molecular dissociation from the $S_{1,n}$ state, a process competing with two-photon ionization via this state. The molecule probably starts to predissociate at energies higher than the energy (24, 140 cm^{-1} in n-Pt of the $S_{1,0}$; see Figure 20), where two-photon ionization still competes effectively with dissociation. As the photon energy increases, direct dissociation seems to predominate, so that the TPI current goes virtually to "zero" at energies of the one-photon resonant transition, at $\sim 27,320$ cm^{-1} (~ 3.39 eV). The excellent agreement between the TPE (curve 2) and the TPI spectra (curve 3) in the spectral region for $\lambda_{exc} < 414$ nm (Figure 19) shows that the observed structure in this spectral range is a result of the vibrational structure of S_1, reached by one-photon, as can also be seen from the spectra in Figure 20.

On the long-wavelength side (Figure 19), the photocurrent reaches asymptotically a "zero level" at $\lambda_{exc} = 551$ nm. This corresponds to an energy for a two-photon transition of 4.50 ± 0.5 eV, which the authors identified with the I_L of fluoranthene in n-Pt. As in the case of pyrene, I_L is smaller than I_G ($I_G - I_L \sim 3$ eV; $I_G = 7.57$ eV).

4.2.2.2.   Experimental Results for Benzene. The first MPI spectrum of liquid (neat) benzene using variable excitation photon energy was investigated by Vaida et al. (1978) over the laser excitation region of ~ 360–450 nm corresponding to a two-photon energy of $\sim 44,400$–55,500 cm^{-1}. A comparison with a similar two-photon thermal lens spectrum led the authors to attribute the observed broad monotonically rising band below $\lambda_{exc} \leq 400$ nm to direct two-photon ionization. This interpretation was also supported by indirect estimations of the ionization threshold of liquid benzene to be around $I_G \approx 6.2$ eV.

The few MPI studies of benzene in the liquid phase that have been reported to date (Vaida et al., 1978; Scott et al., 1979; Scott et al., 1982; Scott and Albrecht, 1982) and discussed above have been limited to neat liquid benzene and have also been at a rather restricted excitation spectral range.

Recently, Siomos and co-workers (Siomos et al., 1984; Faidas and Siomos, 1987, 1988), in an effort to understand the MPI mechanism of benzene in the liquid phase, conducted the first systematic MPI studies of benzene in dilute solutions, using both linearly and circularly polarized light. Such studies provide higher spectral resolution and are free of any effects, such as excimer formation, which traditionally complicate the interpretation and understanding of the photoionization mechanism (see discussion in previous section) and open new possibilities for the utilization of MPI spectroscopy in analytical instrumentation and procedures in liquid environments (see Section 4.2.3).

4.2.2.2.1. Polarization Measurements.

Figures 21 and 22 show the MPI and the TPE spectra of benzene in TMSi and n-Pt as a function of λ_{exc} from 360 to 560 nm and from 440 to 560 nm, respectively. The spectra in Figures 21 and 22 were measured with linearly polarized light. The structure observed for $530 \leq \lambda_{exc} \leq 420$ nm corresponds to the two-photon resonant $^1B_{2u} \leftarrow {}^1A_{1g}$ transitions, while the structure between 400 and 420 nm corresponds to the $^1B_{1u} \leftarrow {}^1A_{1g}$ transitions of benzene.

The strong peak of $\lambda_{exc} \sim 386$ nm (see Figure 21) corresponds to a two-photon allowed transition. This peak observed also in the TPE spectrum (dashed line Figure 21) exhibits a characteristic solvent dependence. Systematic studies (Faidas and Siomos, 1987; Siomos and Faidas, 1988) in a

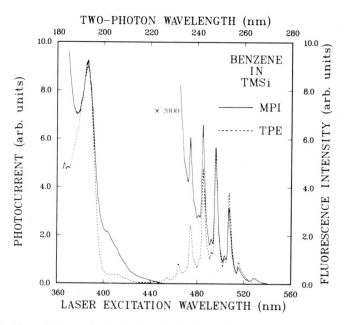

Figure 21. The multiphoton ionization (——) and the two-photon excitation (- - - -) spectra of benzene in tetramethylsilane. Both spectra have been normalized at $\lambda_{exc} = 386$ nm. The portion of the multiphoton ionization spectrum corresponding to $^1B_{2u}$ state of benzene is also shown amplified 2000 times to reach the intensity of the two-photon excitation spectrum.

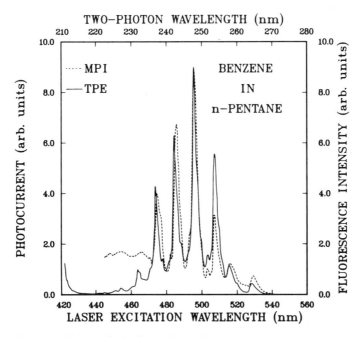

Figure 22. The multiphoton ionization and two-photon excitation of benzene in *n*-pentane.

variety of different hydrocarbon and perfluorocarbon solvents, in conjunction with polarization measurements, showed that the peak at $\lambda_{\text{exc}} \sim 386$ nm exhibits a strong Rydberg-state character and could be attributed to the (3S) $^1E_{1g}$ Rydberg state of benzene.

In Section 2.2 we discussed the importance of polarization studies in the MPI process. We have seen that the polarization ratio Ω ($\Omega = I_{\text{pc}}^{\text{circular}}/I_{\text{pc}}^{\text{linear}}$) of the photoionization signal reflects not only the symmetry properties of the intermediate resonant states involved in the MPI process (for a two-photon resonant transition $\Omega \leq 3/2$) but also the dynamics and symmetry of the post-resonant transitions.

Figure 23a shows the TPI spectra of benzene in TMSi (Siomos et al., 1984; Faidas and Siomos, 1987) using both linearly and circularly polarized light for $460 \leq \lambda_{\text{exc}} \leq 540$ nm. These spectra have been obtained at $I_1 < 120\,\mu$J per pulse. The polarization ratio Ω is shown (solid line) in Figure 23b. For comparison, the corresponding Ω ratio (dashed line) calculated from TPE measurements by Faidas and Siomos (1988) is also presented.

Although the interpretation of the Ω ratios of the MPI of polyatomic molecules in liquids is complicated by the very fast interaction mechanisms between excited molecules and surrounding medium and the ensuing depolar-

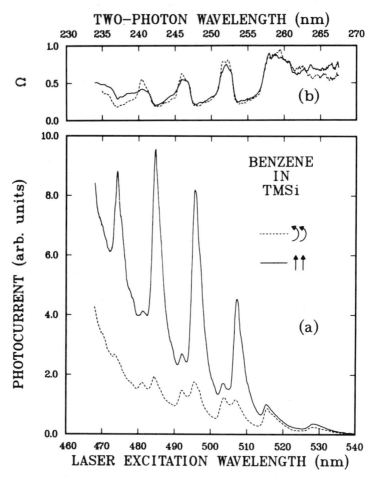

Figure 23. (*a*) The multiphoton ionization spectra of benzene in tetramethylsilane measured with circularly (- - -) and linearly (——) polarized light. (*b*) The polarization ratio Ω of the multiphoton ionization (——) and the two-photon excitation (- - -) spectra (see text).

ization of the excited states, the fact that the MPI and the TPE Ω ratios are very similar (see Figure 23*b*) indicates that for benzene in dilute solutions at room temperature and of low laser light intensities, I_1 is less than 120 μJ per pulse: (i) The final ionizing step is isotropic (i.e., bound to continuum transition), and (ii) the polarization characteristics of the MPI process are dominated (Ω ≤ 3/2) by the symmetry of the intermediate two-photon resonant $^1B_{2u}$ state of benzene.

As the laser light intensity increases, however, the MPI Ω values change drastically. Figure 24 shows the MPI spectra of benzene (Faidas and Siomos,

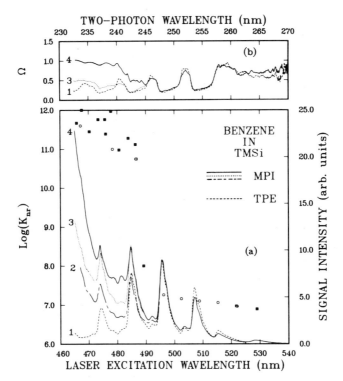

Figure 24. (*a*) The two-photon excitation spectrum of benzene in tetramethylsilane (curve 1) and the multiphoton ionization spectra measured at 70, 120, and 150 μJ/pulse (curves 2, 3, and 4), respectively. The symbols represent the nonradiative relaxation rate constants (K_{nr}) of different vibrational levels of the $^1B_{2u}$ state of benzene in the vapor phase [reproduced, with permission, from Yakovlev and Lukin (1985) and McClain (1971)]. (*b*) The polarization ratios Ω corresponding to the spectra 1, 3, and 4 in part a.

1987) in TMSi measured at successively higher I_l[~ 70 μJ (curve 2); ~ 120 μJ (curve 3); ~ 150 μJ (curve 4)]. Curve 1 in Figure 24*a* represents the TPE spectrum. All the spectra in Figure 24*a* were obtained using linearly polarized light (for the sake of simplicity, the corresponding MPI spectra obtained using circularly polarized light are not shown). The corresponding Ω ratios are shown in Figure 24*b*. The solid and open points in Figure 24*a* represent the nonradiative relaxation rate constants (K_{nr}) of benzene in the gas phase (Farmosinho and da Silva, 1974; Callomon et al., 1972). The sharp increase of K_{nr} at $\lambda_{exc} \sim 490$ nm is associated with the onset of the well known channel-three region. The channel-three mechanism is a very fast nonradiative relaxation mechanism, which sets in at ~ 3000 cm^{-1} above the origin

($38,086$ cm^{-1}) of the S_1 manifold of benzene and increases the nonradiative decay rate by three orders of magnitude.

Although the TPE and MPI spectra coincide very well for $\lambda_{exc} \geq 490$ nm and for low laser light intensities (curves 1 and 2 in Figure 24a), as the laser intensity increases and the λ_{exc} decreases ($\lambda_{exc} \leq 490$ nm), they start deviating substantially from each other and their corresponding Ω ratios approach a value $\Omega = 1$ (see curves 4 in Figure 24a, b). This dramatic increase in the MPI and TPE Ω ratios begins at $\lambda_{exc} \sim 490$ nm, exactly where the onset of the channel-three region is located.

4.2.2.2.2. *Photocurrent Versus Laser-Intensity Measurements.* In view of the aforementioned observations, it is necessary in liquid-phase MPI studies not only to consider carefully the known spectroscopy of the solute and solvent under study but also to establish the order of the MPI process by conducting laser-intensity dependence measurements. Siomos and co-workers (Faidas and Siomos, 1987) conducted a series of photocurrent vs. laser light intensity (I_{pc} vs. I_l) measurements at various λ_{exc}. Figure 25 shows the I_{pc} vs. I_l measurements of benzene in TMSi. These and similar studies conducted in *n*-

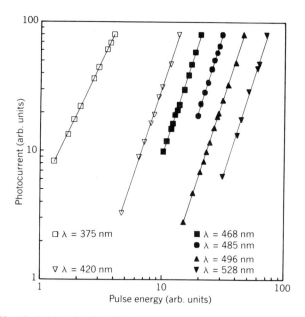

Figure 25. Photoionization signal vs. laser pulse energy measurements of benzene in tetramethylsilane. The slopes are $S = 2$ at $\lambda_{exc} = 375$ nm, and $S = 3$ at $\lambda_{exc} = 420, 468, 485, 496,$ and 528 nm. The slope $S = 3$ represents the lower laser light intensity limit.

Pt show that: (i) For benzene in TMSi, two photons are required to ionize benzene for $\lambda_{exc} \leq 375$ nm, and three photons are needed in the spectral region $420 \leq \lambda_{exc} \leq 530$ nm, whereas for $375 \leq \lambda_{exc} \leq 420$, two and three photons are absorbed, leading to photoionization. (ii) For benzene in n-Pt, although two photons are involved in the photoionization mechanism for $\lambda_{exc} \leq 337$ nm, three and four photons are needed to photoionize benzene in the spectral regions $360 \leq \lambda_{exc} \leq 380$ nm and $420 \leq \lambda_{exc} \leq 530$ nm, respectively.

These measurements have been carried out at the lowest laser light intensity that would yield a measurable photoionization signal. At higher I_1, in general, a larger number of photons were absorbed as shown in Figure 26 for benzene in TMSi. For $I_1 > 100 \, \mu J$ per pulse and for $\lambda_{exc} \sim 490$ nm, as many as nine photons were absorbed, corresponding to a total absorbed energy which exceeds the ionization threshold of benzene ($I_G = 9.24$ eV) in the gas phase by a factor of ~ 3 (see also discussion in Section 3.2) and resulting in an "apparent" high-order ionization process.

These measurements, together with the multiphoton polarization studies, led Siomos and co-workers to propose that the MPI mechanism of benzene in solutions for $\lambda_{exc} \leq 490$ nm is associated with the channel-three region (see

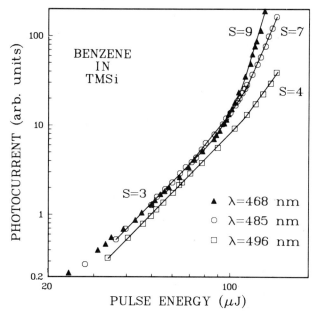

Figure 26. Photoionization signal vs. laser pulse energy measurements of benzene in tetramethylsilane at $\lambda_{exc} = 468$, 485, and 496 nm. Note the increase in the number of the photons absorbed (increase in the S value) with increasing laser light intensity (see text).

Figure 27). The benzene molecules are first excited by two-photon absorption into the higher-vibrational levels of the $^1B_{2u}$ state, from where they enter the vibrational manifold associated with channel three. Although the nature of channel three is not well understood, recent studies (Nakashima and Yoshihara, 1982) indicate that it depopulates the $^1B_{2u}$ state of benzene via the very dense (at these high energies) vibrational manifold of the ground state $^1A_{1g}$.

Subsequently they "oscillate" (Figure 27) in this manifold, absorbing photons and relaxing until they either ionize or relax back to the ground state. This mechanism occurs concomitantly with the three-photon ionization via the $^1B_{2u}$ state and since it is of higher order, it is more pronounced at higher I_1 values.

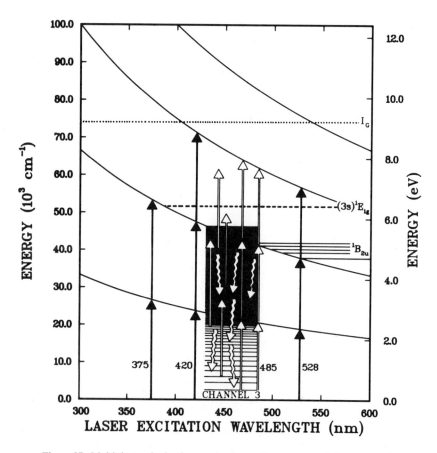

Figure 27. Multiphoton ionization mechanisms of benzene in solution (see text).

4.2.2.2.3.   The photoionization Threshold of Benzene in Solutions. Using fluorescence quantum yield techniques, Fuchs and Voltz (1973) have estimated the I_L of neat liquid benzene to be ~ 7.1 eV. Vaida et al. (1978), however, using MPI techniques, estimated a value $I_L = 6.2$ eV. Scott et al. (1979, 1982) and Scott and Albrecht (1982), in a series of MPI studies in neat liquid benzene (see also discussion in Section 4.2.1), concluded that I_L (benzene) = 7.1 eV.

Faidas and Siomos (1987), in their studies of dilute solutions ($< 10^{-3}$ M) of benzene in nonpolar liquids, where no excimers are likely to be formed, observed a direct two-photon ionization process for benzene in TMSi at $\lambda_{exc} = 375$ nm (6.61 eV); they also observed a three-photon ionization process at $\lambda_{exc} = 540$ nm, where also the onset of the MPI spectrum is located (see Figure 21), corresponding to a three-photon energy of 6.89 eV. At $\lambda_{exc} = 420$ nm, however, three-photon ionization becomes the sole ionization mechanism, setting thus a lower limit for the I_L (benzene) at the corresponding two-photon energy of 5.90 eV.

For benzene in n-Pt, Faidas and Siomos (1987) observed a two-photon ionization process at $\lambda_{exc} = 337$ nm, whereas they found that a three-photon process is supported by the measurements at $\lambda_{exc} = 365$ nm, corresponding to two-photon energies of 7.36 and 6.79 eV, respectively. The four-photon ionization process, on the other hand, observed in their experiments at $\lambda_{exc} = 420$ nm, indicates that the I_L (benzene) in n-Pt should be larger than 7.65 eV, since one photon of energy 2.95 eV in this spectral region cannot reach the ionization continuum from the lowest vibrational level of $^1B_{2u}$ state (4.7 eV); ($I_L = 2.95 + 4.7 = 7.65$ eV). This argumentation, however, is in contradiction to the two-photon ionization process observed at $\lambda_{exc} = 337$ nm (7.36 eV). On the basis of their studies, Siomos and co-workers concluded that for benzene in dilute solutions of TMSi I_L lies between 5.90 and 6.60 eV, whereas while in n-Pt it is less than 7.36 eV. The relatively large difference between the I_L values of benzene in TMSi and n-Pt could be understood on the basis of parameters such as V_0-values, electron mobilities, and solvation power of these liquids (see discussions in Christophorou and Siomos, 1984).

4.2.3.   Multiphoton Ionization as a Highly Sensitive Detection Technique in the Liquid Phase

Currently commercially available spectrometers utilized in liquid-phase spectroscopy use conventional xenon-arc lamps in conjunction with optical spectrometers suitable for the traditional absorption and fluorescence spectroscopical measurements. Although these instruments have been widely utilized for both qualitative and quantitative analysis in the liquid phase, they often lack spectral selectivity, spectral brightness, noise rejection quality, and identification power. They are susceptible to interference effects resulting from either background fluorescence, Raman scattering, or solvent absorption. Their utilization is limited to near UV and visible spectral regions and, most importantly, can be applied only for the detection of compounds that fluoresce appreciably.

MPI techniques, such as the one developed by Siomos and co-workers discussed above, triggered a series of successful attempts toward a unique high-sensitivity MPI spectrometer suitable for liquid-phase analytical studies (Wright, 1982).

Winefordner and co-workers (Voigtman et al., 1981; Voigtman and Wine-fordner, 1982; Winefordner and Voigtman, 1983) developed an MPI detector, based on Siomos' MPI technique for the detection of PAHs and drugs. Using a flow cell detector (see inset in Figure 28), they were able to study a large number of polynuclear aromatic compounds and drugs in solutions (see Table 3), achieving detection sensitivities on the order of 6 ng/ml for anthracene and 9 ng/ml for pyrene.

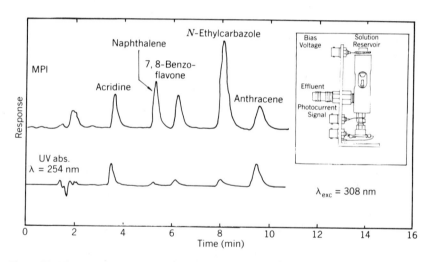

Figure 28. High-performance liquid chromatograms obtained using multiphoton ionization and UV-absorption detectors. *Inset*: Cross-sectional view of a multiphoton ionization flow detector. Reproduced, with permission, from Winefordner and Voigtman (1983).

In similar studies, however, using a stationary photoionization cell, Yamada et al. (1982) were able to detect 0.1 ng/ml of pyrene (see Figure 29) with a signal/noise ratio of 2. This corresponds to 5×10^{-10} M solution of pyrene in *n*-hexane or 10^{11} molecules cm^{-3}.

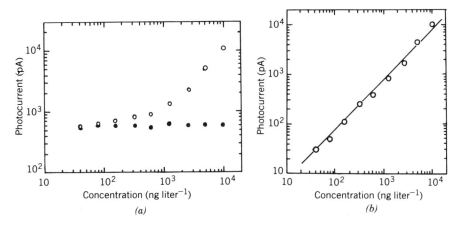

Figure 29. (*a*) The multiphoton ionization signal I_{pc}, of pyrene in *n*-hexane as a function of pyrene concentration (unfilled circles). The filled circles represent measurements of the dark current, I_{dc}, at an applied voltage of 1 kV. (*b*) Actual photoionization signal I_0 ($I_0 = I_{pc} - I_{dc}$) as a function of pyrene concentration. Reproduced, with permission, from Yamada et al. (1982).

In their studies (see Table 3), Winefordner and co-workers (Voigtman and Winefordner, 1982) found that, interestingly enough, there were a number of drugs that could not be detected when excited by an N_2 laser ($\lambda_{exc} = 337$ nm). These results, although easily understood because these compounds (as well as many others) do not absorb at $\lambda_{exc} = 337$ nm appreciably, strongly point to the need for intense light sources frequency tunable from the UV to the red spectral region. The utilization of frequency tunable dye lasers, as applied by Siomos and co-workers, offers a natural choice and a solution to the problem. Wavelengths shorter than 337 nm with sufficient intensity can be easily obtained by doubling the output frequency of a dye laser. Furthermore, frequency tunable dye lasers could be utilized to selectively excite and detect individual drugs, with different ionization threshold values, in a mixture by scanning the frequency of a dye laser, thus utilizing the ability of the MPI technique to measure accurately ionization thresholds of molecules in the liquid phase (Siomos and Christophorou, 1980; Siomos et al., 1981a, b; Faidas and Siomos, 1987).

In an effort to apply MPI detection techniques in the quantitative analysis of complex mixtures of PAHs, drugs, and pesticides, Winefordner and co-workers (Winefordner and Voigtman, 1983) employed a high-performance liquid chromatograph (HPLC), together with a flow-cell MPI detector, to selectively identify and measure a number of organic compounds in a solution of ethanol. Typical results are shown in Figure 28. Figure 28 shows the MPI and UV absorbance chromatograms of a 20-ml ethanol solution of acridine

Table 3. Detection Limits of Polycyclic Aromatic Hydrocarbons (PAHs) and Drugs Using MPI Techniques

Compound	λ_{exc} (nm)	Detection Limits (ng/ml)
PAH[a]		
Anthracene	399	6
1,2-Benzanthracene	385	20
1,12-Benzoperylene	419	60
1,2-Benzopyrene	388	40
3,4-Benzopyrene	403	20
Chrysene	402	100
Coronene	444	200
1,2:5,6-Dibenzanthracene	394	200
Naphthacene	471	1,000
Perylene	438	200
Pyrene	383	9 (0.1)[b]
Rubrene	549	100
Drugs[c]		
Carbamazepine	371	500
Clorazepam	371	2
Diazepam	371	1,000
Hydrochlorthiazide	371	30
Iodochlorhydroxyquin	371	6
Methimazole	371	20
Methyrapone	371	900
Rifampine	371	600
Syrosingapine	371	3,000
Δ^9-THC	371	10,000
Danthron	430	4
Psilocybin	290	10,000

[a] From Voigtman et al. (1981).
[b] From Yamada et al. (1982).
[c] From Voigtman and Winefordner (1982).

(20 g/ml), naphthalene (22 g/ml), 7,8-benzoflavone (20 g/ml, N-ethyl carbazole (22 g/ml), and anthracene (8.0 g/ml). For the MPI detection mode, an excimer laser ($\lambda_{exc} = 308$ nm) was used. In these experiments the authors achieved a detection sensitivity of (a) 0.4 g/ml for acridine, naphthalene, and 7,8-benzoflavone and (b) 0.2 g/ml for N-ethylcarbazole and anthracene.

However, despite, these interesting applications, MPI spectrometry in the liquid phase has not yet reached its highest sensitivity and detection capabilities. Polarization selection rules that are important for MPI processes have not yet been utilized for analytical purposes, although they could provide additional information about the symmetry of the molecules under study and thus improve identification power as has been demonstrated by Siomos and co-workers (Faidas and Siomos, 1987, 1988). Furthermore, the capability of the MPI techniques to determine accurately the ionization thresholds in the liquid phase, in conjunction with the energy tunability offered by the dye lasers, could further improve the identification power of the MPI technique needed for the analysis of complex mixtures of organic polynuclear compounds, drugs, and pesticides.

From the technical point of view, a more careful design of the MPI conductivity detector, by taking into account the parameters that mostly affect both the photoionization signal and dark current, such as (i) photoionization cell geometry, (ii) design and electrode construction, (iii) electrode separation, (iv) RF shielding, and (v) laser beam quality, will lead to an improved MPI detector suitable for liquid-phase studies. In addition, the utilization of modern detection electronics (e.g., boxcar averages, improved charge-sensitive preamplifiers) and computer-aided data analysis could enhance the photoionization signal and improve the signal/noise ratio, thus increasing detection sensitivity and accuracy.

5. MULTIPHOTON IONIZATION DETECTION: FUTURE DEVELOPMENTS AND LIMITATIONS

The future developments of laser MPI spectroscopy of polyatomic molecules both in gaseous and liquid environments will probably follow two directions, depending, on one hand, upon the existing need for a better understanding of the fundamental MPI processes and phenomena observed, especially in the liquid phase, and, on the other hand, upon the obvious desire for a "perfect" detector—a sensitive, selective, and state-specific device with quantitative precision—for the analysis of complex mixtures of organic compounds. In the gas phase and at low densities, there is no theoretical limitation for single-molecule detection, since every molecule could be ionized with unit probability and single electrons can be detected.

In the gas phase, Wessel et al. (1981) have established an absolute MPI sensitivity level for ~ 10 molecules cm^{-3} in a well-controlled environment using a proportional counter. Letokhov and co-workers (Antonov and Letokhov, 1981) have estimated that, in the case of a supersonic molecular beam in conjunction with a TOF mass spectrometer, a sensitivity of a few parts per trillion could be achieved assuming an ionization efficiency of 0.5. Laser light intensities, however, of $10 \, MW \, cm^{-2}$ required for such an ionization efficiency, lead to extensive molecular fragmentation, which, in turn, limits both the sensitivity and selectivity of the MPI process. Furthermore, processes involving rapid internal conversion and intersystem crossing can further reduce the detection sensitivity.

In the liquid phase, Siomos and co-workers were able to show that it is possible to detect as many as 10^4 electrons in the liquid phase under well-defined experimental conditions using extensively purified chemicals.

In view of the aforementioned discussions, it is clear that MPI spectroscopy is a versatile and powerful technique. Its direct bearing on basic photophysical and photochemical processes, as well as its future utilization toward the "perfect" molecular detector, has become evident. More systematic work is needed, however, for it to be fully utilized in analytical methodology.

ACKNOWLEDGMENTS

The author wishes to thank Mr. George Doukas and Mrs. P. Siomos for their invaluable assistance with the manuscript. The author also wishes to thank Mr. George Haralabidis for his assistance in preparing part of the technical figures.

REFERENCES

Abella, I. D. (1962). *Phys. Rev. Lett.*, **9**, 453.

Adreev, S. V., Antonov, V. S., Knyazev, I. N., and Letokhov, V. S. (1977). *Chem. Phys. Lett.*, **45**, 166.

Agostini, P., Barjot, G., Bonual, J. F., Mainfray, G., Manus, C., and Marellec, J. (1968). *IEEE J. Quant. Electron.*, **QE-4**, 667.

Agostini, P., Barjot, G., Mainfray, G., Manus, C., and Theboult, J. (1970). *IEEE J. Quant. Electron.*, **QE-6**, 782.

Alchalal, A., Tamir, M., and Ottolenghi, M. (1972). *J. Chem. Phys.*, **76**, 2229.

Antonov, V. S., Knyazev, I. N., Letokhov, V. S., Matiuk, V. M., Movshev, V. G., and Potapov, V. K. (1978). *Opt. Lett.*, **3**, 37.

Antonov, V. S., Letokhov, V. S., and Shibanov, A. N. (1980). *Zh. Eksp. Teor. Fiz.*, **78**, 2222.

Antonov, V. S., and Letokhov, V. S. (1981). *Appl. Phys.*, **24**, 89.

Arutyunyan, N., Askaryan, G. A., and Pogosyan, V. A. (1970). *Sov. Phys. JETP*, **31**, 548.

Babenko, S. B., Benderskii, V. A., and Gol'danskii, V. I. (1969). *JETP Lett.*, **10**, 129.

Bakos, J. S. (1974). *Adv. Electron. Electron Phys.*, **36**, 57.

Berkowitz, J. (1979). *Photoabsorption, Photoionization and Photoelectron Spectroscopy*, Academic Press, New York.

Boesl, U., Neusser, H. J., and Schlag, E. W. (1978). *Z. Naturforsch.*, **A33**, 1546.

Boesl, U., Neusser, H. J., and Schlag, E. W. (1980). *J. Chem. Phys.*, **72**, 432.

Boesl, U., Neusser, H. J., and Schlag, E. W. (1981). *Chem. Phys.*, **55**, 193.

Born, M. (1919). *Verh. Physikal. Ges.*, **21**, 679.

Born, M. (1920). *Z. Phys.*, **1**, 45.

Boschi, R., and Schmidt, W. (1972). *Tetrahedron Lett.*, **25**, 2577.

Brophy, J. H., and Rettner, C. T. (1979). *Chem. Phys. Lett.*, **67**, 351.

Bunkin, F. V., and Prokhorov, A. M. (1964). *Sov. Phys. JETP*, **19**, 739.

Callomon, J. H., Parkin, J. E., and Lopez-Delgado, R. (1972). *Chem. Phys. Lett.*, **13**, 125.

Christophorou, L. G., and Siomos, K. (1984). In *Electron–Molecule Interactions and Their Applications*, Vol. 2, L. G. Christophorou, Ed. Academic Press, New York, p. 221.

Cooper, C. D., Williamson, A. D., Miller, J. C., and Compton, R. N. (1980). *J. Chem. Phys.*, **73**, 1527.

Dalby, F. W., Petty-Sil, G., Pryce, M. H., and Tai, C. (1977). *Can. J. Phys.*, **55**, 1033.

Delone, N. B. (1965). *Usp. Fiz. Nauk*, **115**, 361–401 [*Sov. Phys. Usp.* **18**, 169–189 (1965)].

Delone, G. A., and Delone, N. B. (1969). *JETP Lett.*, **10**, 25.

Demtroeder, W. (1981). *Laser Spectroscopy*, Springer-Verlag, Berlin, Chapter 7.

Duncan, M. A., Dietz, T. G., Liverman, M. G., and Smolley, R. E. (1981). *J. Chem. Phys.*, **75**, 7.

Eberly, J. H., and Lambropoulos, P., Eds. (1978). *Multiphoton Processes.*, Wiley, New York.

Eland, J. H. D. (1984). *Photoelectron Spectroscopy*, Butterworth, London.

Engelke, F., Hahn, J. H., Henke, W. and Zare R. N. (1987). *Anal. Chem.*, **59**, 909.

Faidas, H., and Siomos, K. (1987). *J. Chem. Phys.*, **87**, 5097.

Faidas, H., and Siomos, K. (1988). *J. Mol. Spectrosc.*, (in press).

Faidas, H., Siomos K., and Christophorou, L. G. (1985). In *Proceedings of the International Conference on Laser Applications*, Vol. 47, p. 93.

Farmosinho, S. J., and da Silva, J. D. (1974). *Mol. Photochem.*, **6**, 409.

Fowler, W. B. (1966). *Phys. Rev.*, **151**, 65.

Franck, J. and Scheibe, G. (1928). *Z. Phys. Chem.*, **A139**, 22.

Franken, P. A., Hill, A. E., Peters, C. W., and Weinreich, G. (1961). *Phys. Rev. Lett.*, **7**, 118.

Fuchs, C., and Voltz, R. (1973). *Chem. Phys. Lett.*, **18**, 394.

Gary, L. P., De Groot, K., and Jarnagin, R. C. (1968). *J. Chem. Phys.*, **49**, 1577.

Georges, A. T., and Lambropoulos, P. (1980). *Adv. Electron. Electron Phys.*, **54**, 191.

Gibbons, W. A., Porter, G., and Savadatti, M. I. (1965). *Nature*, **206**, 1355.

Glownia, J. H., Riley, S. J., Colson, S. D., Miller, J. C., and Compton, R. N. (1982). *J. Chem. Phys.*, **77**, 68.

Goeppert-Mayer, M. (1929). *Naturwissenschaften*, **17**, 932.

Goeppert-Mayer, M. (1931). *Ann. Phys.*, **9**, 273.

Grellmann, K. H., and Watkins, A. R. (1971). *Chem. Phys. Lett.*, **9**, 439.

Haber, F. (1919). *Verh. Physikal. Ges.*, **21**, 750.

Haensch, T. W. (1972). *Appl. Optics*, **11**, 895.

Hall, G. E., and Kenney-Wallace, G. A. (1978). *Chem. Phys.*, **28**, 205.

Hamanoue, K. Hidaku, T., Nakayanu, T., and Teranishi, H. (1981). *Chem. Phys. Lett.*, **82**, 55.

Haseyama, K., and Yoshimura, S. (1965). *J. Phys. Soc. Jpn.*, **20**, 460.

Hauser, N., and Jarnagin, R. C. (1970). *J. Chem. Phys.*, **52**, 1069.

Hurst, G. S., Payne, M. G., Kramer, S. D., and Young, J. P. (1979). *Rev. Mod. Phys.*, **51**, 767.

Ito, M., and Mikami, N. (1980). *Appl. Spectrosc. Rev.*, **16**, 299.

Johnson, P. M. (1975). *J. Chem. Phys.*, **62**, 4562.

Johnson, P. M. (1976a). *J. Chem. Phys.*, **64**, 4143.

Johnson, P. M. (1976b). *J. Chem. Phys.*, **62**, 4638.

Johnson, P. M., Berman, M. R., and Zakheim, D. (1975). *J. Chem. Phys.*, **62**, 2500.

Kaiser, W., and Garrett, C. G. B. (1961). *Phys. Rev. Lett.*, **7**, 229.

Kellmann, A., and Tfibel, F. (1980). *Chem. Phys. Lett.*, **69**, 61.

Klimcak, C. M., and Wessel, J. E. (1980). *Anal. Chem.*, **52**, 1233.

Kourouklis, G. A., Siomos, K., and Christophorou, L. G. (1981). In *Proceedings of the Radiation Research Society*, Academic Press, New York, p. 93.

Kourouklis, G. A., Siomos, K., and Christophorou, L. G. (1982a). *Chem. Phys. Lett.*, **88**, 572.

Kourouklis, G. A., Siomos, K., and Christophorou, L. G. (1982b). *J. Mol. Spectrosc.*, **92**, 127.

Krogh-Jespersen, K., Rava, R. P., and Goodman, L. (1979). *Chem. Phys.*, **64**, 413.

Letokhov, V. S. (1983). *Nonlinear Laser Chemistry*, Springer-Verlag, Berlin.

Letokhov, V. S., and Chebotayev, V. P. (1977). *Nonlinear Laser Spectroscopy*, Springer-Verlag, Berlin.

Leutwyler, S., and Even, U. (1981). *Chem. Phys. Lett.*, **81**, 578.

Levenson, M. D. (1982). *Introduction to Nonlinear Laser Spectroscopy*, Academic Press, New York.

Lin, S. H., Fujimura. Y., Neusser, H. J., and Schlag, E. W. (1984). *Multiphoton Spectroscopy of Molecules*, Academic Press, New York.

Lubman, D. M., and Kronick, M. N. (1982). *Anal. Chem.*, **54**, 660.

Lubman, D. M. (1987). *Anal. Chem.*, **59**, 31A.

Lubman, D. M., Tembreull, R., and Sin, C. H. (1985). *Anal. Chem.*, **57**, 1084.

Lyons, L. E. O. (1957). *J. Chem. Soc. Perkin Trans. 4*, 5001.

Manson, P. R., and McClain, W. M. (1970). *J. Chem. Phys.*, **53**, 29.

McClain, W. M. (1971). *J. Chem. Phys.*, **55**, 2789.

McClain, W. M. (1974). *Acc. Chem. Res.*, **7**, 129.

McDiarmid, R. (1975). *Chem. Phys. Lett.*, **34**, 130.

Miller, J. C., and Compton, R. N. (1981). *J. Chem. Phys.*, **75**, 22.

Miyasaka, H., Masuhara, H., and Mataga, N. (1985). *J. Chem. Phys.*, **89**, 1631.

Mott, N. F., and Littleton, M. L. (1938). *Trans. Faraday Soc.*, **34**, 485.

Nakashima, N., and Yoshihara, K. (1982). *J. Chem. Phys.*, **77**, 6040.

Nakato, Y., Yamamoto, N., and Tsubomura, H. (1967). *Bull. Chem. Soc. Jpn.*, **40**, 2480.

Neusser, H. J., and Schlag, E. W. (1984). In *Multiphoton Spectroscopy of Molecules*, S. H. Lin, Y. Fujimura, H. J. Neusser, and E. W. Schlag, Eds., Academic Press, New York, Chapter 4, p. 71.

Nielsen, T. (1983). *Anal. Chem.*, **55**, 286.

Nieman, C., and Colson, D. (1978). *J. Chem. Phys.*, **68**, 5656.

Nikogosyan, D. N., and Letokhov, V. S. (1983). *Riv. Nuovo Cimento*, **6**, 1.

Parker, D. H. (1983). In *Ultrasensitive Laser Spectroscopy*, D. S. Kliger, Ed. Academic Press, New York.

Parker, D. H., Berg, J. O., and El-Sayed, M. A. (1978). *Chem. Phys. Lett.*, **56**, 197.

Parker, D. H., and El-Sayed, M. A. (1979). *Chem. Phys.*, **42**, 379.

Peticolas, W. L., Goldborough, J. P., and Reichhoff, K. E. (1963). *Phys. Rev. Lett.*, **10**, 43.

Petty, G., Tai, C., and Dalby, F. W. (1975). *Phys. Rev. Lett.*, **34**, 1207.

Piciulo, P. L., and Thomas, J. K. (1978). *J. Chem. Phys.*, **68**, 3260.

Pilloff, H. S., and Albrecht, A. C. (1966). *Nature*, **212**, 499.

Pilloff, H. S., and Albrecht, A. C. (1968). *J. Chem. Phys.*, **49**, 4891.

Raz, B., and Jortner, J. (1969). *Chem. Phys. Lett.*, **4**, 155.

Raz, B., and Jortner, J. (1970). *Chem. Phys. Lett.*, **4**, 511.

Reichardt, H., and Bonhoeffer, K. F. (1931). *Z. Phys.*, **67**, 780.

Reilly, J. P., and Kompa, K. L. (1980). *J. Chem. Phys.*, **73**, 5468.

Rhodes, G., Opsal, R. B., Meek, J. T., and Reilly, J. P. (1983). *Anal. Chem.*, **55**, 280.

Rice, S. A., and Jortner, J. (1966). *Chem. Phys.*, **44**, 4470.

Richards, J. T., West, G., and Thomas, J. K. (1970). *J. Chem. Phys.*, **67**, 4137.

Robin, M. B. (1974). *Higher Excited States of Polyatomic Molecules*, Vol. 1, Academic Press, New York, Chapters I and II.

Robin, M. B. (1980). *Appl. Opt.*, **19**, 3941.

Robin, M. B., and Kuebler, N. A. (1978). *J. Chem. Phys.*, **69**, 806.

Rockwood, S., Reilly, J. P., Hohla, K., and Kompa, K. L. (1979). *Opt. Commun.*, **28**, 175.

Rutherford, E., and Geiger, H. (1908). *Proc. R. Soc. London Ser.*, **A 81**, 141.

Sack, T. M., McCrevy, D. A., and Gross, M. L. (1985). *Anal. Chem.*, **57**, 1290.

Salvi, P. R., Foggi, P., and Castellucci, E. (1983). *Chem. Phys. Lett.*, **98**, 206.

Scott, T. W., and Albrecht, A. C. (1981). *J. Chem. Phys.*, **74**, 3807.

Scott, T. W., and Albrecht, A. C. (1982). In *Advances in Laser Spectroscopy*, Vol. 1, B. A. Garetz and J. R. Lombardi, Eds. Heyden, London, p. 55.

Scott, T. W., Twarowski, A. J., and Albrecht, A. C. (1979). *Chem. Phys. Lett.*, **66**, 1.

Scott, T. W., Brown, C. L., and Albrecht, A. C. (1982). *J. Chem. Phys. Lett.*, **76**, 5195.

Shaefer, F. P., Ed. (1973). *Dye Lasers*, Springer-Verlag, Berlin.

Siomos, K. (1972). Ein Blitzlampen-gepumpter Farbstofflaser mit breiter und schmaller Emissionsbande, Diplom Thesis, Institut fuer Angewandte Physik, University of Heidelberg, Heidelberg, West Germany.

Siomos, K. (1987). In *Proceedings of the Fifth International Symposium on Ultrafast Phenomena in Spectroscopy*, Vilnius, Lithuania, USSR, p. 232.

Siomos, K., Ed. (1988). *Lasers in Medicine*, Pellekanakis Press, Crete, Greece.

Siomos, K., and Christophorou, L. G. (1980). *Chem. Phys. Lett.*, **72**, 43.

Siomos, K., and Christophorou, L. G. (1982a). *Appl. Phys.*, **B28**, 225.

Siomos, K., and Christophorou, L. G. (1982b). *J. Electrost.*, **12**, 147.

Siomos, K., and Faidas, H. (1988) (to be published).

Siomos, K., Kourouklis, G. A., and Christophorou, L. G. (1981a). *Chem. Phys. Lett.*, **80**, 504.

Siomos, K., Kourouklis, G. A., and Christophorou, L. G. (1981b). In *Electron and Ion Swarms*, L. G. Christophorou, Ed., Pergamon Press, New York, p. 139.

Siomos, K., Kourouklis, G. A., Christophorou, L. G., and Carter, J. G. (1981). *Radiat. Phys. Chem.*, **17**, 75.

Siomos, K. Kourouklis, G. A., and Christophorou, L. G. (1983a). In *Proceedings of the Conference on Lasers and Electrooptics*, Optical Society of America, p. 160.

Siomos, K., Kourouklis, G. A., and Christophorou, L. G. (1983b). *In Proceedings of the Seventh International Congress of Radiation Research*, *Vol. A1*, J. J. Broerse, G. W. Barendsen, H. B. Kal, and A. J. van der Kogel, Eds., Martinus Nijhoff Publishers, Amsterdam, p. 35.

Siomos, K., Faidas, H., and Christophorou, L. G. (1984). In *Laser Techniques in the Extreme Ultraviolet*, S. E. Harris and T. B. Lucatorto, Eds., American Institute of Physics, New York, p. 135.

Smith, S. J., and Leuchs, G. (1988). *Adv. At. Mol. Phys.*, **24**, 157.

Speiser, S., and Kimel, S. (1970). *Chem. Phys. Lett.*, **7**, 19.

Speiser, S., and Jortner, J. (1976). *Chem. Phys. Lett.*, **44**, 399.

Takeda, S. S., Hauser, N. E., and Jarnagin, R. C. (1971). *J. Chem. Phys.*, **54**, 3195.

Tamir, M., and Ottolenghi, M. (1970). *Chem. Phys. Lett.*, **6**, 369.

Taniguchi, Y., Nishina, Y., and Mataga, N. (1972). *Bull. Chem. Soc. Jpn.*, **45**, 2923.

Tembreull, R., and Lubman, D. M. (1984). *Anal. Chem.*, **56**, 1962.

Tembreull, R., and Lubman, D. M. (1987). *Anal. Chem.*, **59**, 1003.

Tembreull, R., Sin, C. H., Pang, H. M., and Lubman, D. M. (1985). *Anal. Chem.*, **57**, 2911.

Vaida, V., Robin, M. B., and Kuebler, N. A. (1978). *Chem. Phys. Lett.*, **58,** 557.

Voigtman, E., and Winefordner, J. D. (1982). *Anal. Chem.*, **54,** 1834.

Voigtman, E., Jurgensen, A., and Winefordner, J. D. (1981). *Anal. Chem.*, **53,** 1921.

Wallenstein, R., and Haensch, T. W. (1974). *Appl. Opt.*, **13,** 1625.

Wallenstein, R., and Haensch, T. W. (1975). *Opt. Commun.*, **14,** 353.

Walther, H., and Hall, J. L. (1970). *Appl. Phys. Lett.*, **6,** 239.

Walther, H., Ed. (1976). *Laser Spectroscopy of Atoms and Molecules*, Springer-Verlag, Berlin.

Warburg, E., and Rump, W. (1928). *Z. Phys.*, **47,** 305.

Wessel, J. E., Cooper, D. E., and Klimcak, C. M. (1981). In *Laser Spectroscopy for Sensitive Detection*, J. Gelbwachs, Ed., SPIE 286, Springer-Verlag, Berlin, p. 48.

Wiley, W. C., and McLaren, I. H. (1955). *Rev. Sci. Instrum.*, **26,** 1150.

Williamson, A. D., and Compton, R. N. (1979). *Chem. Phys. Lett.*, **62,** 295.

Winefordner, J. D., and Voigtman, E. (1983). In *New Directions in Molecular Luminescence*, D. Eastwood, Ed., *ASTMSTP* 822, American Society for Test and Matter, p. 17.

Wright, J. C. (1982). In *Applications of Lasers in Analytical Chemistry*, T. R. Evans, Ed., Wiley, New York, p. 36.

Wunsch, L., Metz, F., Neusser, H. J., and Schlag, E. W. (1977). *J. Chem. Phys.*, **66,** 386.

Yakovlev, B. S., and Lukin, L. V. (1985). In *Advances in Chemical Physics*, Vol. LX, I. Prigogine and S. A. Rice, Eds. Wiley, New York, p. 99.

Yamada, A., Kano, K., and Ogawa, T. (1982). *Bunseki Kogaku*, **31,** E247.

Yamamoto, N., Nakato, Y., and Tsubomura, H. (1966). *Bull. Chem. Soc. Jpn.*, **39,** 2603.

Zandee, L., and Bernstein, R. B. (1979a). *J. Chem. Phys.*, **70,** 2574.

Zandee, L., and Bernstein, R. B. (1979b). *J. Chem. Phys.*, **71,** 1359.

Zandee, L., Bernstein, R. B., and Lichtin, D. A. (1978). *J. Chem. Phys.*, **69,** 3427.

CHAPTER

10

MULTIDIMENSIONAL RESONANCE TWO-PHOTON IONIZATION MASS SPECTROMETRIC-BASED ANALYSIS

S. J. WEEKS AND A. P. D'SILVA

Ames Laboratory
Iowa State University
Ames, Iowa

R. L. M. DOBSON

Proctor & Gamble Company
Cincinnati, Ohio

1. INTRODUCTION

Resonance two-photon ionization (R2PI) can be distinguished from resonance-enhanced multiphoton ionization (REMPI) (1); however this distinction is not always observed in the literature. R2PI consists of two one-photon processes, that is, a resonance absorption followed by photoionization. REMPI involves a 'multi' photon absorption through virtual levels. R2PI dominates at lower laser power densities. It has the distinct advantage of much higher absorption cross sections, which translates into significantly lower detectabilities for polycyclic aromatic compounds (PACs). Although R2PI has not been used to extensively characterize real samples for PACs, an R2PI-based instrument can be hybridized with a variety of sampling systems and serial, parallel, and simultaneous detectors producing one instrumental system that may be powerful enough for efficient, accurate determinations of PACs even in the most complex environmental and biological samples.

Ions and electrons created in the R2PI process can be collected with near unity efficiency. A linear time-of-flight mass spectrometer (TOFMS) is the practical detector of choice for R2PI. It has great sensitivity with small throughput losses even when run at maximum resolution and can detect the entire mass spectrum from each ion creation pulse. This last item is of immense significance since pulsed lasers needed to obtain the required laser power densities operate at low repetition rates.

PACs are of obvious interest because they represent a class of the most potent known mutagens and carcinogens. R2PI provides a means of selectively exciting PACs, and mass spectrometry (MS) adds highly selective detection. Utilizing high-resolution chromatography (HRC) adds selectivity in the sample introduction stage through "matrix simplification" that relaxes spectral selectivity requirements for the determination of PACs in complex mixtures. Collecting the laser-induced fluorescence (LIF) radiation inherent from the R2PI process adds complementary detection selectivity. All together, multidimensional HRC–R2PI–MS–LIF technique, when combined with chemometric data reduction, could yield an extremely powerful means of efficiently and accurately characterizing complex biological and environmental samples for trace level PACs. Additional versatility of R2PI–MS instrumentation can be provided by alternative sampling systems such as laser desorption or supercritical fluid sample introduction. Compound-specific excitation can also be provided by utilizing rotationally cooled (RC) supersonic expansions as a means of introducing the sample into the ion source region of the MS. The multidimensional information available from R2PI–MS-based instrumentation provides the analytical chemist with a powerful and versatile tool to achieve trace level determinations of polar and nonpolar PACs in complex biological and environmental sample matrices with the desired selectivity.

An emphasis in this chapter is the complete characterization of samples for PACs, with the greatest confidence in the analytical measurements and in the shortest analysis time. The "PAC problem" will be presented from an analytical point of view. Other techniques are briefly mentioned before introducing the concepts and basic instrumentation for R2PI–MS. The applications have been critically reviewed and describe the variety of hybrids and their potential for determining PACs in complex samples. The significance of R2PI–MS-based systems may simply be stated as informing power.

2. BACKGROUND

2.1. The Analytical PAC Problem

Of the thousands of chemical compounds that have been deemed mutagenic or carcinogenic, it is generally agreed that the polynuclear aromatic compounds (PACs) are among the most potent (2). Within the PAC chemical class, polynuclear aromatic hydrocarbons (PAHs) have been studied most extensively. A number of PAHs and their metabolites have been shown to exhibit a high degree of carcinogenicity and/or mutagenicity in bioassay tests (3–5). With the recent emphasis on the development of alternative energy sources, such as production of solvent refined coal, shale oil, or the use of high-sulfur coal and crude oil, there has been a growing interest in the more polar PAC classes. These include nitrogen (PANH)-, sulfur (PASH)-, and oxygen (PAOH)-containing heterocyclic compounds. Many compounds within each of these classes have also been shown to exhibit strong carcinogenic and/or mutagenic tendencies (6–9).

There are many sources of PACs in the environment, both natural and anthropogenic. Natural sources include, for example, fossil fuels (10), combustion products of forest fires, and even animal pigments (11). Anthropogenic PAC sources are many and varied. PACs have been found in exhausts from internal combustion engines (12), in residue from fuel-rich flames (13), in coal conversion processes (14), and in the environment of the coke production industry (15). Studies even show that PAC concentrations of up to 164 parts per billion may be found in smoked and charcoal-broiled meats (16).

The possible fates of these PACs in the environment are as numerous as their sources. They may, for example, collect in soils and sediments (17, 18) or in fresh-water supplies (19), or they may enter an ecosystem food chain (20). The high chemical stability of most of these PACs further compounds the problem, as their environmental concentrations will continue to increase with time (11). It is, therefore, becoming more and more important to control the level of anthropogenic PACs released into the environment.

Before PAC levels can be judiciously regulated, they must be monitored by appropriate analytical methodology. The many different sources and variety of matrix compositions of PAC-contaminated materials present a serious problem to the analytical chemist. To further complicate the situation, it has been shown that there is a wide range of potency, even among geometric isomers (3, 21, 22) and various substitutional derivatives (3, 6, 23), which often have virtually identical physical, chemical, and spectroscopic properties. Figures 1 and 2 provide several examples of PAC isomers and derivatives characterized by considerably different carcinogenic and mutagenic potencies, respectively. On top of all this, there is evidence that carcinogenicity of certain compounds is enhanced in mixtures through synergistic effects (24). Thus it is not always sufficient to just detect certain "target" compounds. Environmental mixtures containing PACs should therefore be characterized as completely as possible.

The complex problem described above imposes several stringent requirements on analytical methodologies. Not only must the technique be capable of detecting trace levels of both polar and nonpolar PACs in complex environmental matrices, but it must also provide adequate selectivity so as to distinguish between compounds differing only slightly in structure. Other requirements include (a) acceptable accuracy and sample throughput and (b) the capability of detecting nonvolatile, as well as volatile, species.

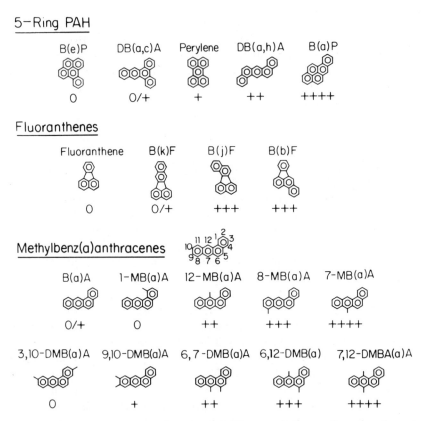

Figure 1. Range of carcinogenic potency among PAH geometric isomers (0 = noncarcinogenic, + + + = highly carcinogenic, as described in ref. 3).

2.2. Conventional Methods

Capillary column gas chromatography combined with mass spectrometry (CGC–MS) has become the most popular and effective analytical method for characterization of complex environmental samples for PACs (25–27). When employed in the conventional manner, the power of this technique is limited by the physical separation capabilities of the CGC, because species that are well-resolved chromatographically are easily identified by the mass spectrometer. To minimize the impact of this limitation, environmental samples must usually undergo an extensive, time-consuming cleanup and chemical class separation procedure prior to analysis by CGC–MS (27, 28). However, for the case of complex samples containing a number of coeluting isomers, it is impossible to provide full characterization when a conventional, nonselective

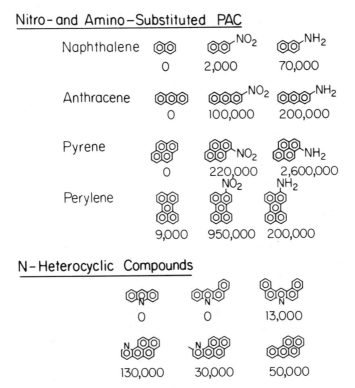

Figure 2. Range of mutagenic potency among PAC substitutional isomers. Mutagenic values have units of revertants per milligram of test chemical (6).

ionization technique is employed prior to mass analysis (28–30). Therein lies the ultimate limitation of CGC–MS for PAC characterization. Recently, novel chemical-ionization mass spectrometric techniques have been utilized to determine specific "targeted" coeluting isomer pairs (31–34); however, these approaches do not represent a general solution to this selectivity problem.

2.3. Laser-Based Techniques

A number of state-of-the-art, laser-based, analytical spectroscopic methodologies have recently been developed to increase the specificity of PAC detection. Generally, these approaches involve prior chromatographic separation and/or low-temperature "spectral simplification."

2.3.1. Solid-State Cryogenic Fluorescence Approaches

Three cryogenic techniques have been utilized to produce narrow-band fluorescence of PACs trapped in appropriate matrices. All of these methods, namely, fluorescence line-narrowing spectroscopy (FLNS) (35), laser-excited Shpol'skii spectroscopy (LESS) (36), and matrix isolation fluorescence spectroscopy (MIFS) (37), have demonstrated highly selective PAC determinations in complex mixtures with minimal sample preparation. Although these techniques show promise, each of them has solvent/matrix restrictions and may not provide sufficient spectral resolution for complete characterization of most real samples for PAC. Also, it is obvious that such fluorescence techniques are not applicable to the determination of "dark" (nonfluorescing) species, and their solid-state nature precludes applications that require on-line monitoring. Additionally, data bases for these techniques are somewhat complex and need to be developed.

2.3.2. Supersonic Expansion Fluorescence Techniques

High spectral selectivity of LIF of molecules seeded in an RC supersonic expansion has been demonstrated (38–42). This low-temperature, gas-phase technique provides great spectral simplification. Discussion of supersonic jet spectroscopy is important here because the concepts involved pertain to ionization as well as to fluorescence detection schemes for gas-phase molecules. In addition to narrowing molecular absorption and emission spectral bandwidths, supersonic expansions translationally cool the molecules, thereby narrowing their kinetic energy distribution. This offers an additional bonus of improved mass resolution when employing a TOFMS ion detection scheme (43). Thus, the supersonic expansion process provides high excitation wavelength selectivity and improved mass resolution of the ions produced. Furthermore, the resulting LIF signal simultaneously offers additional emission wavelength and lifetime detection selectivity. This additional selectivity and complementary information could prove invaluable for efficient and precise determination of PACs in complex matrices. The analytical applications of supersonic expansions have recently been reviewed (43).

The spectral resolution provided by the rotational cooling technique (typically on the order of 1 cm^{-1} for PACs) is more than adequate to be analytically useful. Small and co-workers showed the analytical feasibility of RC–LIF by distinguishing between methyl-substitutional isomers of naphthalene in simple mixtures (44). They also first suggested a gas chromatographic sample introduction scheme to provide quantitative transfer of analyte to an RC–LIF detector (44, 45) and later demonstrated the analytical utility for the direct determination of some naphthalene derivatives in crude oil along with nanogram detectability using GC–RC–LIF (46, 47). This work proved that RC–LIF could be quantitative and sensitive as well as highly selective.

Imasaka et al. have developed a high-temperature (300 °C), small-dead-volume (0.2 ml) pulsed nozzle for capillary-GC–supersonic-jet spectrometry (48). Although the ~ 1.3-ms sample pulse width improved the duty factor of the nanosecond pulsed laser excitation measurement, detectabilities only in the microgram regime were obtained for anthracene derivatives.

Callis and co-workers (49, 50) developed and utilized a CGC-pulsed RC–LIF system for the detection of the geometric isomers of monomethylanthracene in an environmental airborne particulate sample. Low nanogram detection limits were achieved. CGC–RC–LIF allowed unambiguous determination of relative and absolute abundances of each isomer. A small spectral interference from 2-methylanthracene for the 1-methylanthracene signal response was noted. These isomers were not able to be determined by conventional CGC–MS under the relatively fast chromatographic conditions used. More extensive sample cleanup and much longer chromatographic runs would have allowed many components of this environmental sample to be resolved. However, CGC–MS would still be subject to isobaric interferences (from methyl phenanthrenes and possible fragments due to high-molecular-weight aliphatic compounds) that the spectral selectivity of RC–LIF eliminates or minimizes.

Johnston and co-workers (51) achieved a 50-pg detection limit for naphthalene by developing a jet nozzle based on sheath-flow, gas-dynamic focusing. This was found to increase the fluorescence intensity by a factor of 30 over that of an unfocused jet expansion and was also found to be a convenient way to couple capillary GC to a supersonic jet expansion. Spectral selectivity for naphthalene in unleaded gasoline was demonstrated (Figure 3).

There are two general requirements that an analyte must meet when rotationally cooled jet spectroscopy is employed. Obviously, as with other spectroscopic techniques, the analyte must absorb in a wavelength range accessible to the excitation source (typically a laser) employed. Also, the analyte must be volatile enough to produce a partial vapor pressure of greater than 10^{-7} torr (52) at jet operation temperatures. Even et al. have conducted fundamental studies of a variety of porphyrins with their jet system, which may be operated at temperatures approaching 520 °C (53–55). Operation of the jet at high temperatures, however, is not a general solution to this requirement, because many compounds are not chemically stable under these conditions.

High-performance liquid chromatography (HPLC) with RC–LIF (56) has been utilized to detect thermally labile 2-chloroanthracene. The determination of optically active molecules that give completely identical spectra even

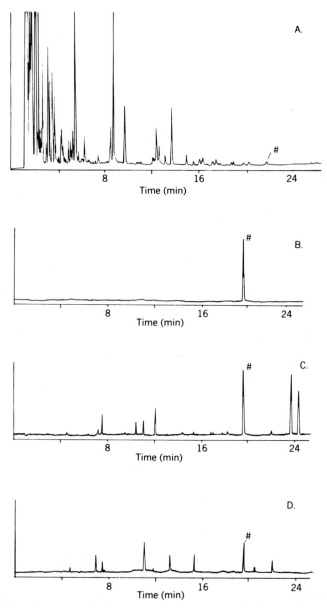

Figure 3. Unleaded gasoline chromatograms, temperature programmed 50–150°C at 4°C min^{-1}. (# denotes naphthalene). (*A*) Flame ionization detector; (*B*) Jet-cooled fluorescence detection, excitation at 308.12 nm, laser beam 60 nozzle diameters downstream; (*C*) Same as part B except excitation at 266 nm; (*D*) Same as part B except laser beam ∼10 nozzle diameters downstream. Reprinted, with permission, from ref. 51.

by supersonic jet spectrometry can also be done using HPLC–RC–LIF. Detection limits in the nanogram range for anthracene were reported using HPLC–RC–LIF. The use of laser desorption and supercritical fluid sampling inlets are also emerging as a means of introducing nonvolatile and thermal labile compounds into both fluorescence and mass spectrometric systems.

In principle, the rotational cooling technique offers significant advantages for the determination of PACs in complex samples. The primary advantage is highly selective (in many cases specific) laser excitation that often results in narrow-line fluorescence emission. This gas-phase scheme is also amenable to on-line, real-time fluorescence emission and ion detection. D'Silva et al. demonstrated the feasibility of employing RC–LIF for on-line monitoring of certain PACs in the effluents from a fluidized bed combustor (57).

A potential selectivity problem, which has not been fully addressed in the context of chemical analysis, is the effect of the vibrational quasi-continuum (VQ) that exists, even in the jet, for all large molecules (58). This VQ probably results from random intramolecular radiative coupling between upper level vibronic states (59) and gives rise to a congested level structure in the molecular absorption spectrum. This structure typically begins above 1000 cm^{-1} on the high-energy side of the first excited singlet state. Thus, the congested level structure does not obstruct the lower-energy, sharp vibronic bands that may be utilized for selective excitation of similar isomers in simple mixtures. However, in very complicated mixtures containing a wide range of compounds, as is found in typical environmental samples, the rotationally cooled narrow-band structure of one isomeric compound group will likely overlap with the congested level structure of other types of species in the mixture. In this case, therefore, true selectivity cannot be achieved in the excitation step alone. If the compounds of interest are fluorescent, the emission spectra or the lifetime decay curve may be useful for selective detection; otherwise an ion detection scheme involving mass spectrometry as an added dimension of selectivity would be required. If these measures do not provide sufficient selectivity, the use of a prior chromatographic separation would also be necessary.

Rotationally cooled jet spectroscopy is still in its infancy as an analytical method. It has shown great potential in a number of applications and, in principle, has appeared to be superior in many cases to the alternative solid-state cryogenic techniques. However, much preliminary work must still be done in several areas, including the development of (a) appropriate sampling systems for particular applications, (b) alternative quantitative methods, and (c) appropriate spectral data bases for PACs. Additionally, cluster formation and decomposition processes need to be studied in the jet. To this point, no one has utilized RC–LIF technique for extensive characterization of PACs environmental and biological samples.

3.   RESONANCE TWO-PHOTON IONIZATION MASS SPECTROMETRY (R2PI–MS)

3.1.   Gas-Phase Laser–Analyte Interactions

When intense monochromatic laser light interacts with a molecule, a variety of processes may occur, depending on the spectral characteristics of the molecule and the wavelength and power density of the laser beam. Such interactions have been discussed at length elsewhere (60–65); therefore, only qualitative consideration of those processes that are relevant here will be given.

Simplified schematic energy-level diagrams for two different molecules are shown in Figure 4. Both molecules have ionization potentials (IP); thus ionization may occur through the successive absorption of two photons. This stepwise photoionization process is termed *multiphoton ionization* (MPI). For molecule A, the first step involves excitation of the molecule to a virtual (nonresonant) state, the lifetime of which is on the order of 10^{-15} (61). If ionization is to occur, a second photon must then be absorbed before the molecule relaxes back into a ground vibronic state. On this time scale, the two photons must be absorbed almost simultaneously, hence requiring an extremely intense light source. It is for this reason that tightly focused laser beams having large power densities are typically employed in such exper-

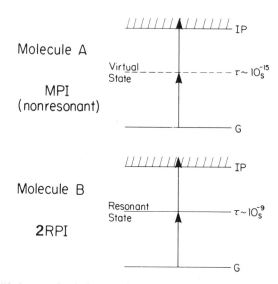

Figure 4. Simplified energy-level diagrams for molecules undergoing photoionization by nonresonant MPI (top) and R2PI (bottom) processes.

iments. Obviously, the MPI process alone provides only limited selectivity because all molecules with sufficiently low IP would be ionized with similar efficiencies.

Spectral selectivity may be achieved through a two-stage process that utilizes a real intermediate (resonant) vibronic molecular state as a "stepping stone" toward ionization. This process is illustrated for molecule B, in which case the laser wavelength has been chosen to provide the precise energy necessary for excitation to a resonant state, the lifetime of which is typically on the order of 10^{-9} (61). This relatively long, intermediate-state lifetime somewhat relaxes the laser power density requirements and results in an increase of up to several orders of magnitude in the overall ionization efficiency. This process, termed *resonance two-photon ionization* (R2PI), provides a great enhancement of molecular ion yield, to a point where the nonresonant MPI process may be negligible by comparison. Thus, by tuning the laser wavelength to selectively populate resonant states in molecules of interest, selective excitation/ionization may be achieved.

The Jablonski diagram in Figure 5, although perhaps still somewhat oversimplified, provides an overview of those dynamic processes associated

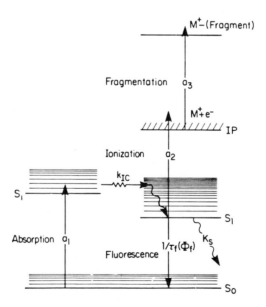

Figure 5. Jablonski diagram (simplified) representing possible energy-transfer pathways of laser–analyte interaction. Absorption cross sections a_1, a_2, and a_3, fluorescence decay lifetime τ_f, fluorescence quantum yield Φ_f, intramolecular conversion rate constant k_{IC}, and total non-radiative S_1 relaxation rate K_s are characteristic of the analyte. Reprinted, with permission, from ref. 105.

with laser–analyte interactions. Typically, a molecule is excited to an upper vibronic state through the absorption of a resonant photon. This state then undergoes "immediate" intramolecular conversion to the congested upper level vibronic bands of the lowest excited electronic singlet (Kasha's Rule), followed by relaxation to the zero-point vibrational level (66, 67). The molecule may then absorb more photons to become ionized and to perhaps undergo subsequent fragmentation. Alternatively, the molecule may relax to a ground vibronic state either through the spontaneous release of a photon (fluorescence) or by a nonradiative decay pathway. The nonresonant MPI contributions are taken to be negligible when the laser beam power densities are sufficiently low. Also, intermolecular deactivation of excited species is essentially nonexistent because the laser–analyte interaction occurs at "collision-free" pressures of 10^{-4} torr or less.

The probability that a molecule will follow a particular energy transfer pathway is dependent on both the intensity of the laser radiation and the spectroscopic properties of the molecule. For instance, at high laser intensities, ionization and fragmentation processes tend to dominate, whereas at lower laser intensities, spontaneous emission and nonradiative decay are more likely to prevail.

The spectroscopic properties of molecules form the primary basis for the selectivity available from this laser–analyte interaction process. The first excitation step is dependent on the absorption cross section, a_1, of the S_0 to S_1 transition (Figure 5). Both the ionization and fluorescence pathways are dependent on the efficiency of this first step and, therefore, are restricted according to the characteristic absorption spectrum of the species of interest. It is important to note that the laser–analyte interaction is very dynamic, since ionization, spontaneous emission, and nonradiative decay are all competing processes. Thus, after the excited molecules settle into the S_1 level, the relative probabilities for each of these processes for a given laser intensity is dictated by a_2 (photoionization cross section of S_1), τ_f (fluorescence decay lifetime), Φ_f (fluorescence quantum yield), and the nonradiative decay probability, respectively (1, 63, 68). Furthermore, an obvious requirement for ionization is that the IP must be low enough to allow one-photon ionization from S_1.

Because the above molecular spectroscopic constants differ from species to species, knowledge of their relative values could be quite helpful in qualitative characterization studies. The precise values for such constants have been determined for only a few PACs; therefore their utility, as applied to analytical characterization experiments, is very limited at present. However, through the simultaneous monitoring of all laser–analyte interaction products, characteristic detector responses may be determined for each species, at a selected laser wavelength and spectral irradiance. These responses are actually the manifestation of the characteristic molecular spectroscopic constants and may equally be utilized in analytical qualitative identification studies, as well as for quantitative determinations.

3.2. Instrumentation

Only a general description of the basic analytical instrumentation will be given here (detailed experimental arrangements can be found in the references cited below). R2PI–MS systems consist of (a) a means of introducing the sample into the ion source region of the MS, (b) a photoexcitation source (usually a laser), and (c) a mass spectrometer with the appropriate detection and signal processing electronics.

3.2.1. Sample Introduction

Capillary column gas chromatography (CGC) is certainly a preferred method for quantitative determination of PACs in complex matrices. CGC has a number of advantages, including: (a) flow rates compatible with MS vacuum systems; (b) quantitative sample delivery; (c) high chromatographic resolving power providing analyte selectivity in the form of a specific retention time (t_r) for each compound, plus the added benefit of matrix simplification in the laser-interrogated "sample cell" volume; and (d) small sample volume (typically 1 μl) required for each analysis. The limitations of CGC are, of course, that the compounds of interest must have sufficient volatility and must not be thermally labile, in addition to the real possibility of coeluting compounds in complex sample analyses.

Although no one to date has reported on supercritical fluid chromatographic (SFC) introduction for R2PI studies, SFC–MS is an active area of research (69–71). With the appropriate interface, SFC–R2PI analysis could be performed on larger, nonvolatile PACs and thermally labile species. However, detectabilities may need to be improved to determine trace-level PACs because of the limited sample loading of capillary SFC. Supercritical fluid injection has, however, been used for R2PI–MS of nonvolatile species (72–74). Laser desorption (LD) has also been utilized for R2PI–MS (72–74). The laser desorption processes considered here are those that either completely separate the desorption and photoionization process or involve surface "soft" ionization approaches. Although R2PI processes may certainly occur in LD–MS (86), where the desorption and ionization process are not separated, this work will not be discussed.

Pulsed valves (43, 48–50, 73, 74, 87, 88) and gas dynamic focusing (51) have been used to improve the experimental duty factor and concentrate the analyte in the sample cell volume, respectively. Both approaches show great promise for improving the detectability of this pulsed laser-based technique.

3.2.2. Lasers

The R2PI process for typical PACs with ionization potentials (IP) of 6–10 eV and resonance absorptions of 3–6 eV (400–180 nm) generally requires UV excitation and sufficiently high power densities to achieve efficient ionization. For this reason, nanosecond pulsed lasers are employed. Fixed-frequency excimer lasers [XeCl* at 308 nm (4.03 eV); KrF* at 249 nm (4.96 eV); KrCl* at 222 nm (5.58 eV); and ArF* at 193 nm (6.42 eV)] and quadrupoled Nd:YAG laser at 266 nm (4.66 eV) have been used in R2PI-MS. For greatly enhanced wavelength selectivity, frequency-doubled tunable pulsed dye lasers are used. Typically, power densities of $\sim 10^6$ W cm^{-2} will provide soft ionization for most PACs at an appropriate excitation wavelength (λ_{ex}).

3.2.3. Mass Spectrometers

Although quadrupole and magnetic sector mass spectrometers have been used in MPI–MS experiments, time-of-flight (TOFMS) and Fourier transform (FTMS) mass spectrometers are the instruments of choice for pulsed laser R2PI–MS experiments. Excellent reviews of TOFMS (89–91) and FTMS (92, 93) are available elsewhere.

The TOFMS is an extremely efficient detector. In principle, every ion that is extracted from the source may be detected and mass analyzed. This is in contrast to magnetic sector and quadrupole mass filter instruments, which must be electronically scanned in order to obtain a full mass spectrum. Such mass spectrometers only transmit a narrow m/q window to the detector at any one time and therefore waste most of the ions. These scanning instruments are adequate for selective ion monitoring (SIM) experiments, but their low efficiency results in relatively poor signal/noise ratios (S/N) when large mass ranges are of interest. This consideration is very important for the case of pulsed R2PI experiments, in which the repetition rate and duty cycle of the ion formation process are presently quite low.

Another shortcoming of scanning mass spectrometers is their susceptibility to problems arising from pulse-to-pulse variations in ion signal. Because such systems are "blind" to all but one m/q signal at a time, there is no way to compensate for signal fluctuations that may occur during the course of a scan. Two general categories of pulse-to-pulse variations in R2PI experiments are laser-related and sample-related processes. The laser-related category includes uncontrollable shot-to-shot laser pulse energy fluctuations, the magnitudes of which are characteristic of the stability of the laser, and controllable variations, intentionally imparted on the system by scanning the excitation wavelength of the laser.

Sample-related variations result from rapid changes in analyte concentration within the ion source of the mass spectrometer. Most, if not all, modes of sample introduction are unable to provide a perfectly constant and uniform concentration of analyte to the ion source. This problem is particularly acute for the case of a CGC sample introduction system, in which a particular effluent may completely pass through the ion source in less than 1 s. Thus, with a 10-Hz laser system, it would be impossible to obtain even a marginal mass spectrum on this time scale when a scanning instrument is utilized.

The relatively high efficiency of TOFMS overcomes the above problems, because an entire mass spectrum may be collected for each and every laser shot. Thus, relative peak heights are invariant to small fluctuations in laser power, and the detector responds immediately to pulse-to-pulse changes in mass spectra because of variations in laser wavelength or fluctuations in analyte concentration. Such response characteristics make the TOFMS an attractive alternative to the traditional scanning instruments, particularly for pulsed R2PI experiments in which chromatographic sample introduction systems are employed. Furthermore, such "real-time" characteristics are highly desirable for applications requiring on-line monitoring of dynamic processes.

A further advantage of the TOFMS is its simple design and operating principles, which allows the cost to remain lower than other mass spectrometers of similar analytical capabilities. Also, this simplicity results in a more durable instrument (94).

The limitation of linear TOFMS instruments is that they have relatively low mass resolving power (200–4000) (78), which restricts their mass spectral specificity and therefore can limit the ability to uniquely identify very heavy molecules. Recently, reflectron time-of-flight mass spectrometers (RE-TOFMS) have been developed and have become commercially available. These instruments have been shown to have mass resolutions of $> 10^5$, and $\sim 10^6$ is possible using picosecond lasers. However, the detection efficiency of RETOFMS is lower than that of a linear TOFMS because the ion drift velocity is usually set at a low value to achieve high mass resolution (82). Therefore, in applications where sensitivity is more important than mass resolution, the simple linear TOFMS would be the MS of choice.

FTMS instruments also offer the multichannel advantage like TOFMS, plus they can achieve mass resolutions of $> 10^6$. Theoretically, the mass resolution is unlimited because it is a function of the time spent observing the circulating ions. However, the cost of an FTMS system is significantly higher than that of the simple linear TOFMS.

4. APPLICATIONS OF R2PI–MS FOR PAC DETERMINATIONS

The applications show the versatility and power of R2PI–MS instrumental systems that demonstrate its potential use as a routine analytical tool. The multidimensionality, excitation and detection selectivity, and limits of detection of the work performed to date are emphasized, together with their novel experimental arrangements. In an HRC–R2PI–MS–LIF system, considerable useful analytical information is obtained from the chromatogram, mass spectrum, and fluorescence spectra. Chromatographic retention time (t_r) is characteristic for each analyte and is a function of the analytes affinity for the stationary phase. A wide variety of stationary (and mobile) phases and chromatographic conditions can be utilized to obtain the desired separation. The use of selective detectors can ease the requirements to make the necessary analytical separation. Here, R2PI plays a significant role. For the determination of PACs, the wavelength of excitation (λ_{ex}) can be adjusted so that aliphatics, alcohols, sulfones, and many other molecules typically present in environmental and biological samples do not undergo resonance photoabsorption and subsequent ionization. This selective excitation by R2PI is a function of molecular spectroscopic constants (e.g., a_i, k_i, IP) and the laser's spectral irradiance (i.e., λ_{ex}, linewidth, and power density). Detection of electrons is a convenient means of measuring R2PI. However, the mass spectra often yields unambiguous identification and quantitation for the analytes of interest. Depending upon the R2PI conditions, molecular weight and structure (via the fragmentation pattern) information are obtained. Fluorescence excitation and emission spectra are often "the forgotten by-products" of the R2PI process. However, these spectra and molecular fluorescence lifetime determinations represent independent and therefore complimentary analytical information which if processed can only serve to increase the confidence of measuring PAC analytes in complex mixtures. Combining λ_{ex}, t_r, m, λ_{em}, and τ_f provides a fivefold qualitative identification of the analyte. The mass and fluorescence intensities are independent measures for quantitation. Thus, an HRC–R2PI–MS–LIF analytical system could provide sufficient information to characterize complex samples for PACs on microliter samples in the time it takes to prepare the sample and run a chromatogram. Attempting to unambiguously measure hundreds of individual components separately would undoubtedly consume more sample and time.

4.1. R2PI–FTMS

McIver and co-workers (95) first reported R2PI–FTMS on PACs. A fixed-frequency excimer laser ($\lambda_{ex} = 222$, 249, or 308 nm) and either a batch inlet with a variable leak value or a direct insertion probe for sample introduction was used. Complete high-resolution mass spectra were obtained for naphthalene, fluoranthene, and triphenylene during each laser pulse. Wavelength selectivity was shown, and R2PI–FTMS signals greater than EI–FTMS signals were achieved even though the laser duty cycle was 10^6 less.

This group (85) has also utilized laser desorption (LD–R2PI–FTMS) to gain ~ 35-fold increase in detectability of naphthalene by R2PI ($\lambda_{ex} = 248$ nm) over that obtained by electron ionization (EI). Naphthalene and benzene were chemisorbed on a platinum single crystal. This work demonstrated that multiphoton ionization of aromatic adsorbates by a UV laser is an efficient process. Most of the naphthalene (98%) was removed by the first laser pulse, which underscored the advantage (and here the necessity) of FTMS to obtain a complete mass spectrum for each laser pulse. Assuming the coverage of one monolayer, it was estimated that ~ 3×10^{-4} of a monolayer of naphthalene could be detected. Naphthalene is also known to undergo thermal decomposition in temperature-programmed reactions. These experiments indicated that laser desorption could remove intact aromatic molecules from the surface for subsequent analysis by mass spectrometry. This group (96) has also utilized the ability of FTMS to store ions in the analyzer cell for subsequent laser-induced photodissociation studies. Using an ArF laser ($\lambda_{ex} = 193$ nm), bromobenzene and oligopeptide ions were efficiently photo-fragmented, demonstrating this technique as a powerful method for elucidating ion structures.

Carlin and Freiser (97) investigated low-molecular-weight aromatic compounds (perdeuteriodiphenyl, azulene, and diethylaniline). They used a fixed-frequency, quadrupoled Nd:YAG laser ($\lambda_{ex} = 266$ nm) having 1–4-mJ pulse energies and a 4–7-ns pulse width. They found the working ranges of pressures and laser power to be highly sample dependent. A mass resolution of 21,000 fwhm at m/z 128 and a mass accuracy of 20 ppm were measured for the molecular ion of azulene at 2×10^{-7} torr. The high resolution and exact mass measurement capability of FTMS, plus the capability of storing ions in a collision-free environment for subsequent structural identification and mixture analysis, were discussed as advantages as compared to TOFMS. They also suggested multiple-photon ionization using a quadrupoled Nd:YAG laser at $\lambda_{ex} = 266$ nm for selective detection of aromatic compounds via CGC–FTMS.

The work to date of most significance, in terms of the characterization of a sample for its PAC content, is that of Gross and co-workers (88). They used a fixed-frequency Nd:YAG operating at 266 nm and a pulsed valve sampling interface maintained at 250°C in a CGC–R2PI–FTMS system. Picogram detection limits for naphthalene, acenaphthene, biphenyl, fluorene, and phenanthrene were obtained at both high and low resolving powers with a

linear dynamic range of 2.5 orders of magnitude. A 5-mm-diameter beam
having 4–15-mJ pulse energies was utilized in order to irradiate a larger
analyte volume, thus yielding better sensitivity and less fragmentation than a
focused beam. Mass measurement accuracies of between 1 and 10 ppm were
obtained. Major differences were observed when the EI and R2PI reconstructed chromatograms of a leaded gasoline sample containing hundreds
of components were compared. As shown in Figure 6, only toluene was
observed in the first 4 min of the R2PI chromatogram, whereas all the light
aliphatic hydrocarbons are seen in the EI chromatogram. This demonstrated
the selective photoabsorption and ionization character of R2PI. Additionally,
the later-eluting compounds ($t_r > 20$ min) that do not appear in the EI trace
are quite evident in the R2PI chromatogram. This indicated the higher
sensitivity capable of being achieved for some compounds, particularly PACs.

4.2. SFI–RC–R2PI–TOFMS

Supercritical fluids allow the use of pressure rather than temperature to
volatilize molecules and therefore are important for analyzing thermally labile
and high-boiling point materials. Lubman and co-workers (73) developed an
ultrahigh-pressure pulsed valve for supercritical fluid injection (SFI) that can
operate at up to 380 atm and 200°C. The pulsed valve (~ 700-μs pulse width)
allowed the use of a larger, 150–200-μm orifice. The on-axis intensity of the
rotationally cooled molecules increased greater than two orders of magnitude
over that of continuous flow operation. With the use of additional cryopum-

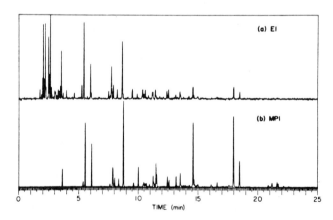

Figure 6. Reconstructed chromatograms for a neat injection of 0.1 μl of commercial leaded
gasoline under (*a*) EI and (*b*) R2PI conditions. The column temperature was programmed from
50°C to 250°C at 5°C min^{-1}. Files were stored at the rate of 1 file s^{-1}. Reprinted, with permission,
from ref. 88.

ping, they have obtained RC–R2PI signals for acenaphthalene, phenanthrene, carbazole, tetracene, pyrene, and triphenylene (72) as well as for several small molecules of biological interest (98). Derivatization methods were used to enhance the solubility of highly polar molecules (e.g., metabolites of catecholamines) in supercritical fluids for SFI–RC–R2PI–TOFMS (74).

4.3. LD–R2PI–TOFMS

Schlag and co-workers (78–81) have combined infrared laser desorption and sample vaporization, pulsed supersonic expansion, R2PI, and a high mass resolution reflection time-of-flight mass spectrometer (RETOFMS) in order to investigate large thermally labile and nonvolatile samples. The separation of the vaporization and ionization processes avoided many of the disadvantages of the simultaneous vaporization–ionization techniques. With this technique, the desorption and ionization step can be individually optimized in terms of laser wavelength and power density. LD–RC–R2PI–RETOFMS provided highly selective ionization, controllable fragmentation, and higher molecular ion yields. Their experimental arrangement minimized or eliminated molecular adduct ion formation. The translational cooling, resulting from the supersonic beam expansion, improved the mass resolution due to the narrow initial velocity distributions of the neutral molecules. The initial potential energy due to the finite spot size of ion generation was efficiently corrected for by the reflectron, thus improving mass resolution. A source of initial analyte ion kinetic energy within the ion source is Coulombic repulsion due to the high ion densities that can be achieved in the efficient R2PI process. The RETOFMS allowed the use of a fairly large laser beam, which minimized Coulombic repulsion effects while maintaining higher signal levels. Mass resolution could be improved by using a more tightly focused laser beam and lower laser intensity, but at the cost of sensitivity. With optimization of laser temporal pulse width and spectral power density, a resolving power of $> 10,000$ was achieved for the molecular ion of p-xylene ($m/z = 106$). The sensitivity was reported to currently be in the femtomole range, with a few orders of magnitude improvement expected (78). Theoretically, the mass range is unlimited. Molecules of biological interest such as chlorophylls, porphyrines, tripeptides, amino acids, and the decapeptide angiotensin I (with mass 1295) have been observed. By controlling the degree of fragmentation, the mass spectra obtained were useful for peptide sequencing.

   This group (99) has also introduced a vapor sample in an effusive beam into their R2PI–RETOFMS instrument to demonstrate wavelength selectivity and mass resolution of 6500 (50% valley) at mass 96 for 2,5-dimethylfuran (96.058 u) and fluorobenzene (96.038 u). Also, simply by removing the blocking filter for the fundamental wavelength in the frequency-doubled dye laser excitation path, they were able to switch from "soft" to "hard" ionization (i.e., from minimal to extensive fragmentation).

Similar work has been performed by Lubman and co-workers (74–77) using a linear TOFMS. They have studied LD–RC–R2PI–TOFMS by investigating a number of important classes of biological molecules, including: catecholamines; indoleamines and their metabolites; vitamins; amino acids; neuroleptic drugs; and small peptides. An excellent discussion of the considerations for LD experiments has been given (74).

Perhaps more significant work, however, is the first demonstration of quantitative analysis of molecules by LD–R2PI–TOFMS. Zare and co-workers showed that 20 primary phenylthioldantoin (PTH) amino acids could be detected and quantitated (100). The key to their methodology was the spatial and temporal separation of desorption and ionization. The laser desorption–ionization source region is shown in Figure 7. A CO_2 laser beam desorbed the sample from a rotating glass cup. The desorbed molecules expanded into the ion source region of a linear TOFMS, where they were interrogated by an Nd:YAG laser (typically 1 mJ, 10 ns) at 266 nm focused to a 0.1–cm-diameter beam waist. The ion signals were linear with PTH amino acid concentration in the picomole to nanomole range. The work showed lower detection limits to be possible. They also demonstrated that within the given concentration range the desorption process is complete.

Reilly and co-workers have developed a prism internal reflection ion source in order to obtain high mass resolution (82, 83). Molecules are laser desorbed from the surface of the prism in order to minimize time broadening due to the spatial width of the ionizing region and the initial ion kinetic energy

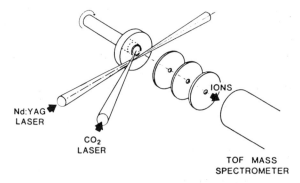

Figure 7. Schematic of the two-step laser-desorption–multiphoton-ionization setup. CO_2 laser desorption is performed at the inner surface of a rotating glass cup in the first electrode, that is, the first acceleration region of a time-of-flight mass spectrometer. A 266-nm laser beam acts as an ionizing laser. Reprinted, with permission from ref. 100.

distribution. A 2-ns-pulse-width frequency-doubled tunable dye laser was used to achieve mass resolutions of 3900 and 11,000 in linear TOFMS and RETOFMS, respectively, for aniline. Mass resolution greater than 1×10^6 was stated to be theoretically possible using picosecond lasers.

Laser desorption does appear to be a viable analytical technique for characterizing complex mixtures for their PAC content. Combining LD–R2PI–MS with instrumentation that allows the chromatographic eluent to be readily "cold trapped" on a rotating disk or moving plate for subsequent analysis would certainly seem to provide a powerful tool for characterizing complex samples for PACs, particularly if it is also combined "on-line" with FTIR.

4.4. CGC–R2PI–TOFMS

Reilly and co-workers (101–103) demonstrated the quantitative analytical utility of CGC–R2PI–TOFMS. An excimer laser was used to investigate several PACs (naphthalene, biphenyl, acenaphthene, fluorene, phenanthrene, anthracene, fluoranthene, and pyrene) in an effusive (ambient temperature) expansion. Detection limits for these molecules ranged from 200 fg to 5 pg with linear dynamic ranges of greater than four orders of magnitude. Selective excitation of chrysene was observed in the presence of its coeluting isomer triphenylene using a XeCl laser. Figure 8 shows chromatograms of 0.2-μl gasoline samples comparing flame ionization detection (FID) with ArF and KrF laser R2PI detection. The aliphatic compounds present in the FID are not detected by KrF and ArF laser ionization because of their high ionization

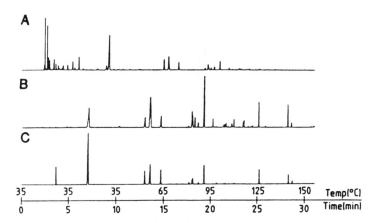

Figure 8. Capillary column gas chromatograms of a 0.2-μl gasoline sample. (*a*) Flame ionization detection (FID). (*B*) ArF-laser-ionization mass-integrated ion yield. (*C*) KrF-laser-ionization mass-integrated ion yield (102).

potentials (IP > 10 eV) and weak UV absorptions at 248 and 193 nm. Efficient and selective ionization of alkylbenzenes, naphthalene, and various alkyl-naphthalenes is apparent, particularly in the later part of the chromatogram. Scale expansion allowed minor components to be readily observed. Selective analysis of nitro- and nitroso-containing compounds using a focused ArF laser has also been demonstrated (103).

Imasaka *et al.* (104) utilized a pulsed valve operated at 200°C in their CGC–RC–R2PI–TOFMS system. They investigated a number of aniline derivatives. Demonstrated limits of detection, however, were only in the microgram range. The coeluting isomers, *N*-methylaniline and *p*-toluidine, were able to be selectively ionized by adjusting the laser wavelength.

4.5. CGC–R2PI–TOFMS–TECD–LIF–FID

The R2PI process provides a unique opportunity to simultaneously collect several types of complementary data for qualitative and quantitative analysis. Dobson *et al.* (105) developed a novel instrument in which cations (M⁺), electrons (e^-), and photons (λ_{Fl}) were simultaneously collected from the ion source region of a TOFMS (see Figure 9). A schematic representation of the laser-based portion of their multidimensional analytical system is shown in Figure 10.

They used a frequency-doubled tunable dye laser (~1 mJ/pulse) in their CGC–R2PI–TOFMS system. Three additional detection systems were in-corporated in their instrumental design. The capillary GC effluent was split to provide for use of a conventional CGC detector in parallel. (A flame

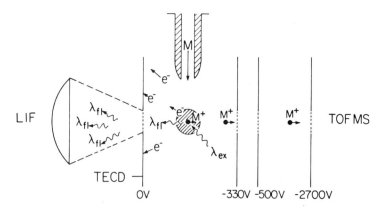

Figure 9. Schematic diagram of a dynamic gas sample cell for monitoring all analytically useful products of laser (λ_{ex})–analyte (M) interaction at low pressure. Interaction products: Cations (M⁺), electrons (e^-), and photons (λ_{fl}). Reprinted, with permission, from ref. 105.

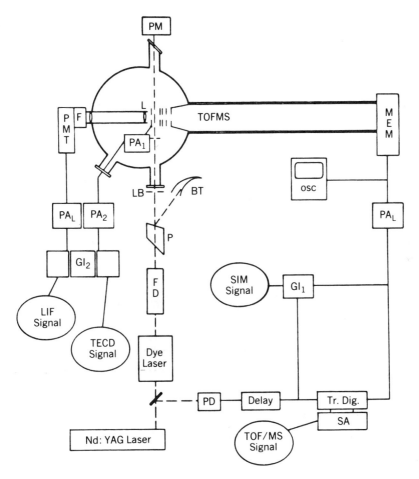

Figure 10. Schematic diagram of CGC–R2PI–TOFMS–LIF–FID analytical instrumentation. MEM, magnetic electron multiplier; PA, preamplifier; GI, gated integrator; Tr. Dig., transient digitizer; SA, signal averager; FD, frequency doubler; PMT, photomultiplier tube; PD, photodiode. Reprinted, with permission, from ref. 105.

ionization detector was used as a universal detector to give information as to the general complexity of the sample matrix). Emission from laser-induced fluorescence (LIF) was collected simultaneously from the gas dyamic sampling cell. Additionally, a novel total-electron-current detector (TECD) was added by attaching a fast preamplifier directly to the backing grid of the nude ion source assembly. This multihyphenated instrumental technique can be flexible in that, depending on the particular application, it may be advantageous to employ only a portion of its capabilities. On the other hand, it is a very powerful analytical instrument that, when utilized to the full extent of its capabilities, should provide sufficient selectivity to solve even the most difficult problems relating to the characterization of samples for PACs.

In order to better interpret the analytical capabilities, the instrumental system was characterized in terms of (a) laser power density effects on R2PI linearity and fragmentation, (b) wavelength selective ionization, (c) limits of detection, and (d) linear dynamic range. Its use as a selective detector for chromatography was also demonstrated. The TECD signal response was shown to be linear for the power density range $\sim 0.2-5 \times 10^6$ W cm^{-2} (i.e., 0.1–2.6 mJ in a 2-mm-diameter laser beam), excitation wavelength, and representative PACs studied. This indicated that the primary (R2PI) ionization mechanism was a stepwise one-photon-limited process! Although the extent of fragmentation increased and was "compound dependent," the PAC analytes largely experienced soft ionization (i.e., the parent ion dominated in the mass spectra). Significant (15–25%) fragmentation was achieved by focusing the laser beam to increase the power density approximately 16-fold. The ability to control fragmentation was noted to be a powerful tool, providing the option of a simpler mass spectrum for coeluting photoionizable PACs or extensive fragmentation of well-resolved chromatographic eluents for the identification of unknown PACs.

Wavelength-selective ionization was also demonstrated for the closely eluting isomers anthracene and phenanthrene. The deuterated analogue of phenanthrene was used in this study to make obvious the identity of the ionized species. It was noted that deuterated analogues should prove to be ideal for use as internal reference standards for quantitation since their physical, chemical, and ambient-temperature spectroscopic properties are nearly identical with those of the parent compound.

The linear dynamic range determined from the TECD peak heights was found to be greater than four orders of magnitude. Picogram limits of detection (LOD for $S/N = 3$) were determined for 21 PACs covering a wide range of molecular weights (see Table 1). These LODs were obtained at a compromise λ_{ex} and pulse energy, which could be optimized for any particular analysis.

The most significant results were graphically demonstrated as shown in Figure 11. These results showed a portion of the analytical information available during a single chromatogram. The FID chromatogram gave information about the complexity of the sample matrix (64 components here). The TECD chromatogram demonstrated that R2PI can provide essentially PAC selective detection for CGC. (A compromise excitation wavelength, λ_{ex}, was selected for the chromatographic runs, and it was noted that the chromatographic peak heights represent relative R2PI responses at a chosen λ_{ex}). The LIF chromatogram obtained under compromise detector gate-

Table 1. TECD and TOF/MS Absolute Limits of Detection for PACs and Other Aromatic Compounds at $\lambda_{ex} = 285$ nm (1 mJ/pulse)

Compound	Structure	LOD (pg)	
		TECD	TOF/MS
p-Cresol	HO—⬡—CH₃	59	7
Naphthalene		130	16
Acenaphthylene		N.D.[a]	
Acenaphthene		7	1
Dibenzofuran		18	2
Fluorene		12	2
p-Phenylphenol		53	7
Dibenzothiophene		260	32
Pentachlorophenol		N.D.	
Phenanthrene		47	6
Anthracene		270	34

Table 1. (*Continued*)

Compound	Structure	LOD (pg)	
		TECD	TOF/MS
Carbazole		6	1
Fluoranthene		N.D.	
Pyrene		33	4
p-Terphenyl		13	2
Triphenylene		43	5
Benz[*a*]anthracene		19	2
Chrysene		24	3
Benzo[*b*]fluoranthene		500	63
Benzo[*j*]fluoranthene		N.D.	

Table 1. (*Continued*)

Compound	Structure	LOD (pg)	
		TECD	TOF/MS
Benzo[*k*]fluoranthene		38	5
Benzo[*e*]pyrene		20	3
Benzo[*a*]pyrene		12	2
Perylene		—[b]	
Indeno[1,2,3-*cd*]pyrene		N.D.	
Dibenz[*a,h*]anthracene		—[b]	
Dibenz[*a,c*]anthracene		37	5
Benzo[*ghi*]perylene		35	4

[a] N.D., not detected.
[b] Detected, not quantitated.

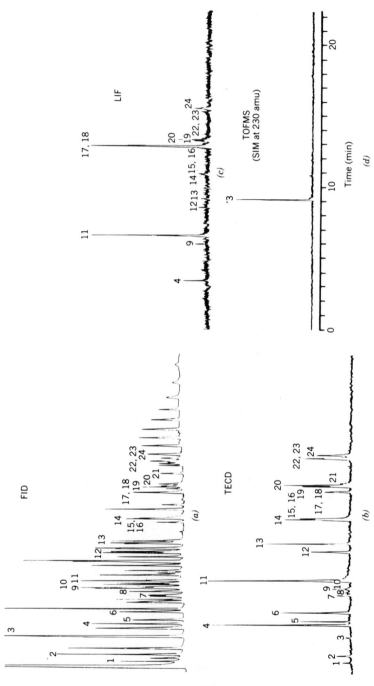

Figure 11. Chromatogram simplification demonstrated for a 64-component mixture by simultaneous (*a*) FID, (*b*) TECD, (*c*) LIF, with WG 320 and WG 335 optical filters, and (*d*) TOFMS, SIM at 230 amu (1-amu window), λ_{ex} 285 nm. Temperature programmed from 100°C to 320°C at 14°C min^{-1} with 1-min initial hold and 6-min final hold. Reprinted, with permission, from ref. 105.

width, delay and emission filtering conditions showed similar selectivity for PACs and provided complementary information. The selective ion monitoring (SIM) chromatogram illustrated, in part, the additional selectivity offered by the incorporation of a mass spectrometer. In this case, only *p*-terphenyl was detected by preselecting only a 1-amu mass window, at 230 amu, for monitoring. When the transient digitizer data processing system is utilized, full advantage of the TOFMS detector could be realized, since an entire mass spectrum may then be collected for each ionizable chromatographic effluent. Thus, all of these chromatograms were simultaneously available for each sample injection.

Many real samples that need to be characterized in terms of their PAC content are considerably more complex than the above-mentioned 64-component mixture. A neutral PAC fraction of a crude shale oil was obtained and analyzed (Figure 12). Under these sample analysis conditions the R2PI process at ambient temperatures did not provide the necessary excitation selectivity to yield a well-resolved CGC–R2PI–TECD chromatogram of the hundreds of different PAC species in this sample under the conditions used. However, the additional detection selectivity provided by the TOFMS allowed for the selective detection of a specific PAC (dibenzothiophene).

5. SUMMARY: R2PI–MS FOR PAC DETERMINATIONS

Multidimensional R2PI–MS-based systems have shown the potential to be more versatile and powerful than conventional CGC–MS for the extensive characterization of real samples for PACs. There is a need, of course, to develop an extensive low-pressure, gas-phase PAC excitation–ionization spectral data base, particularly if rotational cooling is to be employed. More sophisticated data acquisition and reduction techniques, including chemometrics, must be applied in order to take full advantage of the available analytical information.

R2PI–MS systems have been combined with a variety of sample introduction systems. The use of capillary gas chromatography, laser desorption, and supercritical fluid injection allows R2PI–MS to be applied quite conveniently to a wide variety of samples. Rotational cooling and laser-induced fluorescence detection enhance the selectivity and analytical information available in the R2PI process. Clearly, R2PI–MS-based instruments can be applied to a wide variety of analytical problems involving trace PAC characterization in complex mixtures.

Greater confidence in the analytical measurements is a major advantage for multidimensional, R2PI–MS-based instrumentation. Combining high-resolution chromatography, selective ionization, and sensitive mass and fluore-

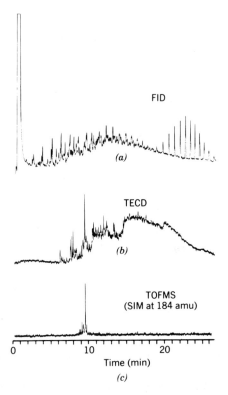

Figure 12. Simultaneous chromatograms of Paraho shale oil (CRM-2): neutral PAC fraction ($\lambda_{ex} = 285$ nm, 1 mJ/pulse). (*a*) FID; (*b*) TECD; (*c*) TOF/MS with SIM at 184 u. Temperature program: column held at 60°C for 1 min, then ramped to 340°C at 12°C min^{-1}, where it was held for 2 min. Reprinted, with permission, from ref. 105.

scence spectrometry yields an instrumental system with high informing power. Complete characterization of complex environmental and biological samples gains extreme importance in light of the synergistic or multistep nature of cancer development (106, 107). Methodology to measure exposure to PACs and their cancer promoters could be developed utilizing HRC–R2PI–TOFMS–LIF instrumentation which would (a) be applicable to a wide range of sample types (b) allow shorter total analysis times, and (c) give the confidence needed in the measurement to allow reasonable assessment of health risks.

REFERENCES

1. D. H. Parker, in *Ultrasensitive Laser Spectroscopy*, D. S. Kliger, Ed., Academic Press, London, Chapter 4 (1983).

2. K. Birdbord and J. G. French, in *Carcinogensis, Vol. 3: Polynuclear Aromatic Hydrocarbons*, P. W. Jones and R. I. Freudenthal, Eds., Raven Press, New York, p. 451 (1978).

3. D. W. Jones and R. S. Mathews, *Prog. Med. Chem.*, **10**, 159 (1974).

4. C. White, in *Polynuclear Aromatic Hydrocarbons, Third International Symposium*, P. Jones, and P. Leber, Ed., Ann Arbor Science, Ann Arbor, Mich. (1979).

5. A. Borgen, H. Darvey, N. Castagnok, T. T. Crocker, R. E. Rassmussen, and I. Y. Wang, *J. Med. Chem.*, **16**, 502 (1974).

6. C.-H. Ho, B. R. Clark, M. R. Guerin, B. D. Barkenbus, T. K. Rao, and J. L. Epler, *Mutat. Res.*, **85**, 335 (1981).

7. M. R. Guerin, I. B. Rubin, T. K. Rao, B. R. Clark, and J. L. Epler, *Fuel*, **60**, 282 (1981).

8. *Chemical Carcinogens, Vols. 1 and 2, 2nd edition*, C. E. Searle, Ed., American Chemical Society, Washington, D.C. (1984).

9. C. Willey, M. Iwao, R. N. Castle, and M. L. Lee, *Anal. Chem.*, **53**, 400 (1981).

10. C. S. Woo, A. P. D'Silva, V. A. Fassel, and J. G. Oestreich, *Environ. Sci. Technol.*, **12**, 173 (1978).

11. M. Blumer, *Sci. Am.* **234**, 34 (1976).

12. D. Hoffman and E. L. Wynder, *Nat. Cancer Inst. Monogr.*, **9**, 91–226 (1962).

13. R. Long, *Studies on Polycyclic Aromatic Hydrocarbons in Flames*, U.S. NTIS No. PB-220151 (1972).

14. D. W. Koppenaal and S. E. Manahan, *Environ. Sci. Technol.*, **10**, 1104 (1976).

15. M. R. Guerin, W. H. Griest, C.-H. Ho, and W. D. Shutts, in *Third ERDA Environmental Protection Conference*, Oak Ridge, Tenn. (September 1975).

16. T. Panalaks, *J. Environ. Sci. Health*, **B11**, 299 (1976).

17. W. Giger and M. Blumor, *Anal. Chem.*, **46**, 1663 (1974).

18. W. Giger and G. Schaffner, *Anal. Chem.*, **50**, 243 (1978).

19. J. Borneff, *Der Landarzt*, **40**, 109 (1964).

20. J. W. Howard, T. Fazio, R. H. White, and B. A. Klimeck, *J. Assoc. Off. Anal. Chem.*, **51**, 122 (1968).

21. W. H. Griest, B. A. Tomkins, J. L. Epler, and T. K. Rao, in *Polynuclear Aromatic Hydrocarbons*, P W. Jones, and P. Leber, Eds., Ann Arbor Science, Ann Arbor, Mich., p. 395 (1979).

22. A. Dripple, in *Chemical Carcinogens*, C. E. Searle, Ed., American Chemical Society, Washington, D.C. p. 245 (1976).

23. A. A. Pelroy and A. Gandolfi, *Mutat. Res.*, **72**, 329 (1980).

24. D. Hoffman and E. L. Wynder, *Cancer*, **27**, 848 (1971).

25. T. Romanowski, W. Funcke, I. Grossmann, J. Konig, and E. Balfanz, *Anal. Chem.*, **55**, 892 (1983).

26. H. Y. Tong and F. W. Karasek, *Anal. Chem.*, **56**, 2129 (1984).

27. G. Grimmer, J. Jacob, K. Naujack, and G. Dettbarn, *Anal. Chem.*, **55**, 892 (1983).

28. D. W. Later, T. G. Andros, and M. L. Lee, *Anal. Chem.*, **55**, 2126 (1983).

29. D. W. Later, M. L. Lee, K. D. Bartle, R. C. Kong, and D. L. Vassilaros, *Anal. Chem.*, **53**, 1612 (1981).

30. R. J. Crowley and S. Siggia, P. C. Uden, *Anal. Chem.*, **52**, 1224 (1980).

31. L. R. Hilpert, G. D. Byrd, and C. R. Vogt, *Anal. Chem.*, **56**, 1842 (1984).

32. W. J. Simonsick, Jr., and R. A. Hites, *Anal. Chem.*, **56**, 2749 (1984).
33. W. J. Simonsick, Jr., and R. S. Hites, *Anal. Chem.*, **58**, 2114 (1986).
34. W. J. Simonsick, Jr., and R. A. Hites, *Anal. Chem.*, **58**, 2121 (1986).
35. J. C. Brown, J. A. Duncanson, Jr., and G. J. Small, *Anal. Chem.*, **52**, 1711 (1980).
36. A. P. D'Silva and V. A. Fassel, *Anal. Chem.*, **56**, 985A (1984).
37. M. B. Perry, E. L. Wehry, and G. Mamantov, *Anal. Chem.*, **55**, 1893 (1983).
38. A. Amirav, U. Even, and J. Jortner, *Anal. Chem.*, **54**, 1666 (1982).
39. T. Imasaka, H. Fukuoka, T. Hayashi, and N. Ishibashi, *Anal. Chim. Acta*, **156**, 111 (1984).
40. M. V. Johnston, *Trends Anal. Chem.*, **3**, 58 (1984).
41. R. E. Smalley, L. Wharton, and D. H. Levy, *Acc. Chem. Res.*, **10**, 139 (1977).
42. J. M. Hayes and G. J. Small, *Anal. Chem.*, **55**, 565A (1983).
43. J. M. Hayes, *Chem. Rev.*, **87**, 745 (1987).
44. J. A. Warren, J. M. Hayes, and G. J. Small, *Anal. Chem.*, **54**, 138 (1982).
45. J. C. Brown, J. M. Hayes, J. A. Warren, and G. J. Small, *Lasers in Chemical Analysis*, G. M. Hieftje, F. E. Lytle, and J. C. Travis, Eds., Humana Press, Clifton, NJ, Chapter 12 (1981).
46. J. M. Hayes, and G. J. Small, *Anal. Chem.*, **54**, 1202 (1982).
47. J. M. Hayes and G. J. Small, *Anal. Chem.*, **55**, 565A (1983).
48. T. Imasaka, T. Okamura, and N. Ishibashi, *Anal. Chem.*, **58**, 2152 (1986).
49. B. V. Pepich, J. B. Callis, J. D. S. Danielson, and M. Gouterman, *Rev. Sci. Instrum.*, **57**(5), 878 (1986).
50. B. V. Pepich, J. B. Callis, D. H. Burns, M. Gouterman, and D. A. Kalman, *Anal. Chem.*, **58**, 2825 (1986).
51. S. W. Stiller and M. V. Johnston, *Anal. Chem.*, **59**, 567 (1987).
52. A. Amirav, U. Even, and J. Jortner, *Anal. Chem.*, **54**, 1666 (1982).
53. U. Even, J. Magen, and J. Jortner, *J. Chem. Phys.*, **77**, 4374 (1982).
54. U. Even, J. Magen, J. Jortner, and J. Friedman, *J. Chem. Phys.*, **77**, 4384 (1982).
55. U. Even and J. Jortner, *J. Chem. Phys.*, **77**, 4391 (1982).
56. T. Imasaka, N. Yamaga, and N. Ishibashi, *Anal. Chem.*, **59**, 419 (1987).
57. A. P. D'Silva, M. Iles, G. Rice, and V. A. Fassel, Ames Laboratory Report IS-4556, Ames, Iowa (April 1984).
58. A. Amirav, U. Even, and J. Jortner, *Opt. Communi.* **32**, 266 (1980).
59. I. Schek and J. Jortner, *J. Chem. Phys.*, **70**, 3016 (1979).
60. P. M. Johnson, *Acc. Chem. Res.*, **13**, 20 (1980).
61. M. B. Robin, *Appl. Opt.*, **19**, 3941 (1980).
62. M. Seaver, J. W. Hudgens, J. J. Decorpo, *Int. J. Mass Spectrom. Ion Phys.*, **34**, 159 (1980).
63. J. H. Brophy and C. T. Rettner, *Opt. Lett.*, **4**, 337 (1979).
64. R. J. Cotter, *Anal. Chem.*, **56**, 485A (1984).
65. J. Grotemeyer, and E. W. Schlag, *Angew. Chem. Int. Ed. Eng.*, **27**, 447 (1988).
66. W. Siebrand, in *Dynamics of Molecular Collisions, Part A.*, W. Miller, Ed., Plenum, New York, Chapter 6 (1976).
67. J. Kasha, *Discuss. Faraday Soc.*, **9**, 14 (1950).
68. R. Botter, I. Dimicoli and J. Lemaire, in *Laser Applications in Chemistry*, D. K. Evans, Ed., Proc. SPIE 669, p. 64 (1986).

69. R. D. Smith and H. R. Udseth, *Anal. Chem.*, **59**, 13 (1987).
70. B. W. Wright, H. T. Kalinoski, H. R. Udseth, and R. D. Smith, *J. High Res. Chromatogr. Chromatogr. Commun.*, **9**, 145 (1986).
71. E. D. Lee, J. D. Henion, R. B. Cody, and J. A. Kinsinger, *Anal. Chem.*, **59**, 1309 (1987).
72. C. H. Sin, H. M. Pang, D. M. Lubman, and J. Zorn, *Anal. Chem.*, **58**, 487 (1986).
73. H. M. Pang, C. H. Sin, D. M. Lubman, and J. Zorn, *Anal. Chem.*, **58**, 1581 (1986).
74. D. M. Lubman, *Prog. Analyt. Spectrosc.*, **10**, 529 (1987).
75. R. Trembreull and D. M. Lubman, *Anal. Chem.*, **59**, 1299 (1987).
76. R. Trembreull and D. M. Lubman, *Anal. Chem.*, **59**, 1082 (1987).
77. D. M. Lubman and R. Trembreull, *Anal. Instrum.*, **16**, 117 (1987).
78. U. Boesl, J. Grotemeyer, K. Walker, and E. W. Schlag, *Anal. Instrum.*, **16**, 151 (1987).
79. J. Grotemeyer, U. Boesl, K. Walter, and E. W. Schlag, *J. Am. Chem. Soc.*, **108**, 4233 (1986).
80. J. Grotemeyer, U. Boesl, K. Walter, and E. W. Schlag, *Org. Mass. Spectrom.*, **21**, 595 (1986).
81. H. V. Weyssenhoff, H. L. Selzle, and E. W. Schlag, *Z. Naturforsch.*, **40a**, 674 (1985).
82. M. Yang and J. P. Reilly, *Anal. Instrum.*, **16**, 133 (1987).
83. M. Yang and J. P. Reilly, *Int. J. Mass Spectrom. Ion Proc.*, **75**, 209 (1987).
84. L. Q. Huang, R. J. Conzemius, G. A. Junk, and R. S. Houk, *Anal. Chem.*, **60**, 1490 (1988).
85. M. G. Sherman, J. R. Kingsley, J. C. Hemminger, and R. T. McIver, *Anal. Chim. Acta*, **178**, 79 (1985).
86. K. Balasanmugam, S. K. Viswanadham, and D. M. Hercules, *Anal. Chem.*, **58**, 1102 (1986).
87. D. M. Lubman and R. M. Jordan, *Rev. Sci. Instrum.*, **56**, 373 (1985).
88. T. M. Sack, D. A. McCrery, and M. L. Gross, *Anal. Chem.*, **57**, 1290 (1985).
89. W. C. Wiley and I. H. McLaren, *Rev. Sci. Instrum.*, **26**, 1150 (1955).
90. D. Price and J. E. Williams, Eds., *Time-of-Flight Mass Spectrometry*, Pergamon Press, New York (1969).
91. J. E. Campana, Ed., *Advances in Time-of Flight Mass Spectrometry*, Analytical Instrumentation, Vol. 16, Marcel Dekker, Inc., New York (1987).
92. M. L. Gross and D. L. Rempel, *Science*, **226**, 261 (1984).
93. C. L. Wilkins and M. L. Gross, *Anal. Chem.*, **53**, 1661A (1981).
94. B. Y. Tailliez and S. H. Hume, in *Dynamic Mass Spectrometry*, Vol. 6, D. Price and J. F. J. Todd, Eds., Heyden & Son, London, Chapter 15 (1981).
95. M. P. Irion, W. D. Bowers, R. L. Hunter, F. S. Rowland, and R. T. McIver, Jr., *Chem. Phys. Lett.*, **93**, 375 (1982).
96. W. D. Bowers, S. S. Delbert, and R. T. McIver, Jr., *Anal. Chem.*, **58**, 969 (1986).
97. T. J. Carlin and B. S. Freiser, *Anal. Chem.*, **55**, 955 (1983).
98. C. H. Sin, H. M. Pang, and D. M. Lubman, *Anal. Instrum.*, **17**, 87 (1988).
99. R. Frey, G. Weiss, and H. Z. Kaminski, *Z. Naturforsch*, **40a**, 1349 (1985).
100. F. Engelke, J. H. Hahn, W. Henke, and R. N. Zare, *Anal. Chem.*, **59**, 909 (1987).
101. G. Rhodes, R. B. Opsal, J. T. Meek, and J. P. Reilly, *Anal. Chem.*, **55**, 280 (1983).

102. R. B. Opsal and J. P. Reilly, *Opt. News*, **12**(6), 18 (1986).
103. R. B. Opsal and J. P. Reilly, *Anal. Chem.*, **58**, 2919 (1986).
104. T. Imasaka, K. Tashiro, and N. Ishibashi, *Anal. Chem.*, **58**, 3242 (1986).
105. R. L. M. Dobson, A. P. D'Silva, S. J. Weeks, and V. A. Fassel, *Anal. Chem.*, **58**, 2129 (1986).
106. L. D. Tomei, I. Noyes, D. Blocker, J. Holliday, and R. Glaser, *Nature*, **329**, 73 (1987).
107. E. Farber, *Cancer Res.*, **44**, 4217 (1984).

CHAPTER

11

PHOTOTHERMAL SPECTROSCOPY

MICHAEL D. MORRIS

Department of Chemistry
University of Michigan
Ann Arbor, Michigan

1. INTRODUCTION

Light absorption measurements can be made indirectly, by measuring the heat evolved as excited molecules relax back to the ground state. While photoacoustic spectroscopy is the most familiar indirect absorption technology, photothermal methods provide experimental advantages in many liquid-phase and solid-phase systems. Like photoacoustic spectroscopy, photothermal spectroscopy, in principle, provides for nonfluorescent molecules many of the advantages traditionally associated with fluorimetry. In particular, as the concentration of absorbing molecules approaches zero, the heat evolved also goes to zero. It is easier to measure small signals against a nominally zero background than to measure small differences in large signals. Photothermal absorbance measurements should be a favorable measurement case.

Photothermal spectroscopy can provide high spatial resolution. Lasers are usually used as the light sources and can be easily focused to near the diffraction limit with simple optics. Laser sources make photothermal spectroscopy well-suited to short-path or restricted-volume liquid-phase measurements. For solids, they allow sampling of a small area with high signal/noise ratio.

The more familiar photoacoustic spectroscopy has been the subject of several recent books (1, 2) and review articles (4, 5). The recent review literature on photothermal spectroscopy (5–9) is no less extensive.

Photothermal spectroscopy is the measurement of the refractive index changes caused by light absorption and subsequent heat deposition. The refractive index change may be probed directly in the absorbing medium or, if the medium is an opaque or highly scattering solid, in a coupling fluid in contact with it.

Many experimental configurations have been used to probe these two kinds of systems. The principles of four popular approaches are outlined in Figure 1.

361

The refractive index distribution generated by each experiment approximates a conventional optical element. The element is shown next to each configuration.

By far, the most common effects used are transverse photothermal deflection (also called the mirage effect) and the thermal lens effect. Transverse photothermal deflection probes the refractive index change of the medium in thermal contact with the sample, whereas the thermal lens technique probes the change of refractive index of the sample directly. The thermal lens can be observed as the defocusing of the laser that heats the solution or as the defocusing of a second weak probe beam. Variants in which the lens is observed with a nonparaxial probe laser are increasingly common. Depending on the details of the experiments, as well as on the laboratories reporting them, these techniques are called collinear photothermal deflection or crossed-beam thermal lens measurements.

It is possible to generate refractive index gradients that behave approximately as circular lens, cylinder lens, prisms, and gratings. In liquids and gases, the circular lens configuration is the most popular, usually under the

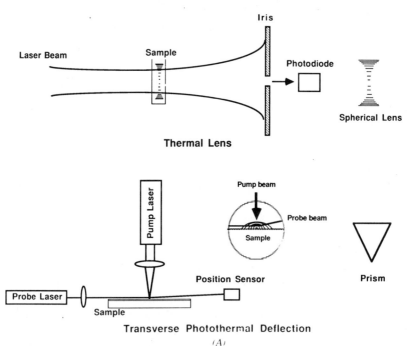

Figure 1. The principles of four photothermal techniques. The experimental configuration (*A*) and the discrete optical element that approximates the refractive index distribution (*B*) are shown. Reproduced, with permission, from ref. 7.

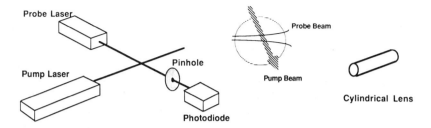

Photothermal Refraction

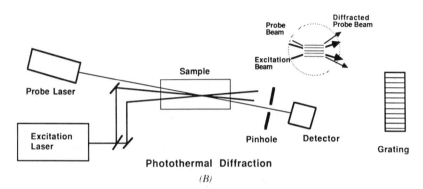

Photothermal Diffraction

(B)

Figure 1 *(Continued)*

name thermal lens. In two-phase media, the prism is the common element. Techniques employing this element are usually called transverse photo-thermal deflection or mirage effect measurements.

The thermal lens (Figure 1) is simply the result of the refractive index that forms in a heated medium. Most materials expand when heated, so that, in practice, dn/dT is negative. If the material is heated with a laser beam or other source with cylindrical symmetry, the thermal lens itself will be negative, or divergent, lens. A completely rigorous treatment of the thermal lens leads to intractable equations (10, 11). Simplifying assumptions are necessary, so that actual lens behavior usually deviates slightly from the approximate governing equations.

2. THERMAL LENS THEORY

2.1. General Aspects

Gordon et al. (10) developed the first theoretical treatment of thermal lens theory. Their treatment was used by Hu and Whinnery (11) to describe the simplest one-laser thermal lens measurement systems. Later workers have extended this theory to other experimental configurations. Occasionally, a different set of simplifying assumptions is used.

Gordon et al. assumed that the sample is a liquid which is heated by a TEM_{00} Gaussian source, whose radius, w_0, is constant throughout the length of the sample cell. If conduction is the only mode of heat transfer in the liquid, the temperature distribution as a function of radial distance from the beam center and time from the start of the laser irradiation, $\Delta T(r, t)$, is given by

$$\Delta T(r, t) = \frac{A\pi w_0^2}{8k} \left[Ei\left(\frac{2r^2}{w_0^2}\right) - Ei\left(\frac{2r^2}{8Dt + w_0^2}\right) \right] \tag{1}$$

In equation (1), A is the heat density of the laser beam; that is, $A = 0.24P$, where P is the power density in watts per square centimeter. D is the thermal diffusivity and is defined by $D = (k/\rho)C_p$, where k is the thermal conductivity, ρ is the density, and C_p is the specific heat of the liquid.

Equation (1) is complicated by the assumption that the laser beam is brought to a focus and diverges, so that its radius changes through the length of the sample. Consequently, a constant power density is assumed over the entire length, b, of the sample.

There are two other problems with equation (1). The assumption of an infinite radius for the medium generates a steady-state infinite temperature gradient. Definition of a finite boundary much larger than the laser beam removes this mathematical artifact. Second, the exponential integral is a transcendental function that is not easily employed in equations relating temperature gradient to the focal length of the thermal lens.

Gordon et al. solve the second problem by approximating the exponential integral as the series defined by equation (1a).

$$Ei(-x) = \gamma nrx - x + \frac{x^2}{4}, \dots, \qquad \gamma = 1.781 \dots \tag{1a}$$

They retain only the terms through quadratic. This parabolic approximation is valid for small values of r. Thus, near the axis of the laser beam, the following equation is a good approximation to the temperature distribution:

$$\Delta T(r, t) \cong \frac{0.06bP}{\pi k} \left[\ln\left(1 + \frac{8Dt}{w_0^2}\right) - \left(\frac{16Dt}{w_0^2 + 8Dt}\right)\frac{r^2}{w_0^2} \right] \tag{2}$$

where b is the length of the illuminated region of liquid.

2.2. The Single-Beam Thermal Lens

A parabolic refractive index distribution [equation (2)] functions as a thermal lens whose focal length is given by

$$F(t) = \frac{k\pi n_0 k w^2(w^2 + 8Dt)}{0.24bPl(dn/dT)8Dt} \tag{3}$$

$$= F_\infty \left(1 + \frac{t_c}{2t}\right)$$

The steady-state focal length F_∞ and critical time are given by

$$F_\infty = \left(\frac{k\pi n_0 w^2}{0.24bPl}\right) \Big/ \left(\frac{dn}{dT}\right) \tag{4}$$

$$t_c = \frac{w^2}{1+D} \tag{4a}$$

The thermal lens is detected by measuring its effect on a laser beam. The probe laser may be the one used to form the lens, or it may be an independent laser. If an independent probe laser is used, it must have much lower power than the heating laser, so that it will not perturb the thermal lens properties of the solution.

The assumption that w is constant throughout the length of the sample is a thin lens assumption. The heating laser need not be collimated. However, its divergence must be small enough to maintain a constant radius through the length of the sample. In the simplest experimental configuration, maximum sensitivity will be achieved if the sample is not placed exactly at beam focus, but about 1 confocal parameter beyond.

Figure 1A (top) shows the basic thermal lens experimental configuration. The maximum sensitivity is obtained if the thermal lens is generated at a distance z from the focal length, which generates the maximum fractional change w/w in beam radius observed at some distance Z in the far field.

The effect of a thermal lens, in the thin lens approximation, is to change the radius of curvature of the Gaussian laser beam at the lens. If the radius of curvature of the beam before the lens is R_0, then after the lens it is R.

Equation (5) describes the thermal lens system, or any thin lens system:

$$\frac{1}{F(t)} = \frac{1}{R_0} - \frac{1}{R} \tag{5}$$

Since the defocused laser beam is a Gaussian beam, it is described by a beam waist w_0' and its radius of curvature, according to equations (6) and (7):

$$w_0'^2 = \frac{w^2}{1+(\pi w^2/\lambda R)^2} \tag{6}$$

$$z' = \frac{R}{1+(\lambda R/\pi w^2)^2} \tag{7}$$

Because $1/wR = -1/F$, equation (8) follows:

$$\frac{\Delta w}{w} = \left(-\frac{\theta}{1+t_c/2t}\right)\left\{\left(\frac{b'}{w}\right)\left[2\left(\frac{w_0^2}{w^2}\right)-1\right]\pm\frac{w_0}{w}\left[1-\left(\frac{w_0^2}{w^2}\right)\right]^{1/2}\right\} \tag{8}$$

$$\theta \equiv \frac{P_{abs}(dn/dT)}{\lambda k} \tag{9}$$

P_{abs} is the absorbed laser power, which is proportional to the incident power P and the absorption coefficient α.

The first term in curly brackets in equation (8) will be negligible compared to the second, since $w \gg w_0'$. Equation (10) is an adequate approximation:

$$\frac{\Delta w}{w} = \pm\left(\frac{-\theta}{1+t_c/2t}\right)\left(\frac{w_0}{w}\right)\left[1-\left(\frac{w_0^2}{w^2}\right)\right]^{1/2} \tag{10}$$

Equation (10) maximizes if $w = 2w_0$, that is, if the lens is generated one confocal distance beyond the waist of the input beam. Equation (10) minimizes if the lens is generated one confocal parameter before the focus of the beam. Equation (10) predicts that the thermal lens cannot be observed by its effect on the input laser beam if the sample is placed at the focus of the input beam.

The thermal lens is measured by placing an aperture in the beam at some distance from the sample cell. The change in the laser intensity passed through the aperture is a measure of the change in the square of the beam radius. It can be shown that equation (11), essentially the square of equation (10), applies:

$$\frac{I(t)}{I(0)} = \left(1-\frac{\theta}{1+(t_c/2t)}+\frac{\theta^2}{2[1+(t_c/2t)^2]}\right)^{-1} \tag{11}$$

$$t_c = \frac{w_0^2}{4D} \tag{11a}$$

Equation (11) relates the thermal lens behavior to the sample absorbance and to its thermal and optical properties. At sufficiently low absorbance and laser power, only the linear term is significant and the time-dependent intensity through the aperture is directly proportional to absorbance. However, if either the laser power or absorbance is sufficiently large, the full quadratic dependence becomes important.

In many thermal lens experiments, a linear concentration/signal relation is assumed, or a component of the signal which contains only the linear response is measured. However, quadratic response is inherent in any experiment which depends on heat conduction.

The quantity $\theta/\alpha = (dn/dT)/\lambda k$ is often called E, the enhancement factor in the analytical literature (5). The enhancement factor is just the relative thermal lens response per unit of laser power. Table 1 lists enhancement factors for several common solvents. From the table it is clear that water is about the least suitable solvent for a thermal lens experiment, whereas nonpolar solvents such as carbon tetrachloride are quite good. The relatively poor sensitivity of water results from both a large thermal conductivity and a small refractive index temperature coefficient. The methanol–water mixtures commonly used as reverse-phase liquid chromatographic solvents will typically have enhancement factors in the range 0.5–1 per milliwatt of laser power.

Table 1. Thermo-optical Constants and Enhancement Factors for Common Solvents

Solvent	k (mW/cm · K)	n	$10^4 dn/dT$ (K^{-1})	E (mW^{-1})
H_2O	6.11	1.334	-0.8	0.09
CH_3OH	2.01	1.311	-3.9	1.33
$(CH_3)_2CO$	1.60	1.362	-5.0	2.16
C_6H_6	1.44	1.504	-6.5	3.05
CCl_4	1.02	1.46	-5.8	3.88

The thermal lens model proposed by Gordon et al. (10) assumes that the refractive index distribution in the laser beam is parabolic. This assumption leads, ultimately, to simple equations such as equation (11), as well as analogous equations for other experiments. A laser beam has a Gaussian intensity profile, not a parabolic profile. The parabolic expression is used for mathematical convenience and, ultimately, because it is satisfactory except at very high absorbances.

The Gaussian intensity profile can be retained, and approximations can be introduced elsewhere in the model, describing the aberration in the thermal lens more fully. The somewhat more complicated equations describe a strong thermal lens better than the basic parabolic lens approximation (12). An empirical equation, which inserts some of the features of the aberrant lens into the parabolic model equations, describes strong thermal lens behavior better than either one alone (13). For applications to chemical analysis, the simpler parabolic approximation is adequate.

2.3. The Dual-Beam Thermal Lens

The most common experimental configuration is the "pump and probe" technique, introduced by Long et al. (14). The CW pump laser is modulated at a low frequency, usually in the range 5–100 Hz, to form a modulated thermal lens. A simple mechanical chopper is usually employed. The thermal lens is detected by its effect on a much weaker CW probe laser beam, which propagates collinearly with the pump laser. Figure 2 shows a typical apparatus. The thermal lens signal is extracted with a lock-in amplifier, which is referenced to the pump beam.

The theory of the repetitively chopped dual-beam thermal lens has been derived by Swofford and Morrell (15) as an extension to the theory of the time-

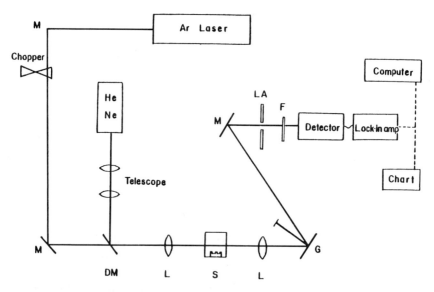

Figure 2. Pump–probe thermal lens apparatus. M, beam steering mirrors; DM, dichroic mirror; L, lens; S, sample cell; G, diffraction grating; LA, limiting aperture; F, filter. Reproduced, with permission, from ref. 55.

dependent single-beam lens. For the chopped beam case, it is necessary to account for the "memory" effect of previous chopper cycles. That is, the lens from earlier cycles has not completely died away during a given cycle. Thermal lens effects caused by the probe laser are assumed to be negligible. This condition can be satisfied by judicious choice of probe wavelength and by keeping probe laser power small compared to pump laser power.

For a chopper with a 50% duty cycle, operating at frequency f, equation (12) applies:

$$\frac{\Delta I}{I_0} = 2z' \frac{\partial n}{\partial T} \frac{\alpha P}{\pi_1^2 Jk} \left(\frac{-1}{1 + ft_c} \right)$$ (12)

Equation (12) predicts that the lock-in amplifier output signal will be proportional to the absorbance of the sample and to the laser power. At low frequency a limiting signal is expected, whereas at frequencies greater than $1/t_c$ an inverse dependence on chopper frequency is predicted.

Carter and Harris (16) point out that equation (12) neglects the quadratic component of the concentration dependence of the thermal lens. This dependence generates a signal at the second harmonic of the chopper frequency. The nonlinearity is less than 3% below absorbance 0.001 and increases to 31% at absorbance 0.05. Consequently, working curves are nonlinear above absorbance 0.001 and actually go through a maximum at sufficiently high absorbance. The linear dynamic range of the pump–probe technique can be increased by decreasing the pump laser power. However, the dynamic range is obtained at the cost of decreased sensitivity.

Equation (12) describes only the thermal lens formed by the heating laser. No assumptions are made about the character of the probe beam. The probe should be well-matched to the pump beam. Maximum sensitivity will occur if the probe beam has the same diameter and divergence—or equivalently, the same focused diameter and confocal parameter—as the heating beam. This condition is rarely completely satisfied.

Carter and Harris (16) have derived a similar approximate description of the pump–probe system, explicitly including the effects of the probe laser. They show that sensitivity, relative to a single laser experiment, is given by equation (13). The subscripts s and p refer to the heating and probe lasers, respectively. Because $\partial n / \partial T$ is a slowly varying function of wavelength, the magnitude of the observed signal is inversely proportional to probe laser wavelength:

$$\frac{E_p}{E_s} = \frac{(\partial n / \partial T)_p \lambda_s}{(\partial n / \partial T)_s \lambda_p}$$ (13)

Although short-wavelength probe lasers are preferred in principle, small He–Ne lasers are almost universally used as probe lasers. Only if the sample itself absorbs at 633 nm is another probe laser, usually argon ion, employed.

Fang and Swofford (6) point out that the strongest thermal lens is formed if the heating beam is focused into the sample. The maximum sensitivity to change occurs if the probe is focused one confocal distance before the heating laser. This condition can be met only by independent focusing of probe and heating lasers. However, the sensitivity increase is only a factor of 2 (21). Few workers appear to have bothered with the added complication of independent beam focusing to achieve this relatively small increase.

A larger thermal lens signal can be obtained in longer cells (18). A limiting signal is achieved in a cell longer than about 10 confocal distances. The reason is that the lens becomes weaker as the heating beam becomes larger before and after focus. Note, however, that the existence of a distance-independent signal is useful. This behavior means that strong signals can be achieved in short-path cells if tightly focused lasers are used.

2.4.　The Pulsed Thermal Lens

The thermal lens can also be formed with a pulsed laser. An independent probe laser is necessary. The lasers are assumed to have identical waists and divergences. Absorption by the probe beam is assumed to be negligible.

The thermal lens can be described by simple equations (19) if one further assumes that the formation time of the lens is negligible compared to its decay time and that the laser pulse itself is infinitely short. The lasers used in thermal lens measurements have pulse durations of 1–20 ns. The rise time, t_r, of the thermal lens is the time it takes an acoustic pulse to travel the radius of the focused laser at the speed of sound. The rise time is given by equation (14), where v_s is the speed of sound in the medium:

$$t_r = v_s/w_0 \tag{14}$$

Because the lasers are focused to a radius of 10–100 μm, and the speed of sound in most liquids is in the range of 900–1700 m/s, the rise time is typically 10–100 ns. The decay time is t_c, about 1–10 ms. The assumptions are valid.

From these assumptions, Twaroski and Kliger (19) derive the focal length of the pulsed thermal lens:

$$\frac{1}{f(t)} = \frac{1}{F_0}\left[\left(1 + \frac{2t}{t_c}\right)^{-2} \right] \tag{15}$$

$$\frac{1}{F_0} = \frac{\alpha(Q/0.5t_c)l(\partial n/\partial T)}{Jkw_0^2} \tag{16}$$

In equations (15) and (16), Q is the energy in the laser pulse and F_0 is the initial focal length of the thermal lens, at the time the laser pulse heats the sample.

The initially strong lens decays with a time dependence that is ultimately t^{-2}. Note that the initial focal length is the same as that produced in a steady-state thermal lens experiment by a CW beam of power $P = Q/0.5t_c$.

For the thin lens approximation the optimum position for the cell is approximately 0.707 confocal parameter beyond the focal point of the focusing lens, at a distance $z' \simeq 0.707b$. In this case, the ratio of the change in beam intensity observed through a limiting aperture, relative to the unperturbed $(t \to \infty)$ intensity, is given by

$$\frac{\Delta I_{bc}}{I_{bc,t \to \infty}} \simeq \frac{-2z'}{F(t)} \qquad (17)$$

The value of t_c is generally a few milliseconds. If the pulse repetition rate is more than 100–200 pulses per second, the system is effectively a CW system. Small thyratron-triggered excimer or nitrogen lasers can be pulsed that rapidly, and some metal vapor lasers operate at pulse rates of 10^3–10^4 per second. However, most pulsed lasers actually employed in thermal lens systems do not behave as pseudo-CW devices.

2.5. Path-Length Dependence of the Thermal Lens

In almost all photothermal measurements, the signal is generated by using a focused laser beam. Because the signal depends on the power density in the laser beam, photothermal signals do not increase linearly with cell length (18). This behavior is generally observed for any phenomenon that depends upon power density.

The length dependence is given by equation (18), where z_a is the beam diameter at the entrance of the cell of length b and where $z_c = w_0^2 \lambda/n$ is the confocal parameter:

$$\frac{\Delta I}{I} = \frac{\theta z_c}{b} \ln\left[\frac{z_c^2 + (z_b + b)^2}{z_c^2 + z_a^2}\right] + \frac{\theta^2 z_c}{b} \arctan\left[\frac{bz_c}{z_c^2 + z_a^2 + bz_a}\right] \qquad (18)$$

Equation (18) is valid only in the limit of a weak thermal lens. Generalized equations do not yield a closed-form solution, although a numerical approximation provides a good fit to experiment (18).

In general, a limiting response is obtained for a cell whose length is many confocal parameters long. Maximum sensitivity is obtained when the experiment is operated under thin lens conditions, with a focusing lens that gives a confocal parameter that is long compared to the cell path length.

Carter and Harris (20) have considered the implications of focused beams for small volume measurement. They suggest that a reasonable maximum path length of b_{max} is less than $z_c/3$. The critical time for the system depends on w^2. For this value of b_{max}, the beam diameter remains close to the minimum and the critical time varies over a range of no more than $\pm 15\%$. This condition provides a single time constant for time-resolved measurements and a constant relative contribution from linear and quadratic components of the thermal lens response.

Assuming that the beam remains truly cylindrical, the boundaries of the region generating the thermal lens can be calculated. A cell length of $z_c/3$ can be accommodated in a volume defined by a cylinder of radius $w = 2w_0$. The minimum sample volume is given by

$$V_{min} = \pi(4w_0)^2 b = 16z_c \lambda b \geq 48 \lambda b^2 \tag{19}$$

The implication of equation (19) is that the tighter focusing improves the mass sensitivity in a tightly focused short-path cell system, relative to a loosely focused long-path cell system, because the volume increases quadratically with path length.

Absorbance detection limits should be independent of volume, because absorbance detection limits are really heat-detection limits. In practice, Carter and Harris find that absorbance detection limits decrease by a factor of 2 when the cell volume is reduced from 2.6 μl to 29 nl. They suggest that the improvement results from two factors. First the tightly focused beam diverges more rapidly. The detector can be placed close to the sample and is less sensitive to pointing errors and misalignment. Second, the tightly focused beam should be less sensitive to thermal or refractive index gradients in the sample.

These properties of photothermal detectors are the basis of their utility for LC detectors—electrophoresis densitometers, for example. Strong signals can be obtained for tightly focused beams in these short-path-length, low-volume systems.

2.6. Transverse Photothermal Deflection

The probe beam angular deflection of the PDS signal in the medium over a sample of length L is given by equation (21).

$$\theta = \frac{L}{n} \frac{dn}{dT} \frac{dT}{dx} \tag{20}$$

In this equation, n is the refractive index of the medium where the monitoring beam is deflected, and dT/dx is the temperature gradient in the same medium.

The observed signal intensity at the detector, for small deflection angles, is

$$S = \Psi l_d \theta \tag{21}$$

where Ψ is a constant depending on the detector sensitivity and gain, and l_d is the distance from the sample center to the detector (lever arm).

A rigorous calculation of the beam deflection for PDS requires calculation of the temperature distribution within the sample and the resulting temperature distribution in the adjacent fluid. The equations describing the beam propagation through this inhomogeneous medium must then be solved.

Murphy and Aamodt (21) have presented an exact theory for the transverse PDS, assuming a homogeneous sample. Their theory describes both thermal and acoustic contributions to the observed response. Acoustic waves are important only at high modulation frequencies.

Mandelis has developed a simple one-dimensional theoretical model for the transverse PDS effect which emphasizes the relationship between the photothermal signal and the optical absorption coefficient of the solid sample under investigation (22). In this model, acoustic waves are not considered, so the derived expressions are simpler than in the Murphy–Aamodt treatment.

Briefly, the model considers the thermal diffusion equations in the coupling gas, sample, and backing. The thermal diffusion equations are

$$\frac{\partial^2}{\partial y^2} T_i(y, t) - \frac{1}{\alpha_i} \frac{\partial}{\partial y} T_i(y, t) = \begin{cases} \dfrac{I(y, t)}{\kappa_s}, & \text{sample} \\ 0, & \text{gas, backing} \end{cases} \tag{22}$$

where α_i is the thermal diffusivity [$i = g$(gas), s(sample), b(backing)]. $I(y, t)$ is the laser intensity at point y.

The heating rate of the sample assuming a quantum efficiency of 1 for nonradiative relaxation is

$$I(y, t) = \tfrac{1}{2} I_0 \beta e^{-\beta y} \operatorname{Re}(1 + e^{i\omega\tau}) \tag{23}$$

where β is the sample absorptivity, I_0 is the intensity of the high incident in the sample, and ω is the modulation frequency. The ac component of the temperature in the gas region is given by

$$T_g(y, t) = \frac{A}{\beta^2 - \sigma_s^2}$$

$$\frac{(r-1)(b+1)e^{\sigma_s L_s} - (r+1)(b-1)e^{-\sigma_s L_s} + 2(b-r)e^{-\beta L_s} + 1}{(\eta+1)(b+1)e^{\sigma_s L_s} - (\eta-1)(b-1)e^{-\sigma_s L_s}} e^{-\sigma_g y + i\omega t} \tag{24}$$

where $b = 1/\xi$; $A = I_0\beta/2\kappa_s$; $\sigma_i = (1 + i)a_i$, where $a_i = (\delta/2\alpha_i)\ 1/u_i$ is the thermal diffusion coefficient of the material i; $r = (1 - g)(\beta/2a_s)$; $g = a_g k_s/a_s k_s$, where $a_s = (jw/\alpha_s)^{1/2}$; $\mu_s = a_s^{-1}$; α_i is the diffusivity [i = g(gas), s(sample), b(backing)]; where L = pump beam and sample surface cross-section diameters; k_i = thermal conductivity [i = g(gas), s(sample), b(backing)]; L_s = sample thickness; and $t = a_g/c_0^2$, where c_0 is the acoustic velocity in the gas.

The trajectory of the probe beam through the refractive index gradient in the gas and for small angles is given by

$$n_0 \frac{\partial^2}{\partial x^2} = \frac{\partial}{\partial y} n(y, t) \tag{25}$$

Integrating equation (25) and combining with equations (23) and (24) yields the ac component of the deflection angle for a thermally thick sample equation (26):

$$\theta_{ac}(y_0, t) = \frac{LI_0\beta}{2T_0\kappa_s} \text{Re} \left\{ \frac{\sigma_g(r-1)}{(\beta^2 - \sigma^2)(\eta + 1)} e^{-\sigma_g y_0 + i\omega t} \right\} \tag{26}$$

This equation contains the probe beam offset from the solid–gas interface.

Using polar coordinate rotation, both the phase (ψ) and the amplitude (θ) of the photothermal signal can be written as a function of the product $\beta\mu_s$, which is a material property:

$$\psi(y_0, \beta\mu_s) \simeq \alpha_\sigma y_0 - \tan^{-1}\left(\frac{2}{\beta\mu_s}\right)^2 - \tan^{-1}\left(\frac{\beta\mu_s}{2} - \beta\mu_s\right) \tag{27}$$

$$\theta_{ac}(y_0, \beta\mu_s) = \frac{LI_0}{T_0\kappa_s}\left(\frac{\alpha_s}{\alpha_g}\right)^{1/2}\left(\frac{1}{2}\frac{(\beta\mu_s)^2 - \beta\mu_s + 1}{(\beta\mu_s)^2 + 4/(\beta\mu_s)^2}\right)^{1/2} e^{-\alpha_g y_0} \tag{28}$$

For an optically thin, thermally thick sample, equation (28) simplifies to

$$\theta_{ac} = \frac{LI_0}{T_0 k_s}\left(\frac{\alpha_s}{\alpha_g}\right)^{1/2}\frac{1}{2}\beta\mu_s l^{-\alpha_g y_0} \tag{29}$$

2.7. Spatially Multiplexed Transverse Photothermal Deflection

Pulsed laser photothermal techniques are difficult to employ with solid samples. Gross thermal decomposition is common, especially with high-peak power pulsed lasers. Less obviously, photochemical transformations, such as cis–trans isomerization, may be quite facile under laser illumination. Since photochemical reaction may cause absorbance changes without obvious changes in color, systematic errors from that source can go completely undetected.

It is important to reconcile the need for high laser power or energy to maximize signal to noise with the need for gentle irradiation to minimize unwanted sample transformations. This problem has been recognized by several groups working in both photoacoustic and transverse photothermal deflection spectroscopies.

Distribution of the laser energy over the total sample surface is a practical way to minimize or avoid the thermal damage and retain the inherent sensitivity of PDS. Coufal et al., have demonstrated the feasibility of power distribution in photoacoustic spectroscopy using both Hadamard (23) and Fourier (24) coding of the expanded excitation beam. Hadamard masks are simpler to construct, whereas Fourier (sinusoidal) masks allow variable resolution from a single mask.

Fotiou and Morris introduced the use of Hadamard coding for transverse photothermal densitometry, employing both pulsed (25) and CW (26) lasers as light sources. The principle is shown in Figure 3. In their work, an encoded laser beam was line-focused and used to irradiate samples on a stationary TLC plate to create refractive index gradients for transverse photothermal deflection measurements.

Hadamard transform photothermal deflection spectroscopy exploits the property of the photothermal deflection signal that the signal generated along a line is the sum of the incremental signals generated at each point along the line. The excitation beam is passed through a mask that encodes it. The encoded beam is line focused along the direction of the probe beam. Therefore, the signal is sum of signals from each illuminated resolution element along the line of the probe beam. A resolution element is defined by the image of the unit aperture of the mask.

If a suitable sequence of masks is chosen, it is possible to make a series of measurements from which the signals at each resolution element can be recovered. When a mask consisting of opaque and transparent slits intercepts the pump beam, the measured signal generated by mask j is given by

$$y_i = \sum_{i=1}^{n} s_{ij} x_i \qquad (30)$$

where y_j is the signal from the jth mask, $s_{ij} = (s_{1j}, s_{2j}, \ldots, s_{nj})$ is the matrix of mask elements of the jth mask, and x_i is the signal at the position i on the sample. The element s_{ij} has the value 1 if the corresponding ith resolution

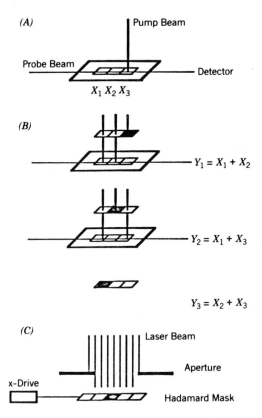

Figure 3. Hadamard transform transverse photothermal deflection imaging. (*A*) Conventional technique sampling element X_3; (*B*) Hadamard encoding with a line-focused pump beam. (*C*) Hadamard encoding with a cyclic mask. Reproduced, with permission, from ref. 56.

element is illuminated and 0 if it is not illuminated. If the n masks are properly chosen, they define a system of n linear independent equations that completely describe the system:

$$y_1 = \sum_{i=1}^{n} s_{i1} x_i \tag{31a}$$

$$y_2 = \sum_{i=1}^{n} s_{i2} x_i \tag{31b}$$

$$y_n = \sum_{i=1}^{n} s_{in} x_i \tag{31c}$$

In matrix notation, the set of equations (31) may be written as

$$\mathbf{Y} = \mathbf{S} \cdot \mathbf{X} \tag{32}$$

The system can be solved by calculating the inverse of the matrix $\mathbf{S}$, according to

$$\mathbf{X} = \mathbf{S}^{-1} \cdot \mathbf{Y} \tag{33}$$

The inverse $\mathbf{S}^{-1}$ is computationally simple to generate and is given by

$$\mathbf{S}^{-1} = \frac{2}{n+1} \cdot \mathbf{W} \tag{34}$$

Here $\mathbf{W}$ is a matrix that has -1's where $\mathbf{S}^T$ has 0's and $+1$'s where $\mathbf{S}^T$ has -1's (T stands for transpose).

There are three known constructions for S-matrices that yield cyclic matrices (27). A cyclic matrix has the property that each row is generated from the previous one by shifting its elements one position to the left (or right) and placing the overflow in the position of the element that was first shifted. The advantage of using a cyclic matrix is that a single mask of $2n - 1$ slits can be used to generate the configurations of all n individual masks. This mask is called a *Hadamard mask*. In operation, the Hadamard mask is shifted incrementally underneath a limiting aperture (or frame) a distance of one slit width. Each shift of the mask generates another row of the S-matrix. Shifting the mask $n - 1$ times generates the entire S-matrix sequentially.

The modulation transfer function (MTF) of a Hadamard encoded photo-thermal system has been investigated by Fotiou and Morris (28). They demonstrated that if the pump laser beam is adequately collimated, photo-thermal measurements are made in the near field. In this case, a mask behaves as a system of independent apertures of unit width rather than as a far-field ensemble of apertures. For masks with coarse unit aperture ($d \geq 20 \ \mu m$), the largest contribution to the transfer function is convolution with the aperture width, which causes a linear decrease in contrast with spatial frequency. If the mask is continuously moved rather than incrementally shifted, then a blur contribution to the MTF becomes important, and the transfer function becomes parabolic in spatial frequency. Fotiou and Morris were unable to observe a thermal diffusion contribution to the MTF. Under their experimental conditions, thermal diffusion should be a small contributor to the MTF.

Like any multiplexing technique, Hadamard masking has the property that A/D converter resolution is divided between representation of multiple channels and representation of signal intensities. If a 2^j element mask is employed, and the signal dynamic range in each channel is 2^k, and one bit is used to represent noise, then equation (35) describes b, the number of bits required to digitize the Hadarmard-encoded photothermal image (28):

$$b = j + k - 4 \tag{35}$$

Equation (35) has been verified for photothermal signals obtained from 63-, 127-, and 255-element mask systems. The A/D converter resolution requirements of multiplexed Hadamard encoded signals are surprisingly small. Dynamic range compression of the signals results from the fact that high resolution is required only to encode small changes in the shape of the signals. Limiting the A/D converter range usually smoothes the signal somewhat and may lead to loss of resolution if conditions of equation (35) are violated.

In practice, the signal dynamic range rarely exceeds 4 or 5 bits. A standard 12-bit converter is adequate to handle signals from a 4095-element mask. A 16-bit converter would be needed to handle masks containing up to 16 K elements with sufficient data dynamic range to be useful. It is unlikely that any transducer now available has sufficient linear dynamic range to allow greater resolution than 16 bits.

3. APPLICATIONS OF PHOTOTHERMAL TECHNIQUES

3.1. Liquid Chromatography Detectors

Photothermal detectors for liquid chromatography are the most highly developed analytical application. Photothermal techniques provide low detection limits at moderate cost combined with ready adaptability to submicroliter flow cells. Applications to liquid chromatography have been recently reviewed (8).

Absorbance cells for liquid chromatography deliberately include provision for introducing some nonuniformity into mobile-phase flow. Typically, the liquid entering the cell is directed onto one window, the exit flow is at an angle to the other window. This "Z" configuration ensures that there are no stagnant regions of liquid into the cell to degrade resolution. Sample mixing is not actually complete during the residence time of the cell (29). The mixing action flushes the cell more rapidly than unperturbed laminar flow but makes local concentration a strong function of position. This behavior can lead to systematic errors in thermal lens measurements, which probe a small fraction of the cell volume. There are other adverse consequences as well. The sudden changes in flow direction introduce high-frequency components in several directions into the flow velocity to induce mixing. The mixing process breaks up thermal lenses and becomes a source of excess noise.

These problems were recognized early in the development of photothermal and photoacoustic detectors. Many early reports in both fields are devoted to attempts to describe and circumvent the problems. Oda and Sawada (30) studied the modulation frequency dependence of photoacoustic signals and signal/noise ratio in a liquid chromatographic detector. Their cell, which employed a "U" flow geometry, was noisy at low frequencies but was quiet above 2 kHz, when pumped with a piston pump.

The problem can be alleviated with pumps that produce smooth, rather than pulsating, flow. Buffett and Morris (31) found that the combination of a Kratos SFA-234 flow cell (which employs a "Z" flow geometry) and a dual piston pump with pulse dampener could not be successfully employed in thermal lens detectors. The same cell performed well when used with a syringe pump, which produces pulseless flow. Because they minimize problems with flow fluctuations, syringe pumps are most commonly employed with thermal lens detectors.

Pang and Morris (32) demonstrated that flow-related noise could also be reduced by using matched sample and mobile-phase blank flows in a series differential thermal lens configuration. To the extent that flow patterns in the cells are matched, their effects can be made to cancel. Imprecise measures of flow matching were used, but noise reduction by a factor of 2.5–4 was observed. However, this approach doubles solvent consumption and would only be practical for use with microbore or capillary columns, in which flow rates are 50 μl/min or less.

Mobile-phase flow is equivalent to an increase in thermal conductivity of the system (33). Over the limited flow range encountered in liquid chromatographic practice, the change in thermal lens enhancement factor with flow velocity, dE/dv, is constant. Flow-rate fluctuations are a source of noise. At low frequencies, the noise in a flowing thermal lens system can be described as the sum of a static contribution and an independent contribution which depends on flow-rate fluctuations:

$$\sigma^2 = \sigma_0^2 + A^2 (dE/dv)^2 \sigma_v^2 \tag{36}$$

The standard deviation in the flow system is σ, the component from the stationary thermal lens is σ_0, and the component from flow-rate fluctuations is σ_v. E is the thermal lens enhancement factor, A is the absorbance, and v is the mobile-phase flow velocity. Equation (36) is an incomplete description of the thermal lens in a typical liquid chromatography cell, since it accounts for pumping fluctuations only and does not account for the perturbations caused by the flow patterns within the cell.

Leach and Harris (34) were the first to use the thermal lens effect for liquid chromatography detection. They employed a time-resolved single-beam measurement with an argon ion laser (190 mW) and an 18-μl 10-mm-path-length flow cell. Using a reverse-phase system (50:50 methanol–water) and a 250- × 4.6-mm column packed with ODS on 5-μm silica, they demonstrated noise levels of about 1.5×10^{-5} absorbance, with a 5-s time constant. These early results were about an order of magnitude better than absorbance detectors of the same period.

Buffett and Morris (35) introduced the two-laser (pump–probe) configuration to LC detection. They also employed a conventional reverse-phase system (LiChrosorb RP-18 on 10-μm silica, 80:20 methanol) and a 10-mm-path-length flow cell. They were able to obtain absorbance noise of about 2×10^{-6} with a 1-s time constant. Their performance improvement over the Leach–Harris system can be attributed to two factors. First, the use of an 80:20 methanol mobile phase provided a larger enhancement factor. Second, with the pump–probe system, they were able to use synchronous demodulation to recover the thermal lens signal. Leach and Harris were forced to use a simplified data extraction algorithm that achieved real-time performance in their time-resolved system at the expense of signal/noise ratio.

Buffett and Morris (31) demonstrated that high laser power was not needed for low-noise liquid chromatography detection. They were able to reduce the pump beam power from 100 mW to 7.5 mW with no degradation in S/N, if they maintained a pump/probe ratio of 30:1 or higher, as shown in Figure 4.

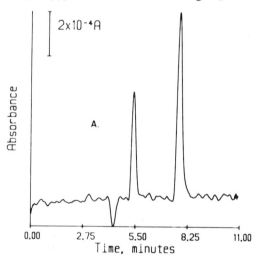

Figure 4. Chromatograms of O-nitroaniline (0.28 ng) and N,N-dimethyl-3-nitroaniline (1.7 ng). (A) 75-mW pump beam; (B) 7.5-mW pump beam; (C) 0.75-mW pump beam. Reproduced, with permission, from ref. 31.

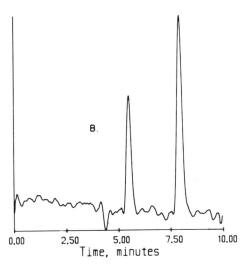

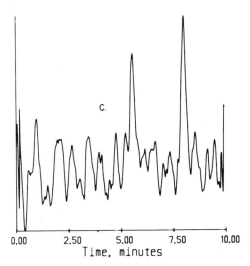

Figure 4 (*Continued*)

Since the demonstration of the advantages of the pump–probe configuration for liquid chromatographic detection, most workers have employed
variants on this technique. In addition to thermal lens measurements,
photothermal deflection with a pump–probe beam crossed at a small angle
(36) or at right angles (37) have been employed in liquid chromatographic
measurements. Yang and Ho (38) have demonstrated that these techniques
give essentially similar performances.

Several groups have attempted single CW laser variants on the pump–probe configuration. Typically a small fraction of the laser beam is used as the probe. Polarization encoding (39, 40) and counterpropagating beams (41) are the most common versions. Alternatively, the thermal lens signal can be detected on the modulated pump laser itself at the second harmonic of the modulation frequency (42). These configurations eliminate the alignment instability problems of the two laser designs. Unless the beams are crossed or counterpropagating, most of these schemes place extraordinary demands on the rejection properties of polarizing prisms or lock-in amplifiers. They provide somewhat worse performance than the traditional configurations.

Skogerboe and Yeung (43) have used high-frequency (150 kHz) acousto-optic modulation to direct a single laser beam alternately through and around a flow cell to the same photodiode, to achieve a pseudo-steady-state measurement by lock-in detection, without the disadvantages associated with some of the earlier single-laser designs.

Early photothermal detectors achieved noise levels equivalent to 10^{-5}–10^{-6} absorbance (34, 35). More recent designs have demonstrated performance approaching 10^{-7} absorbance (36, 37). The improvement results largely from careful choice of pump and flow cells to minimize noise associated with these components, rather than from improvements in the photothermal experiment itself.

Most research on photothermal detectors has focused on detector design rather than application. Nitroanilines have been common substrates for these studies (34, 35). Nitroanilines absorb fairly strongly at the blue wavelengths of argon ion and helium–cadmium lasers, although their absorption maxima are actually in the 350–400-nm range.

Nolan et al. (37) have shown that DABSYL derivatives of amino acids can be detected at the femtomole level, using a microbore (1-mm i.d.) C-18 column and 60% methanol as the mobile phase, along with argon ion 488 nm to generate the thermal lens. Detection limits range from about 5 fmol for glycine to 300 fmol for methionine. The wide range of detection limits results from differences in the reactivity of the amino acids to the chromogenic reagent and differences in extraction efficiency. The absorption coefficients of the amino acid DABSYL derivatives at 488 nm vary by only a factor of 2.

Collette et al. (36) have used photothermal detection to quantify acid sulfonate rug dyes with about 500-fg (1–2 fmol) detection limits. A C-18 microbore column and 65% methanol were employed. They also successfully employed gradient elution to improve the separation. At 488 nm the solvent background is sufficiently low that the change in background is small compared to the size of the signals observed. However, detection limits are increased about an order of magnitude.

Although the need for UV photothermal detectors has been appreciated since the first applications, virtually all of the development has been carried out with visible lasers operating in the mid-visible range. Most researchers have preferred to use convenient CW lasers, primarily the argon ion laser, rather than the pulsed systems that are required for flexible access to the UV range.

Kettler and Sepaniak (44) have successfully used an excimer laser pumped dye laser in a photothermal refraction detector for capillary liquid chromatography. Their apparatus is shown in Figure 5. Measurements were made in a short section of 200-μm-i.d. silica tubing cemented to the end of their 50-μm-i.d. column. The dye laser was operated with 400-nm 30-μJ pulses at 23 pulses per second.

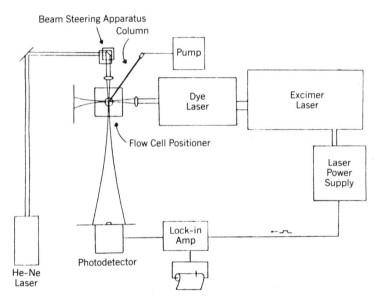

Figure 5. Apparatus for pulsed laser photothermal refraction detection for capillary liquid chromatography. Reproduced, with permission, from ref. 44.

Figure 6 shows the performance of the system to an extract of coal fly ash which had been coated with pyrene and exposed to NO_2. The linear dynamic range for 1-nitropyrene, the major component, is 4×10^{-6} to 6×10^{-4} M.

The absorbance detection limits for pulsed laser photothermal refraction are not particularly low—about 1×10^{-4} absorbance in these experiments. The pulsed laser is focused tightly to generate a good thermal lens. This, in turn, means that the laser must be used at low power to avoid burning the

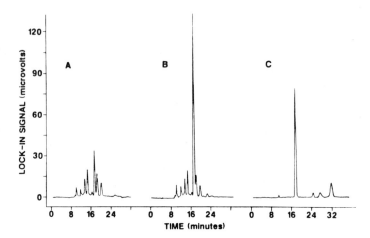

Figure 6. Chromatograms of synthetic sample coal ash extracts (*A* and *B*) and 1-nitropyrene and dinitropyrene isomer mixture (*C*), obtained with the apparatus of Figure 5. Reproduced, with permission, from ref. 44.

quartz capillary. The 30-μJ pulse energy is about 1% of the available output. The average power is only about 0.5 mW. At these levels, photothermal spectroscopies are not more sensitive than direct absorbance measurements.

A higher pulse rate would increase the average power without exposing the capillary to greater risk of damage. Recently available dye lasers operate satisfactorily at 200 pulses per second. More durable sample cells would also improve the performance of the system. Indeed, thick-wall capillaries do allow operation at somewhat higher power levels, with a 10-fold decrease in detection limits (44). The proposed sheathed-flow cells (44) are an interesting alternative.

3.2. Thin-Layer Chromatography Detectors

Morris and Chen (45) have used transverse photothermal deflection to quantify compounds separated by thin-layer chromatography. Detection limits in the picogram range were achieved for intensely absorbing derivatives of several quinones. Their system employed an argon ion laser to generate the mirage and mechanical scanning of the thin-layer chromatography plate in a stationary optical system. As with most mirage effect systems, satisfactory operation required shielding from laboratory air currents, which degraded signal/noise ratios by a factor of 10.

Pump–probe beam shaping has been shown to reduce detection limits about one order of magnitude, by allowing integration over local irregularities in the chromatogram (46). Using a one-dimensional $2\times$ beam expander to increase the probe area parallel to the plate, along with a defocused pump laser beam to match the probe cross section, Chen and Morris achieved a 1.3-ng detection limit for phenanthrenequinone, compared to 31 ng with an unexpanded beam. A similar improvement was obtained for 1,2-naphthoquinone. However, detection limits for α-ionone were unchanged, apparently because the limiting factor is partial decomposition of the ionone in the sulfuric acid reagent. The performance of this system is shown in Figure 7.

Peck et al. (47) have shown that careful optimization of experimental parameters, including (a) the angle between the probe laser beam and the TLC plate, (b) the probe beam offset, and (c) the pump beam modulation frequency,

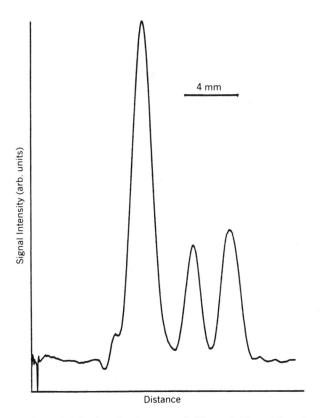

Figure 7. Photothermal deflection densitometry of 3.8 ng of 1,2-naphthoquinone, 46 ng of phenanthrenequinone, and 75 pg of α-ionone, visualized by sulfuric acid oxidation. Reproduced, with permission, from ref. 57.

are required to obtain maximum sensitivity from the photothermal densitometer. Using only 21-mW pump beam power, they have demonstrated that their system gives five times lower detection limits than commercial reflectance or transmission densitometers. They have also identified wobble in the mechanical stage as a serious source of excess noise, and they estimate that a further fourfold reduction in detection limits could be obtained with smooth stage travel.

Masujima et al. (48) have described a mirage effect densitometer using an intracavity probe arrangement. They obtained detection limits 20 times lower with the intracavity configuration than with a conventional extracavity design. Although the observed detection limits are similar to those of Peck et al. (47), the work suggests that a completely optimized intracavity probe design would produce extremely low detection limits.

Fotiou and Morris (49) have developed a densitometer based on a pulsed laser, which provides access to the UV range. They used a 351-nm XeF excimer laser in the mirage effect configuration, with gated integrator processing of the transient thermal refractive index gradient.

To avoid thermal damage, the laser pulse energy delivered to the sample was limited to 1–2 mJ. With this system they obtained a subnanogram detection limit and three-decade linearity for 2,4-dinitroaniline. In contrast to CW photothermal deflection, the pulsed system was found to be only moderately susceptible to laboratory air currents. Operation of the system without its enclosure increased noise signals by no more than fourfold. Even with millijoule pulse energies, detection limits were set by absorption by residual impurities on the plate and possibly by the organic binder on the plate. The background from these sources obscured small signals.

Pulsed PDS with UV lasers is sensitive and experimentally simple. The low power requirements demonstrate that the technique can be useful with relatively inexpensive pulsed lasers. Lower detection limits can be reached by careful pre-elution of plates to remove impurities. It may also be necessary to use plates with binders selected for lowest absorption at the wavelengths of the experiment. Even further improvements are possible by increasing the incident power to the sample. However, power increases must be made with caution, because photochemical transformations or photodecomposition are common. High average power is best achieved with low-pulse-energy–high-rep-rate lasers or by the use of distributed power, as described below.

3.3. Photothermal Detectors for FTIR

Photothermal deflection does not require a laser light source. Only the refractive index gradient, not its curvature, is probed. The intensity of the light source is important, but its spatial profile is not. Transverse photothermal deflection with an arc lamp source is widely used to study band gap energies of semiconductors, for example.

Low and co-workers (50, 51) introduced transverse photothermal deflection as an alternative to photoacoustic (PAS) detection for Fourier transform infrared (FTIR) spectroscopic measurement of irregular solids. Photothermal deflection has both advantages and drawbacks as compared to PAS.

Perhaps the major advantage of photothermal deflection is that it makes few demands on the physical form of the sample. The technique is capable of handling large irregular solids as well as finely divided powders. Simple mechanical stages can be used to position almost any sample at the focus of the spectrometer beam. PAS, on the other hand, requires that the sample and detector be completely enclosed in the cell; furthermore, PAS is limited to small samples. An elegant example is the simple cell shown in Figure 8, which can be evacuated and heated to 1000°C. Spectra for pyrolysis studies using this cell are shown in Figure 9.

For powdered samples, photothermal deflection is about a factor of 3 less sensitive than PAS, at least with the present state of the instrumentation (52). The major reason is that the photothermal signal is partially dissipated within the porous sample. This problem affects the PAS sample less. Despite this

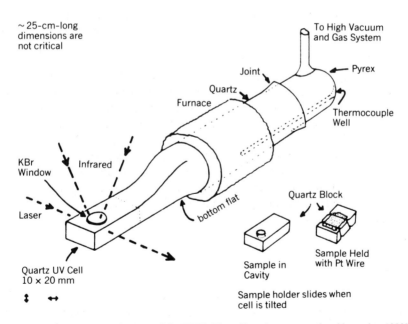

Figure 8. Photothermal deflection cell for FTIR. The cell can be evacuated and heated to 1000°C. Reproduced, with permission, ref. 54.

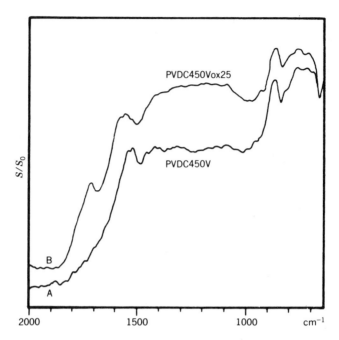

Figure 9. FTIR spectra following a pyrolysis of polyvinylidene chloride to 450°C (curve A) and following pyrolysis in apparatus of Figure 8 (curve B). Reproduced, with permission, from ref. 54.

problem, finely ground powders produce better photothermal signals than do coarsely ground powders or undivided solids (53). The powders present the largest surface area for absorption of radiation and generation of heat.

Photothermal deflection is also susceptible to noise pick-up from air movement, mechanical vibrations, and dust (54). These noise sources affect PAS measurements as well, of course. The problems have been minimized by careful engineering of sample cells. It is likely that the performance of photothermal deflection FTIR will also benefit by engineering attention to cell design. The technique may ultimately equal or surpass the sensitivity of photoacoustic detection.

4. SUMMARY

Photothermal detectors are a class of simple, but sensitive, absorbance detectors. They are ideally suited to low absorbance measurements on samples of limited volume or area. After several years of intensive development, the strengths and weaknesses of the devices are well understood.

Widespread adoption of photothermal techniques has been hindered by the cost, size, and inflexibility of many laser systems. The recent commercialization of diode laser pumped solid-state lasers—and their extension into the visible—suggests that the size problem is almost behind us. While these devices are still no less expensive than ion lasers, there is a good probability of major cost reductions in the near future. With a solution to the cost problem, we may expect a rapid increase in the popularity of photothermal spectroscopy.

REFERENCES

1. A. Rosencwaig, *Photoacoustics and Photoacoustic Spectroscopy*, Wiley, New York (1980).
2. I. Letokhov, *Laser Optoacoustic Spectroscopy*, Springer-Verlag, New York (1986).
3. C. K. N. Patel and A. C. Tam, *Rev. Mod. Phys.*, **53**, 517 (1981).
4. A. C. Tam, *Ultrasensitive Laser Spectroscopy*, in D. Kliger, Ed., Academic Press, New York (1983).
5. J. M. Harris and N. T. Dovichi, *Anal Chem.*, **52**, 692A (1980).
6. H. L. Fang and R. L. Swofford, in *Ultrasensitive Laser Spectroscopy*, D. Kliger, Ed., Academic Press, New York (1983).
7. M. D. Morris and K. Peck, *Anal. Chem.*, **58**, 811A (1986).
8. M. D. Morris, in *Detectors for Liquid Chromatography*, E. S. Yeung, Ed., Wiley, New York, pp. 105–147 (1986).
9. N. J. Dovichi, *CRC Crit. Rev. Anal. Chem.*, **17**, 357 (1987).
10. J. P. Gordon, R. C. C. Leite, R. S. Moore, S. P. S. Porto, and J. R. Whinnery, *J. Appl. Phys.*, **36**, 3 (1965).
11. C. Hu and J. R. Whinnery, *Appl. Opt.*, **12**, 72 (1973).
12. S. J. Sheldon, L. V. Knight, and J. M. Thorne, *Appl. Opt.*, **21**, 1663 (1982).
13. C. A. Carter and J. M. Harris, *Appl. Opt*, **23**, 476 (1984).
14. M. E. Long, R. L. Swofford, and A. C. Albrecht, *Science*, **191**, 183 (1976).
15. R. L. Swofford and J. A. Morrell, *J. Appl. Phys.*, **49**, 3667 (1978).
16. C. A. Carter and J. M. Harris, *Anal. Chem.*, **55**, 1256 (1983).
17. T. Berthoud, N. Delorme and P. Mauchien, *Anal. Chem.*, **57**, 1216 (1985).
18. C. A. Carter and J. M. Harris, *Appl. Spectrosc.*, **37**, 166 (1983).
19. A. J. Twaroski and D. S. Kliger, *Chem. Phys.*, **20**, 253 (1977).
20. C. A. Carter and J. M. Harris, *Anal. Chem.*, **56**, 922 (1984).
21. J. C. Murphy and L. C. Aamodt, *J. Appl. Phys.*, **51**, 4580 (1980).
22. A. J. Mandelis, *J. Appl. Phys.*, **54**, 3404 (1983).
23. H. Coufal, U. Moller, and S. Schneider, *Appl. Opt.*, **21**, 116 (1982).
24. H. Coufal, U. Moller, and S. Schneider, *Appl. Opt.*, **21**, 2339 (1982).
25. F. K. Fotiou and M. D. Morris, *Appl. Spectrosc.*, **40**, 704 (1986).
26. F. K. Fotiou and M. D. Morris, *Anal. Chem.*, **59**, 185 (1987).
27. M. Harwit and N. A. Sloane, *Hadamard Transform Optics*, Academic Press, New York (1979).

28. F. K. Fotiou and M. D. Morris, *Anal. Chem.*, **59**, 1446 (1987).
29. K. Peck and M. D. Morris, *J. Chromatogr.*, in press.
30. J. Oda and J. Sawada, *Anal. Chem.*, **53**, 471 (1981).
31. C. E. Buffett and M. D. Morris, *Anal. Chem.*, **55**, 376 (1983).
32. T.-K. J. Pang and M. D. Morris, *Anal. Chem.*, **57**, 2158 (1985).
33. N. J. Dovichi and J. M. Harris, *Anal. Chem.* **53**, 689 (1981).
34. R. A. Leach and J. M. Harris, *J. Chromatogr.*, **218**, 15 (1981).
35. C. E. Buffett and M. D. Morris, *Anal. Chem.*, **54**, 1821 (1982).
36. T. W. Collette, N. J. Parekh, J. H. Griffin, L. A. Carreira, and L. B. Rogers, *Appl. Spectrosc.*, **40**, 164 (1986).
37. T. G. Nolan, B. K. Hart, and N. J. Dovichi, *Anal. Chem.*, **57**, 2703 (1985).
38. Y. Yang and T. V. Ho, *Appl. Spectrosc.*, **41**, 583 (1987).
39. Y. Yang, *Anal. Chem.*, **56**, 2336 (1984).
40. T.-K. J. Pang and M. D. Morris, *Appl. Spectrosc.*, **39**, 90 (1985).
41. Y. Yang, S. C. Hall, and M. S. DeLa Cruz, *Anal. Chem.*, **58**, 758 (1986).
42. T.-K. J. Pang and M. D. Morris, *Anal. Chem.*, **56**, 1467 (1984).
43. K. J. Skogerkoe and E. S. Yeung, *Anal. Chem.*, **58**, 1014 (1986).
44. C. N. Kettler and M. J. Sepaniak, *Anal. Chem.*, **59**, 1733 (1987).
45. M. D. Morris and T. I. Chen, *Anal. Chem.*, **56**, 19 (1984).
46. T. I. Chen and M. D. Morris, *Anal. Chem.*, **57**, 1359 (1985).
47. K. Peck, F. K. Fotiou, and M. D. Morris, *Anal. Chem.*, **57**, 1359 (1985).
48. T. Masujima, A. N. Sharda, L. B. Lloyd, J. M. Harris, and E. M. Eyring, *Anal. Chem.*, **56**, 2975 (1984).
49. F. K. Fotiou and M. D. Morris, *Appl. Spectrosc.*, **40**, 700 (1986).
50. M. J. D. Low, C. Morterra, and A. G. Severdia, *Spectrosc. Lett.*, **15**, 415 (1982).
51. M. J. D. Low and M. LaCroix, *Infrared Phys.*, **22**, 139 (1982).
52. P. G. Varlashkin, M. J. D. Low, G. A. Parodi, and C. Morterra, *Appl. Spectrosc.*, **40**, 636 (1986).
53. M. J. D. Low and C. Morterra, *Appl. Spectrosc.*, **41**, 280 (1987).
54. M. J. D. Low, *Appl. Spectrosc.*, **40**, 1011 (1986).
55. J. P. Haushalter, Ph.D. thesis, University of Michigan (1981).
56. F. K. Fotiou, Ph.D. thesis, University of Michigan (1987).
57. T. I. Chen and M. D. Morris, *Anal. Chem.*, **56**, 1674 (1984).

IMMUNOLOGICAL METHODS FOR THE DETECTION AND QUANTITATION OF EXPOSURE TO AROMATIC HYDROCARBONS

REGINA M. SANTELLA and MARINA STEFANIDIS

Comprehensive Cancer Center and Division of Environmental Sciences
School of Public Health
Columbia University
New York

1. INTRODUCTION

Immunoassays have gained increasing acceptance as a viable method for quickly and easily quantitating an array of substances. They make use of the unique nature and high affinity of antibody–antigen reactions. Antibodies have been generated to a number of antigens, including hormones, bacterial proteins, and cell surface antigens as well as organic chemicals of environmental interest. The chemicals to which antibodies have been developed include dioxin (1, 2), aflatoxin (3–6), T-2 toxin (7–11), and aromatic hydrocarbons such as benzo[*a*]pyrene (12–15). These antibodies can be used in sensitive immunoassays for quantitating femtomole levels of these compounds in various biological samples. In addition, the ease with which these assays can be run makes them ideal for use on large numbers of samples and have helped in the expansion of the field of molecular epidemiology (16).

2. ANTIBODY DEVELOPMENT

Development of an immunoassay for detection and quantitation of a chemical starts with immunization of animals with the proper antigen. Low-molecular-weight organic chemicals are not immunogenic in themselves and must be coupled to a carrier protein. Proteins and polypeptides with molecular weights greater than 10,000 are generally good immunogens and can induce an antibody response. The immunogenicity of compounds below 5000 (called *haptens*) must be enhanced via chemical coupling to a carrier protein. Various proteins have been used as carriers, including bovine serum albumin (BSA), bovine gamma globulin, and keyhole limpet hemocyanin (KLH). Coupling of the chemical can be carried out by a number of different techniques; specific examples are given below.

Soluble antigens can be rapidly absorbed, circulated, and metabolized in animals. Adjuvants are used to increase the antigen lifetime and thus make it more immunogenic. The most common method involves emulsification of the antigen with an oil mixture called *incomplete Freund's adjuvant*. Complete Freund's adjuvant also contains killed mycobacteria, which makes the emulsion even more immunogenic. These oil emulsions ensure the slow absorption of the antigen over a period of weeks. Other adjuvants include hydrated aluminum hydroxide, inorganic colloidal suspensions, clays (bentonite), and charcoal (17).

Mice, rats, rabbits, and goats have all been used for antibody production and are immunized over the course of several weeks to months. Although there is no ideal protocol, potential advantages of different sites of injection and immunization schedules have been discussed (17). Either polyclonal or monoclonal antibodies are produced. For polyclonal antibody production, rabbits or goats are immunized and sera are collected. This polyclonal sera contains a number of antibody proteins recognizing different parts of the antigen (epitopes) with different affinities.

For monoclonal antibody production (18), mice or rats are immunized and sera are tested for specific antibody (see below). If specific antibodies are present, a cell suspension is made from the spleen of the immunized animal. These cells are then fused with special mutant myeloma cells deficient in the enzyme hypoxanthine guanine phosphoribosyl transferase (HGPRT), which is needed in the salvage pathway of purine biosynthesis. Since aminopterine inhibits normal purine synthesis, the myeloma cells are killed by selection in HAT (hypoxanthine, aminopterine, and thymine) media. Only hybrid cells that have received the normal HGPRT gene from the spleen cells, during the fusion process, will survive in HAT media. Unfused spleen cells die after several days in culture, whereas hybrids become immortal by fusion with the transformed myeloma cells. Some hybrids will have retained the capacity to produce the specific antibodies found in the animal sera. They must be identified by careful screening then cloned to obtain a pure population of cells. These monoclonal cells, resulting from the fusion of a single spleen cell, produce a single antibody protein. Cells can be cultured to obtain large amounts of antibody and can be frozen and regrown years later. Even larger quantities of antibody can be obtained by growing the cells as ascites tumors in the peritoneal cavity of the same genetic strain of animals from which the spleen cells were originally derived. As much as 10 mg/ml of antibody can be obtained by this procedure as compared to the approximately 10–100 μg/ml present in the culture media. Thus, a major advantage of monoclonal antibody production is the limitless amount of a uniform antibody that can be obtained.

3. IMMUNOASSAYS

A number of different types of immunoassays have been used both for the characterization of antibodies and quantitation of antigen. The two most commonly used assays are radioimmunoassay (RIA) (19) and enzyme-linked immunosorbent assay (ELISA) (20, 21). The general protocol for RIA is given in Figure 1. The antigen is radiolabeled and serves as the tracer. Non-radiolabeled antigen serves as the known inhibitor. A standard curve is generated by keeping the antibody and tracer constant and by adding increasing amounts of inhibitor. With increasing levels of nonlabeled antigen, decreasing levels of labeled antigen are bound by the antibody. To separate the bound tracer from the free tracer, the antibody is precipitated with a second antibody. For example, if the antigen-specific antibody is from a

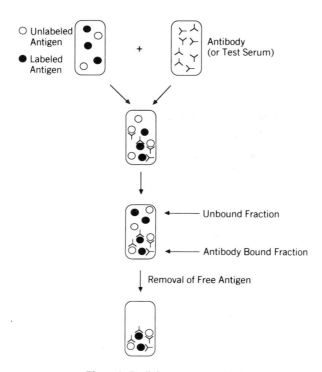

Figure 1. Radioimmunoassay (RIA).

rabbit, the second antibody could be a goat anti-rabbit IgG. Alternatively, ammonium sulfate, which precipitates antigen–antibody complexes, can be used. After centrifugation, the amount of radioactivity in the supernatant or precipitate is measured. Unknowns are mixed with the same amount of antibody and tracer and are quantitated from inhibition of tracer binding and the standard curve.

The first step in a competitive ELISA (Figure 2) is binding of the antigen to the wells of a 96-microwell plate. Since low-molecular-weight compounds cannot bind reproducibly, the hapten–carrier complex is normally used. These protein complexes can be adsorbed to plastic by simply incubating high-pH carbonate solutions of the protein in the wells. Competitive mixtures, made by mixing diluted antibody with serially diluted standards, are then added to the wells. Antibody not bound by antigen in solution, can bind to the antigen on the plate. After washing off nonbound antibody, the amount of bound antibody is quantitated by use of a second antibody coupled to an enzyme. For example, if the first antibody is a mouse monoclonal antibody, the second can be goat anti-mouse IgG antibody coupled to alkaline

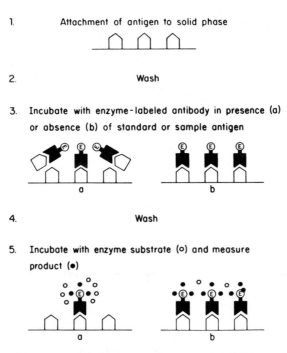

Figure 2. Competitive enzyme-linked immunosorbent assay (ELISA). Reproduced, with permission, from ref. 21.

phosphatase, peroxidase, or beta-glucuronidase. Finally, substrates that are colorless but become colored after enzyme cleavage are added to the well. For alkaline phosphatase, the substrate normally used is *p*-nitrophenyl phosphate, which is converted to *p*-nitrophenol after phosphate cleavage. With increasing amounts of antigen in the competitive mixture, decreasing amounts of antibody bind to the well, and lower levels of enzyme are present. The standard curve gives percent inhibition of antibody binding to the plate as a function of antigen concentration (Figure 3). Antigen levels in unknowns are determined from their percent inhibition of antibody binding and the standard curve. ELISAs are generally much more sensitive than RIAs, and various methods have been developed to further improve the sensitivity of the standard color assay. For example, it is possible to use substrates that become fluorescent after enzyme cleavage. With alkaline phosphatase, methylumbelliferyl phosphate, which becomes fluorescent after phosphate hydrolysis, is used (Figure 3) (22, 23). Alternate approaches use radiolabeled substrates, but hydrolyzed material must then be isolated for counting (24). The biotin–avidin technology has also been applied to immunoassays to increase sensitivities (25, 26).

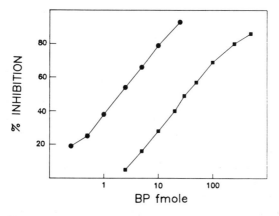

Figure 3. Competitive inhibition of polyclonal antibody binding to BPDE-I–DNA in an ELISA. Enzyme level is detected by substrates giving colored (■) or fluorescent (●) products.

4. ADVANTAGES

The low cost and ease of RIA and ELISA make them ideal for use on large numbers of samples. ELISA has the advantage over RIA that the use of radiolabeled compounds with their necessary safeguards are eliminated. Automated equipment is also available to wash 96-microwell plates, add reagents, and measure both color and fluorescence. A major consideration in the development of any assay is the sensitivity and specificity. High sensitivities can be obtained with immunoassays. Specific examples are discussed in detail below, but, in general, femtomole (10^{-15}) levels of detection are readily obtainable. A major area of difficulty frequently encountered is cross-reactivity of the antibody with structurally related compounds. This is also discussed in detail below and can make absolute quantitation of chemicals present in mixtures difficult. Total antibody binding to the sample will be the result of binding to multiple compounds with different affinities.

Two structurally similar compounds were quantitated with two cross-reacting antibodies, each of which preferentially reacted with one of the compounds (11). However, it is unlikely that this approach will work with complex mixtures. Alternatively, specificity can be increased by chromatographically separating the compounds before quantitation in the immunoassay. Although sufficient quantities are usually not available for UV or fluorescence detection, fractions can be collected blind, based on previous runs with standard compounds. The fractions can then be quantitated in the immunoassay. This technique has been used successfully in several DNA adduct studies (27, 28).

5. BENZO[a]PYRENE

Benzo[a]pyrene (BP) (Figure 4, I) is a ubiquitous environmental pollutant that is found in cigarette smoke (20–50 ng per cigarette), urban air (up to 100 ng/m^3), and smoked foods (100 μg/kg). In addition, occupational exposures are particularly high for coke oven workers, asphalters, and foundry workers. The reactive intermediate of BP responsible for its carcinogenic and mutagenic activity has been identified as 7,8-dihydroxy-9,10-epoxy-7,8,9,10-tetrahydrobenzo[a]pyrene (BPDE-I) (Figure 4, II). This diol epoxide is produced by enzyme activation of the parent compound in several steps as outlined in Figure 4. It is highly reactive and can bind to cellular macromolecules, including DNA, RNA, and protein (for review see ref. 29). Reaction with DNA is preferentially at the N-2 position of guanine (Figure 4, IV). This adduct has been identified as the major adduct in a number of cell culture and tissue culture system as well as in animals treated with BP (30–32).

Polyclonal and monoclonal antibodies have been developed to monitor levels of BP in human fluids and tissues. Polyclonal antibodies were developed in animals immunized with BP coupled to bovine serum albumin (12). BP containing an isocyanate group at the 6 position was conjugated to BSA via a carbamido linkage. An RIA, using 6-[^{3}H]BP as the tracer, could detect 0.4 pmol (0.1 ng) of BP. Significant cross-reactivity was seen with a number of

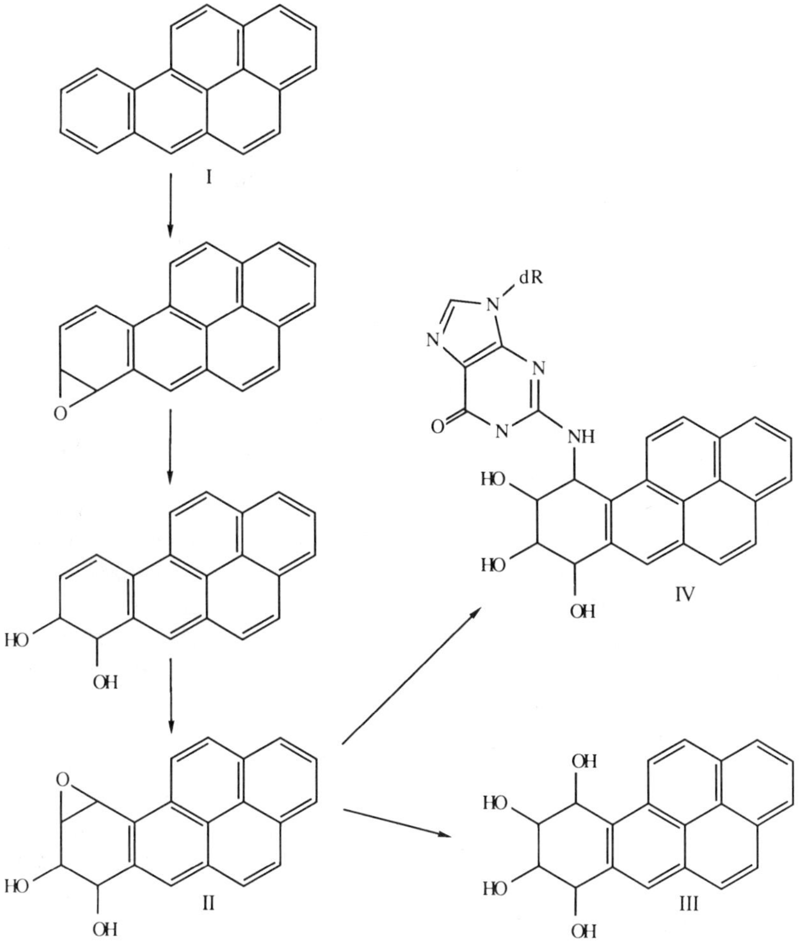

Figure 4. Chemical structure of benzo[a]pyrene (**I**), benzo[a]pyrene diol epoxide, BPDE-I (**II**), BPDE-I–tetraols (**III**), and the N-2 guanine adduct of BDPE-I on DNA (**IV**).

structurally related compounds, including 3-OH-BP, benz[a]anthracene, pyrene, phenanthrene, and anthracene. Although this antibody could be used to measure BP in a complex mixture without purification, absolute quantitation would be impossible because of this cross-reactivity. Total antigenicity of the sample would be the sum of antibody binding to a number of polycyclic aromatic hydrocarbons (PAHs), but with different affinities. Another group used a similar approach to produce polyclonal antibodies to BP by immunization with BP coupled to horse serum albumin, but this antibody has not been used in an RIA or ELISA (14).

A new approach to measurement of BP in solution takes advantage of fiber optics technology (15). Specific antibodies to BP were coupled to a fused silica optical fiber which was then immersed in the test solution. Bound BP was then excited with a helium–cadmium laser, and fluorescence, transmitted back through the fiber, was measured. Detection of as little as 1 fmol of BP in 5 μl of solution was possible.

6. DNA ADDUCTS

Methods have recently been developed to measure carcinogens bound to cellular macromolecules, including DNA, RNA, and protein (reviewed in ref. 33). Measurement of DNA adducts has been suggested to be a more relevant marker of carcinogen exposure than measurement of the carcinogen itself or its metabolites (16). It takes into account individual differences in absorption, metabolism, and excretion of the compound. DNA binding is considered the critical first step in the carcinogenic process, and measurement of DNA adducts measures the biologically effective dose, that is, the amount that reacts with critical cellular targets. Immunologic methods have been developed to measure carcinogen–DNA adducts. Specific monoclonal and polyclonal antibodies have been produced which recognize a number of different adducts (reviewed in refs. 34 and 35).

To quantitate levels of BP bound to DNA, in collaboration with Dr. M. Poirier at the National Cancer Institute, we developed a polyclonal antibody to DNA modified by BPDE-I (36). Rabbits were immunized with BPDE-I–DNA electrostatically complexed to methylated bovine serum albumin (mBSA) to increase antigenicity. These antibodies were originally characterized in an RIA that could detect 2 pmol of adduct. They have also been used in highly sensitive ELISAs, with color or fluorescence endpoints, detecting as little as 1 fmol of adduct per well (Figure 3).

More recently, a panel of monoclonal antibodies has been developed in our laboratory. Five stable clones producing antibody specific for BPDE-I–DNA have been isolated and characterized (37, 38). Four clones designated 5D11, 5D2, 1D7, and 4C2 were from the spleen cells of animals immunized with BPDE-I–DNA electrostatically complexed with methylated BSA. The other clone, 8E11, was obtained from an animal immunized with the monomeric form of the adduct coupled to BSA, namely, BPDE-I–G–BSA. The ribose form of guanosine was used so that the adjacent hydroxides could be used to couple the adduct to protein with a periodate oxidation procedure (39). Competitive ELISAs were employed to determine the sensitivity and relative

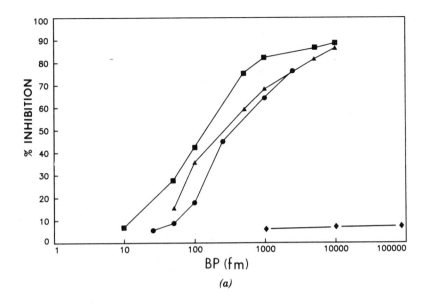

(a)

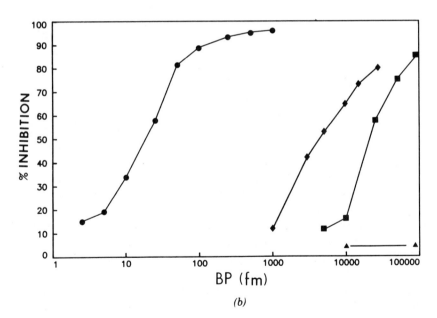

(b)

Figure 5. Competitive inhibition of monoclonal antibody binding to BPDE-I–DNA. The competitors were BPDE-I–DNA (●), BPDE-I–dG (■), BPDE-I–tetraols (▲), and BPDE-II–DNA (◆). (*a*) Antibody 8E11; (*b*) antibody 5D11. Reproduced, with permission, from ref. 38.

specificity of the antibodies. This information, expressed as the amount required for 50% inhibition of antibody binding, is shown in Figure 5 and summarized in Table 1. Antibody 8E11 reacts better with the deoxyribose form of the original antigen, BPDE-I–dG (50% inhibition at 145 fmol), than with BPDE-I–DNA (50% inhibition at 350 fmol). This antibody also cross-reacts with BPDE-I–tetraols (Figure 4, III), the hydrolysis products of BPDE-I (50% inhibition at 250 fmol). These results indicate that antibody 8E11 mainly recognizes determinants present on the BP ring. It does not react with DNA modified by BPDE-II, a stereoisomer of BPDE-I. Recently, we have tested several additional BP metabolites with this antibody. Significant cross-reactivity was detected with both the 7,8- and 9,10-diols of BP.

Table 1. Competitive Inhibition of Monoclonal Antibody Binding to BPDE-I–DNA

	Femtomole Level Causing 50% Inhibition				
Antibody	8E11	5D11	5D2	4C2	1D7
BPDE-I–DNA	350	19	17	160	370
BPDE-I–dG	145	21,000	$>10^5$	$>10^5$	40,000
BPDE-I–tetraols	250	$>10^5$	90,000	64,000	1,200
BPDE-I–RNA	580	18,000	20,000	2,100	900
BPDE-II–DNA	$>3 \times 10^4$	4,400	$>10^5$	$>10^5$	8,800
BPDE-II–dG	$>10^5$	$>10^5$	$>10^5$	$>10^5$	6,200
AAF–DNA	$>10^5$	$>10^5$	$>10^5$	$>10^5$	$>10^5$
AP–DNA	$>1.5 \times 10^4$	$>1.5 \times 10^4$	$>10^5$	$>10^5$	$>10^5$

Antibody 5D11, derived from an animal immunized with BPDE-I–DNA, has better sensitivity for the original antigen, BPDE-I–DNA (50% inhibition at 19 fmol), than for the monoadduct BPDE-I–dG (50% inhibition at 21,000 fmol) (Figure 5b). There is no cross-reactivity with BPDE-I–tetraols or BPDE-II–dG, but there is some low cross-reactivity with BPDE-II–DNA (50% inhibition at 4400 fmol). The other three clones produced from animals immunized with BPDE-I–DNA had similar characteristics. The higher reactivity with modified DNA as compared to that with the monoadduct suggests that these antibodies recognize antigenic determinants present in the surrounding DNA structure as well as in the adduct itself. The values given above for reactivity with DNA are for denatured, single-stranded DNA. There is lower reactivity with double-stranded DNA perhaps because of lowered accessibility of the antibody to the adduct.

Other carcinogen-modified DNAs were tested for cross-reactivity with the antibodies. Even at high concentrations, neither N,2-acetylaminofluorene nor

1-aminopyrene-modified DNA (AAF–DNA or 1-AP–DNA) showed any cross-reactivity with the antibodies. Unlike BPDE-I, which binds at the N-2 position of guanine, these carcinogens produce adducts at the C-8 position of guanine. In contrast, significant cross-reactivity was observed with DNAs modified by several other PAH diol epoxides. These included benz[a]anthracene trans-3,4-diol anti-1,2-epoxide, benz[a]anthracene trans-8,9-diol anti-10,11-epoxide, or chrysene trans-1,2-diol anti-3,4-epoxide, all of which are stereochemically similar to BPDE-I. Similar cross-reactivities were obtained with the polyclonal antibodies to BPDE-I–DNA (40, 41). Both the monoclonal and polyclonal antibodies therefore recognize N-2 guanine adducts that are formed from trans-diol epoxides with stereochemistry similar to that of BPDE-I. As in other cases with cross-reacting antibodies, absolute adduct determination in a sample with multiple diol epoxide adducts will be impossible.

Polyclonal and monoclonal antibodies that react with BPDE-I–DNA have been developed by several other groups. Wallin et al. developed monoclonal antibodies to 6-amino-BP coupled to BSA with carbodiimide (13). An ELISA using plates coated with BP coupled to ovalbumin was carried out to quantitate levels of BP bound to DNA, RNA, and protein. The lower detection limit of the assay was 1 pmol of BP bound to DNA or RNA and 5 pmol bound to protein. A highly sensitive ELISA with femtomole sensitivity has been developed by Schooten et al. (42) using polyclonal and monoclonal antibodies against BPDE-I–DNA. These antibodies were also shown to cross-react with chrysene-diol-epoxide-modified DNA.

The amount of BPDE-I–dG adduct present in animal or human samples can be quantitated by competitive ELISA. DNA, isolated from the sample of interest, is sonicated to make a homogeneous solution and is then denatured before analysis. With femtomole levels of sensitivity and the ability to test $50\,\mu g$ of DNA per microwell, one adduct in 10^8 normal nucleotides can be measured. This level of sensitivity is particularly important for analysis of the low level of adducts present in human samples. Humans are normally exposed to complex mixtures of environmental pollutants, including BP and a number of structurally related PAHs. Measured antigenicity could result from antibody binding, with variable affinity, to multiple diol epoxide adducts. Since BPDE-I–DNA is used in the standard curve, PAH adduct levels are expressed as femtomole equivalents of BPDE-I–DNA which would cause similar inhibition in the ELISA.

This approach has been used to monitor BP exposure in humans. Most studies have measured DNA adducts in lymphocytes since blood samples are readily and repeatedly available, but adducts have also been measured in placental and lung tissues. The first study used a polyclonal antibody to BPDE-I–DNA to determine adduct levels in lung tissue and lymphocyte

DNA of lung cancer patients (43). Adducts were detectable in a small number of the tissue samples, thereby demonstrating the feasibility of this approach. Another group has used the same polyclonal antibody to BPDE-I–DNA to quantitate adducts in lymphocytes from exposed worker populations. Detectable adducts were present in 20–40% of foundry workers and roofers (44) and about 30% of coke-oven workers (45, 46). In addition, antibodies to the adducts were seen in the sera of a number of workers and may serve as an alternate marker of exposure (45, 46). Adducts have also been detected in placental DNA of smokers and nonsmokers (47). Although adduct levels were higher in the smokers compared to the nonsmokers, this difference was not statistically significant. We have recently measured adducts in lymphocyte DNA of smokers and nonsmokers (48). The number of samples with detectable adducts is much lower for lymphocyte DNA as compared to placental DNA and did not differ in smokers and nonsmokers. Among positive samples, however, smokers had higher adduct levels than nonsmokers. Studies on BP adducts are complicated by the ubiquitous nature of this environmental contaminant making control unexposed populations hard to identify. Nonsmokers may be exposed to significant amounts of BP through environmental and dietary exposures. We have also measured adduct levels in the lymphocyte DNA of a group of Finnish foundry workers. Exposure to BP, as a measure of total hydrocarbon exposure, was classified as high, medium, or low based on an industrial hygienist evaluation of job description. All workers with high and medium exposure had detectable adduct levels, whereas only 2 of 10 controls had detectable adducts. When the mean of positive samples was compared, a dose-related increase in adducts was found (Table 2).

Table 2. PAH–DNA Antigenicity in Lymphocyte DNA from Finnish Foundry Workers and Controls

Exposure Group (μg BP/m^3)	Fraction Positive	BPDE-I–dG Equivalents[a] per 10^7 Nucleotides
High (>0.2)	4/4	5.0
Medium (0.05–0.2)	13/13	2.0
Low (<0.05)	13/18	1.1
Control	2/10	0.72

[a] Mean of positive samples.

7. IMMUNOFLUORESCENCE DETECTION

In addition to quantitation of adducts in isolated DNA, antibodies can be used to localize adducts in specific cell types and tissues. Immunofluorescence detection of adducts involves the incubation of fixed cells or tissue sections with the specific antibody. This is followed by incubation with a second antibody, such as goat anti-rabbit IgG, coupled to fluorescein or rhodamine. The polyclonal antibodies to BPDE-I–DNA discussed above have been used to monitor adduct levels in cultured mouse keratinocytes treated with BPDE-I (49). Specific nuclear staining was seen in treated cells but not in control cells. Treatment with as little as 10^{-7} M BPDE-I resulted in positive immunofluorescence staining. DNA was also extracted from the treated cells, and adduct levels were determined by competitive ELISA. The lowest level of adduct formation which could be detected by the immunofluorescence technique was 1.5 adducts per 10^5 nucleotides. Conventional immunofluorescence is thus less sensitive than competitive ELISA and is limited to detection of adducts in the range of one in 10^5 to 10^6 nucleotides. More recently, video-enhanced microscopy systems have been used to increase sensitivities up to a reported two adducts per 10^8 nucleotides (50).

BPDE-I–DNA adduct-specific antibodies have been used in electron microscopy studies to visualize and quantitate BP adducts in modified DNA (51). In these studies, a ferritin-labeled second antibody was used and good agreement was obtained between ELISA quantitation of adducts and visualization of antibodies bound to adducts by electron microscopy. Antibodies have also been used to visualize selective binding of carcinogens to specific genomic regions (52). Treatment of *Chironomus* polytene chromosomes with BPDE-I or larvae with BP resulted in nonrandom binding of the carcinogen. Another study took advantage of the selective binding of protein A to immunoglobulins to visualize antibody binding. Autoradiography following ^{125}I-protein A binding to anti-BPDE-I–DNA antibodies was used to study effects of DNA sequence and chromatin structure on carcinogen binding and indicated a three fold higher binding in linker DNA than in nucleosomal core DNA (53).

8. PROTEIN ADDUCTS

Even though DNA is believed to be the critical target in chemical carcinogenesis, most carcinogens also bind to RNA and proteins. Thus, the use of carcinogen binding to hemoglobin has been suggested as a surrogate for DNA binding in human dose monitoring studies (54, 55). There are several advantages to this approach, including the fact that (a) large quantities of hemoglobin are readily available in small amounts of blood, (b) there is no repair system for modified hemoglobin, and (c) chronic low levels of exposure may be quantifiable because of accumulation over the lifetime of the erythrocyte (about 4 months in humans). To develop a sensitive immunoassay for quantitating levels of BP on human hemoglobin, we developed monoclonal antibodies that specifically recognize BPDE-I–modified protein (56). These antibodies as well as 8E11, developed against BPDE-I–G, were found to have lower affinity for the BP chromophore when it was bound to protein (as compared to DNA) or free in solution as BPDE-I–tetraols (Table 3). This lowered reactivity with the BP protein adduct may be due to burying of the BP chromophore in hydrophobic regions of the protein and made quantitation of adducts on intact protein difficult. However, tetraols can be released from BPDE-I-modified proteins with acid treatment and can be quantitated by ELISA (30, 57, 58). Since the antibodies recognize BDPE-I–tetraols with tenfold higher sensitivity than BP bound to protein, a much more sensitive assay is possible. In addition, tetraols can be released from large amounts of protein before analysis, which also increases sensitivity. The applicability of this method for human hemoglobin samples is currently limited by the uncertainty of the structure of BP globin adducts formed *in vivo* (59).

Table 3. Competitive Inhibition of Antibody Binding to
BPDE-I–Rabbit Serum Albumin (RSA)

Antibody	Femtomole Level Causing 50% Inhibition		
	8E11	2B6	2G4
BPDE-I–RSA	3,000	10,000	6,400
BPDE-I–DNA	350	1,150	250
BPDE-I–Tetraols	250	1,400	870
BPDE-I–dG	145	1,075	505
BPDE-II–RSA	2,325	3,400	4,400

9. NITROAROMATICS

Nitroarenes, including a number of nitro-substituted fluoranthrenes and pyrenes, have been found in the particulate fraction of diesel exhaust. Monoclonal antibodies have been developed to a mixture of nitrated fluoranthrenes coupled to KLH (60). Although there were differences seen with the individual clones, all antibodies cross-reacted with other compounds structurally related to the original antigen. These included 1-nitropyrene, fluoranthrene, and 1- and 2-nitronaphthalene. One of the most sensitive clones was able to detect 40 ng of 3-nitrofluoranthrene but unfortunately had almost equal cross-reactivity with both 1-nitropyrene (1-NP) and BP. These compounds could also be detected in spiked urine down to a level of about 50 ng/ml by RIA. Because of significant cross-reactivity with related nitroaromatics and PAHs, this technique will not provide quantitative data on particular compounds but only a qualitative measure of general exposure.

1-NP is responsible for a significant portion of the mutagenic activity of diesel exhaust particulates. The structure of the DNA adduct formed when both *Salmonella typhimurium* and rats are treated with 1-NP has been identified as N-(deoxyguanosine-8-yl)-1-aminopyrene (61, 62). To quantitate levels of this adduct in human and animal samples, we developed a panel of monoclonal antibodies specific for 1-aminopyrene–modified DNA(1-AP–DNA) (63). The most sensitive clone, 11H2, has been completely characterized by competitive ELISA (Table 4). It reacts best with AP–modified denatured DNA (50% inhibition at 18 fmol), whereas with double-stranded modified DNA the 50% inhibition value is 250 fmol. These results are similar to those seen with the BPDE-I–DNA antibodies and are probably due to the adduct being more accessible to antibody binding when present in single-stranded

Table 4. Competitive Inhibition of Monoclonal Antibody Binding to Denatured AP–DNA

Antibody	Femtomole Level Causing 50% Inhibition		
	11H2	5H5	8B10
1-AP–d-DNA[a]	18	23	25
1-AP–n-DNA[b]	250	380	200
BPDE-I–d-DNA	3,000	4,000	3,000
BPDE-I–n-DNA	51,000	58,000	64,000
AAF–n-DNA	2,400	3,600	4,400
AAF–d-DNA	2,200	N.D.[c]	N.D.
8-nitro-1-AP–n-DNA	250	N.D.	N.D
6-nitro-1-AP–n-DNA	17,000	N.D.	N.D.
1-Nitropyrene	$>10^5$	$>10^5$	$>10^5$
1-Aminopyrene	$>10^5$	$>10^5$	$>10^5$

[a] AP–modified denatured DNA.
[b] AP–modified native DNA.
[c] N.D., not determined.

DNA. Antibody 11H2 also cross-reacts with other C-8 of guanine–modified DNAs. With denatured 8-nitro-1-AP–DNA, 50% inhibition is at 34 fmol, whereas with denatured 6-nitro-1-AP–DNA it is at about 3200 fmol. Denatured AAF–modified DNA cross-reacts with a 50% inhibition at 2200 fmol. Surprisingly, denatured BPDE-I–DNA also cross-reacts with the antibody (50% inhibition at 3000 fmol) even though this adduct is at the N-2 position of guanine. The assay was validated by quantitating adducts in *S. typhimurium* treated with [^{3}H]-1-NP and has been used to monitor adducts in animals treated with 1-NP (64).

10. FUTURE APPLICATIONS

Studies on exposure to aflatoxin B$_1$ (AFL–B$_1$) can serve as a model for potential applications of immunoassays for aromatic hydrocarbons. The major DNA adduct of AFL–B$_1$ is at the N-7 position of guanine. This adduct is chemically unstable and either depurinates (releases from DNA) or opens in the guanine ring (65). Animals treated with AFL–B$_1$ have detectable levels of released AFL–guanine adduct in their urine (66). A number of antibodies have been developed against AFL–B$_1$ and, have been used to quantitate aflatoxin levels in foods, urine, milk, and serum (reviewed in ref. 67). These antibodies have been used in sensitive immunoassays to detect as little as 300 fmol of AFL–B$_1$. Antibodies have also been developed against AFL-B$_1$–modified DNA (68–70).

A number of studies monitoring aflatoxin levels in urine (71) or sera have been carried out in regions of high exposure (72). Urine is partially purified by passage through a C18 Sep-Pak cartridge. The retained AFL metabolites and the excised AFL–N,7-guanine adduct are then released for quantitation by immunoassay or HPLC (71, 73). Another approach has used an affinity column of anti-aflatoxin antibodies to pull metabolites out of urine (70). The retained material can then be released and quantitated by immunoassay. The advantage of this approach is that the antibody columns are reusable and can rapidly purify material from a large number of samples. Because of cross-reactivity of the antibodies with the various metabolites and excised adduct, absolute quantitation may be difficult. As with the studies on aromatic hydrocarbon–DNA adducts, metabolites with lower affinity for the antibody will be underestimated.

A similar approach can be used to study exposure to BP. The presence of an unstable adduct of BP at the N-7 position of guanine has been reported in the urine of rats treated with BP and has been tentatively identified in the urine of a smoker (74). If confirmed, measurement of this adduct could serve as an alternate marker of exposure to BP. In addition, the predominant *N,2-*

guanine adduct of BP has recently been detected in the feces of mice treated with [^{3}H]-BP (75). These authors estimated that as much as 0.3 fmol of BP adduct may be excreted in the feces of humans who smoke and may be detectable by immunologic methods.

We are currently developing immunoaffinity methods, using antibody 8E11, to isolate and quantitate BP adducts and metabolites from urine. Antibody 8E11 reacts with a number of BP metabolites, in addition to the DNA adducts. Further purification of material retained by the affinity column will be necessary before the adducts and metabolites can be quantitated. Measurement of total BP urinary metabolites would be one useful marker of exposure to BP. However, it may not be as relevant to human risk as measurement of BP–DNA adducts.

11. SUMMARY

Immunoassays have potential application to a wide variety of exposure studies. The number of antibodies available to particular chemicals or DNA and protein adducts is rapidly increasing. These antibodies can be used in highly sensitive immunoassays to quantitate antigen levels. Because of their ease, large numbers of samples can be assayed, making them ideal for molecular epidemiology studies.

ACKNOWLEDGMENTS

Research in the authors' laboratory was supported by the Council for Tobacco Research, number Inc. #1483, NIH CA21111 and NIEHS ESO3881. The authors thank R. Yang for secretarial assistance.

REFERENCES

1. S. J. Kennel, C. Jason, P. W. Albro, G. Mason, and S. H. Safe, *Toxicol. Appl. Pharmacol.*, **82**, 256–263 (1986).
2. M. I. Luster, P. W. Albro, G. Clark, K. Chae, S. K. Chaudhary, L. D. Lawson, J. T. Corbett, and J. D. McKinney, *Toxicol. Appl. Pharmacol.*, **50**, 147–155 (1979).
3. J. J. Pestka and F. S. Chu, *Appl. Environ. Microbiol.*, **47**, 472–477 (1984).
4. T. S. L. Fan, G. S. Zhang, and F. S. Chu, *Appl. Environ Microbiol.*, **47**, 526–532 (1984).
5. P. Sizatet, C. Malaveille, R. Montesano, and C. Frayssinet, *J. Nat. Cancer Inst.*, **69**, 1375–1381 (1982).
6. J. J. Langone and H. Van Vunakis, *J. Nat. Cancer Inst.*, **56**, 591–595 (1976).

7. F. S. Chu, S. Grossman, R. D. Wei, and C. J. Mirocha, *Appl. Environ. Microbiol.*, **37**, 104–108 (1979).

8. J. J. Pestka, S. C. Lee, H. P. Lau, and F. S. Chu, *J. Assoc. Off. Anal. Chem.*, **64**, 940a–944a (1981).

9. E. H. Glendloff, J. J. Pestka, S. P. Swanson, and L. P. Hart, *Appl. Environ. Microbiol*, **47**, 1161–1163 (1984).

10. J. F. Hewetson, J. D. Pace, and J. E. Beheler, *J. Assoc. Off. Anal. Chem.*, **70**, 654–656 (1987).

11. T. S. L. Fan, Y. Xu, and F. S. Chu, *J. Assoc. Off. Anal. Chem.*, **70**, 657–670 (1987).

12. N. Y. Kado and E. T. Wei, *J. Nat. Cancer Inst.*, **61**, 221–225 (1978).

13. H. Wallin, C. A. K. Borrebaeck, C. Glad, B. Mattiasson, and B. Jergil, *Cancer Lett.*, **22**, 163–170 (1984).

14. A. Tompa, G. Curtis, W. Ryan, C. Kuszynski, and R. Langenbach, *Cancer Lett.*, **7**, 163–169 (1979).

15. T. Vo-Dinh, B. J. Tromberg, K. R. Griffin, M. J. Ambrose, M. J. Sepaniak, and E. M. Gardenhire, *Appl. Spectrosc.*, **41**, 735–738 (1987).

16. F. P. Perera and I. B. Weinstein, *J. Chronic Dis.*, **35**, 581–600 (1982).

17. D. M. Weir, Ed., *Handbook of Experimental Immunology*, Vol. 3, Blackwell Scientific Publications, St. Louis (1978).

18. G. Galfre and C. Milstein, in *Methods Enzymol.*, **73**, 3–46 (1981).

19. D. Parratt, H. McKenzie, K. H. Nielsen, and S. J. Cobb, *Radioimmunoassay of Antibody*, Wiley, New York (1982).

20. A. Voller, A. Bartlett, and D. E. Bidwell, *J. Clin. Pathol.*, **31**, 507–520 (1978).

21. E. T. Maggio, *Enzyme-Immunoassay*, CRC Press, Boca Raton, Fla. (1983).

22. J. Nakayama, S. H. Yuspa, and M. C. Poirier, *Cancer Res.*, **44**, 4087–4095 (1984).

23. X. Y. Yang, V. DeLeo, and R. M. Santella, *Cancer Res.*, **47**, 2451–2455 (1987).

24. I. C. Hsu, M. C. Poirier, S. H. Yuspa, D. Grunberger, I. B. Weinstein, R. H. Yolken, and C. C. Harris, *Cancer Res.*, **41**, 1090–1095 (1981).

25. B. Leipold and W. Remy, *J. Immunol. Methods*, **66**, 227–234 (1984).

26. A. M. Shamsuddin and C. C. Harris, *Arch. Pathol. Lab. Med.*, **107**, 514–517 (1983).

27. D. Umbenhauer, C. P. Wild, R. Montesano, R. Saffhill, J. M. Boyle, N. Huh, U. Kristein, and M. F. Rajewsky, *Int. J. Cancer*, **36**, 661–665 (1985).

28. A. M. J. Fichtinger-Schepman, A. T. van Oosterom, P. H. M. Lohman, and F. Berends, *Cancer Res.*, **47**, 3000–3004 (1987).

29. A. M. Jeffrey, T. Kinoshita, R. M. Santella, and I. B. Weinstein, in *Carcinogenesis: Fundamental Mechanisms & Environmental Effects*, P. O. P. Ts'o and H. Gelboin Eds., D. Reidel, Boston, p. 565 (1980).

30. M. Koreeda, P. D. Moore, P. G. Wislocki, W. Levin, A. H. Conney, H. Yagi, and D. M. Jerina, *Science*, **199**, 778–781 (1978).

31. V. Ivanovic, N. Geacintov, H. Yamasaki, and I. B. Weinstein, *Biochemistry*, **17**, 1597–1603 (1978).

32. K. Shinohara and P. A. Cerutti, *Proc. Nat. Acad. Sci. USA*, **74**, 979–983 (1977).

33. R. M. Santella, *Mutat. Res.* **205**, 271–282 (1988).

34. P. T. Strickland and J. M. Boyle, in *Progress in Nucleic Acid Research & Molecular Biology*, Vol. 31, W. E. Cohn, Ed., Academic Press, New York, pp. 1–58 (1984).

35. M. C. Poirier, *Environ. Mutagens*, **6**, 879–887 (1984).

36. M. C. Poirier, R. Santella, I. B. Weinstein, D. Grunberger, and S. H. Yuspa, *Cancer Res.*, **40,** 412–416 (1980).

37. R. M. Santella, L. L. Hseih, C. D. Lin, S. Viet, and I. B. Weinstein, *Environ. Health Perspect.*, **62,** 95–100 (1985).

38. R. M. Santella, C. D. Lin, W. L. Cleveland, and I. B. Weinstein, *Carcinogenesis*, **5,** 373–377 (1984).

39. B. R. Erlanger and S. M. Beiser, *Proc. Nat. Acad. Sci. USA*, **52,** 68–74 (1964).

40. R. M. Santella, F. Gasparro, and L. Hsieh, *Prog. Exp. Tumor Res.*, **31,** 63–75 (1987).

41. A. Weston, G. Trivers, K. Vahakangas, M. Newman, and M. Rowe, *Prog. Exp. Tumor Res.*, **31,** 76–85 (1987).

42. F. J. van Schooten, E. Kriek, M. S. T. Steenwinkel, H. P. J. M. Noteborn, M. J. X. Hillebrand, and F. E. V. Leeuwen, *Carcinogenesis*, **8,** 1263–1269 (1987).

43. F. P. Perera, M. C. Poirier, S. H. Yuspa, J. Nakayama, A. Jaretzki, M. M. Curnen, D. M. Knowles, and I. B. Weinstein, *Carcinogenesis*, **3,** 1405–1410 (1982).

44. A. K. M. Shamsuddin, N. T. Sinopoli, K. Hemminki, R. B. Boesch, and C. C. Harris, *Cancer Res.*, **45,** 66–68 (1985).

45. C. C. Harris, K. Vahakangas, J. M. Newman, G. E. Trivers, A. Shamsuddin, N. Sinopoli, D. L. Mann, and W. E. Wright, *Proc. Nat. Acad. Sci. USA*, **82,** 6672–6676 (1985).

46. A. Haugen, G. Becher, C. Benestad, K. Vahakangas, G. E. Trivers, M. J. Newman, and C. C. Harris, *Cancer Res.*, **46,** 4178–4183 (1986).

47. R. B. Everson, E. Randerath, R. M. Santella, R. C. Cefalo, T. A. Avitts, and K. Randerath, *Science*, **231,** 54–57 (1986).

48. F. P. Perera, R. M. Santella, D. Brenner, M. C. Poirier, A. A. Munshi, H. K. Fischman, and J. Vanryzin, *J. Nat. Cancer Inst.*, **79,** 449–456 (1987).

49. M. C. Poirier, J. R. Stanley, J. B. Beckwith, I. B. Weinstein, and S. H. Yuspa, *Carcinogenesis*, **3,** 345–348 (1982).

50. J. Adamkiewicz, G. Eberle, N. Huh, P. Nehls, and M. F. Rajewsky, *Environ. Health Perspect.*, **62,** 49–55 (1985).

51. R. S. Paules, M. C. Poirier, M. J. Mass, S. H. Yuspa, and D. G. Kaufman, *Carcinogenesis*, **6,** 193–198 (1985).

52. P. D. Kurth and M. Bustin, *Proc. Nat. Acad. Sci. USA*, **82,** 7076–7080 (1985).

53. M. Seidman, H. Mizusawa, H. Slor, and M. Bustin, *Cancer Res.*, **43,** 743–748 (1983).

54. M. Tornqvist, S. Osterman-Golkar, S. Kautiainen, S. Jensen, P. B. Farmer, and I. Ehrenberg, *Carcinogenesis*, **7,** 1519–1521 (1986).

55. C. J. Calleman, L. Ehrenberg, B. Jansson, S. Osterman-Golker, D. Segerback, K. Svensson, and C. A. Wachteneister, *J. Environ. Pathol. Toxicol.*, **2,** 427–442 (1978).

56. R. M. Santella, C. D. Lin, and N. Dharmaraja, *Carcinogenesis*, **7,** 441–444 (1986).

57. L. Shugart, *Toxicology*, **34,** 211–220 (1985).

58. L. Shugart, *Analytical Biochem.*, **152,** 365–369 (1986).

59. H. Wallin, A. M. Jeffrey, and R. M. Santella, *Cancer Lett.*, **35,** 139–146 (1987).

60. R. Hass, Unpublished studies presented at Air Pollution Control Association, New York (1987).

61. P. C. Howard, R. H. Heflich, F. E. Evans, and F. A. Beland, *Cancer Res.*, **43,** 2052–2058 (1983).

62. C. A. Stanton, F. L. Chow, D. H. Phillips, P. L. Grover, R. C. Garner, and C. N. Martin, *Carcinogenesis*, **6**, 535–538 (1985).
63. L. L Hsieh, A. M. Jeffrey, and R. M. Santella, *Carcinogenesis*, **6**, 1289–1293 (1985).
64. L. L. Hsieh, D. Wong, V. Heisig, R. M. Santella, J. L. Mauderly, C. E. Mitchell, R. K. Wolff, and A. M. Jeffrey, in *Carcinogenic and Mutagenic Effects of Diesel Engine Exhaust*, N. Ishinishi, A. Koizumi, R. O. McClellan, and W. Stober, Eds., Elsevier, New York, pp. 223–232 (1986).
65. W. F. Busby and G. N. Wogan, in C. D. Searle, Ed., *Chemical Carcinogens*, Vol. 182, ACS, Washington, D.C., pp. 945–1136 (1984).
66. R. A. Bennett, J. M. Essigmann, and G. N. Wogan, *Cancer Res.*, **41**, 650–654 (1981).
67. C. Garner, R. Ryder, and R. Monesano, *Cancer Res.*, **45**, 922–928 (1985).
68. P. J. Hertzog, A. Shaw, J. R. Lindsay Smith, and R. C. Garner, *J. Immunol. Methods*, **62**, 49–58 (1983).
69. J. D. Groopman, A. Hauten, G. R. Goodrich, G. N. Wogan, and C. C. Harris, *Cancer Res.*, **42**, 3120–3124 (1982).
70. J. D. Groopman, L. J. Trudel, P. R. Donahue, A. Marshak-Rothstein, and G. N. Wogan, *Proc. Nat. Acad. Sci. USA*, **81**, 7728–7731 (1984).
71. H. Autrup, T. Seremet, J. Wakhisi, and A. Wasunna, *Cancer Res.*, **47**, 3430–3433 (1987).
72. S. Tsuboi, T. Nakagawa, M. Tomita, T. Seo, H. Ono, K. Kawamura, and N. Iwamura, *Cancer Res.*, **44**, 1231–1234 (1984).
73. H. Autrup, K. A. Bradley, A. K. M. Shumsuddin, J. Wakhisi, and Q. Wasunna, *Carcinogenesis*, **4**, 1193–1195 (1983).
74. H. Autrup and T. Seremet, *Chem.–Biol. Interact.*, **60**, 217–226 (1986).
75. B. Tierney, C. N. Martin, and R. C. Garner, *Carcinogenesis*, **8**, 1189–1192 (1987).

CHAPTER

13

RECENT ADVANCES IN INFRARED ANALYSIS OF POLYCYCLIC AROMATIC COMPOUNDS

PHIL STOUT* AND GLEB MAMANTOV

Department of Chemistry
University of Tennessee
Knoxville, Tennessee

1. INTRODUCTION

Conventional (dispersive) infrared (IR) spectroscopy has not been a popular technique in the analysis of polycyclic aromatic compounds (PACs). In fact, in the early 1970s a publication of the National Academy of Sciences stated (1) that

No special techniques have been developed for IR analysis of polycyclic compounds . . . . The disadvantages of vibrational spectroscopy are the relatively weak bands, the fact that IR band strengths are not proportional to concentration, the requirement for a vibrationally transparent medium, and the lack of unique polycyclic structural features. The disadvantages far outweigh the advantages of these techniques.

In recent years, several advances have made IR spectroscopy a much more useful method for the analysis of PACs. These advances include:

1. The development of Fourier transform infrared (FTIR) spectroscopy and the availability of sophisticated and readily available FTIR spectrometers.
2. The application of matrix isolation (MI) techniques to qualitative and quantitative analysis of PACs by FTIR spectroscopy.
3. The coupling of chromatographic techniques, particularly gas chromatography (GC), with FTIR spectroscopy.
4. The development of specialized techniques such as (a) photoacoustic spectroscopy in the IR region and (b) excited triplet-state spectroscopy.

---

* Present address: Bio-Rad Digilab Division, Cambridge, Massachusetts.

411

All PACs exhibit a number of IR absorptions. The most intense of these arise from C–H stretching in the region 3200–3000 cm^{-1} and C–H out-of-plane bending vibrations in the region 900–675 cm^{-1} (2). Other absorptions may be caused by vibrations involving heteroatoms in the ring structure or substituent groups on the ring. Environmental samples will typically contain a mixture of PACs. Because of the similar structures of many of these compounds, overlap or near overlap of IR bands is likely. Any useful analytical technique for individual PACs must first address the problem of overlapping bands.

There are two strategies to be considered in addressing this problem. The first of these strategies is to perform some kind of physical separation of the individual PACs prior to IR analysis, thereby eliminating or minimizing the problem of band overlap. This is the underlying philosophy for all chromatographic schemes involving an IR detector. The second strategy is to try to achieve optical isolation by observing the bands of an individual PAC without interference of bands from other compounds. This can be accomplished by affecting the separation of bands in a spectrum of a mixture with some line-narrowing technique. All successful IR techniques for the analysis of PACs employ one of these two strategies or employ a combination of the two.

2. MI–FTIR SPECTROSCOPY

In MI a gaseous sample is mixed with a large excess of an inert diluent gas (the matrix); the mixture is then deposited on an optical surface for spectroscopic observation. Although MI was initially described in 1954 (3), wide application of MI–FTIR spectroscopy to analysis occurred only after FTIR spectrometers became readily available and after the development of closed-cycle refrigerators, which can attain the temperatures required for MI (<25 K) without the use of liquid cryogens. The analytical applications of MI–FTIR spectroscopy were reviewed in 1982 (4) and 1987 (5). The 1982 paper contains an extensive compilation of IR band positions of matrix-isolated polycyclic aromatic hydrocarbons (PAHs) [see Table 1 (4)]. A companion paper (6) contains a similar compilation for matrix-isolated heteroaromatics (see Table 2). A very recent paper (7) contains data for a number of nitro PAHs (see Table 3).

3. CHROMATOGRAPHIC TECHNIQUES

Environmental samples of PACs are normally obtained by extraction of particulate matter from some combustion process. These samples may contain a large number of PACs. In order to determine the identity of the individual PAC, it is often necessary to perform a chromatographic separation of the extract. Gas chromatography/mass spectrometry (GC/MS) has been used extensively for the identification of PACs in environmental samples. There are limitations associated with this technique. Positive characterization of all compounds is difficult because of the limited fragmentation of aromatic molecules when subjected to electron-impact ionization. For this reason, isomers frequently cannot be distinguished by MS. The importance of distinguishing between isomers of PACs cannot be overstated. The mutagenic and carcinogenic properties of a group of isomers may vary greatly within the group (1, 8–10).

Table 1. Infrared Band Positions of Matrix-Isolated Polycyclic Aromatic Hydrocarbons[a]

Compound	Frequencies $(cm^{-1})^b$
Aromatic hydrocarbons containing three or more rings[c]	
Triphenylene	1502, 1438, *746*
Chrysene	1365, 1234, (819, 816), *765*
Benz(*a*)anthracene	1504, 1365, 886, 809, 785, *752*
Pyrene	1436, 1185, (*850, 846*), 715, 713
Anthracene	881, (*733, 730*), 603
Phenanthrene	1504, 1462, 870, 818, *740.5*, 617
Tetracene	*900.5*, (747, 744.5), 551.5
Fluoranthene	1461, 1455.5, 1444, 1138, 830, (*782, 778*), 750, 620
Acenaphthene	(1431, 1422), 1374, 1017.5, 843, *792*, 750
Perylene	1498, 1385, *818*, (776, 774)
Benzo(*b*)fluorene	957, 872, *775*, 764, 728, 570, 475
Benzo[*a*]fluorene	1223, 1020, (828, 826), (*763, 760.5*)
1,3,5-Triphenylbenzene	1580, 1558, 1503, 1417, 1080, 1032, 884, *758.5*, 702, 623, (*614, 611*), 509
Benzo[*a*]pyrene	1184, (888, 886), 855, 843, 831, *765*, 747, 694
1,2,5,6-Dibenzanthracene	1513, 1458, 892, (*813, 809*), (750, 745), 667, 649, 529
Benzo[*e*]pyrene	1480, 1444, 1415, 883, (*755, 753*)
Naphthalenes	
Naphthalene	1140, (1131, 1127), 1014, 1012, 964, 959.5, (*790.5, 787, 785*)
1-Methylnaphthalene	(1517, 1513), 1401, 1383, 1022, (*797, 795*), (777, 775)

Table 1. (*Continued*)

Compound	Frequencies (cm^{-1})^b
2-Methylnaphthalene	1513, 1460, 1430, 1382, 1355, (1007, 1004), 887, (854, 852), (*817, 815, 813*), (745.5, 741.5)
1,3-Dimethylnaphthalene	1467, 1464, 1445, 1415 (886.5, 864.5, 861.5), (850, 848), 775.5, (*751, 749.5, 747*)
1,4-Dimethylnaphthalene	1467, 1445, 1425, 1398, (827, 824), (*759, 756, 755*)
1,5-Dimethylnaphthalene	(1426, 1424), (*794, 787*)
2,3-Dimethylnaphthalene	(1468, 1465), 1453, 1434, 1384, 1376, 1365, 1355, 1273, 1235, 1222, 1098, *872*, (748, 745)
2,6-Dimethylnaphthalene	1510, 1451, *875*, (815, 813.5, 812)
Biphenyls^c	
Biphenyl	(*741, 739, 738*), 701
2-Methylbiphenyl	(779, 778, 775), *751*, 728, 704
3-Methylbiphenyl	795, *758*, 700
4-Methylbiphenyl	(827, 824), *760*, (700, 696)
Methylchrysenes^c	
1-Methylchrysene	822, (797, 795.5), *774*
2-Methylchrysene	880, 818, 805, *754.5*
3-Methylchrysene	(865.5, 862), (838, 836.5), 827.5, 774, *754.5*
4-Methylchrysene	*828*, (800, 798), 767.5, *756*
5-Methylchrysene	(892, 883), (867, 864), 824, 795.5, *765*, 748
6-Methylchrysene	823.5, *764.5*
Hydrogenated polycyclic aromatic hydrocarbons	
9,10-Dihydrophenanthrene	1308, (1265, 1263), (1210, 1209), 1094, 1050, (1008, 1006) (944, 943), (776.5, 774), (*751, 748.5*)
1,4-Dihydronaphthalene	1303, 1287, 1283, 1272, (1205, 1202), 1188, 1115, 1045, (1015, 1013), 1000, (929, 928), 922, (791, 787.5, 785.5), (*756, 752, 750.5, 749, 745*)
1,2,3,4-Tetrahydrofluoranthene	1300, 1262, (1185, 1184), 1138, 1106, 1037, 1020, (831, 829.5), (*782, 778.5*), (758.5, 757.5), (752, 750)
1,2-Dihydropyrene	1309, 1251, 1240, (1197, 1193), (1166, 1160), (850, 846.5), (*837.5, 834, 833, 831*), (821, 819, 816), (785.5, 784, 783, 782), (760, 759, 758, 757), (738, 735, 734), 727, (715.5, 714)
1,2-Dihydronaphthalene	1287, 1282, 1089, 1040, 1032, (1015, 1012), 889, (807, 805.5), 798, 795, (*787, 785*), (777.5, 775.5), (750, 748, 747, 745), (696, 694)

Table 1. (*Continued*)

Compound	Frequencies (cm^{-1})^b
1,2,6,7-Tetrahydropyrene	1309, 1300, 1251, 1240, (1187, 1184), (1169, 1167), (904, 903), (872, 870.5), (837, 836, 833), (821, 818), (786, 785, 783, 782), (760, 757), 735, 727, (716, 713.5)
Dodecahydrotriphenylene	(1465, 1464), (1457, 1454), (*1439, 1437*), 1258
9,10-Dihydroanthracene	(1488, 1485, 1481), (1459, 1457), 1332, (1324, 1323), 1312, 1289, 1263, 940, 933 (815, 812, 811), (*769, 764, 763*), 756, 751, (733, 730), (705, 703)
Decahydronaphthalene	1479, 1463, (*1457, 1453, 1451, 1447*), 1379, 1351, 1310, 1306, 1264, 1256, 1014, 981, 974, 971, 926, (858, 855), 841, 826
Tetrahydronaphthalene	1499, 1441, 1287, 1114, (807, 805.5), (*750.5, 748, 745*)

^{a} From ref. 4.
^{b} Resolution = 1 cm^{-1}; parentheses indicate multiplets; strongest bands are in italics.
^{c} Resolution = 2 cm^{-1}.

The use of an FTIR spectrometer as a detector for GC has been studied for the analysis of PACs. In one study, Chiu et al. demonstrated the complementary nature of information obtained from the parallel analyses of two mixtures of PACs by GC/MS and GC/FTIR (11). The mixtures, one synthetic containing 20 compounds, the other a coal combustion extract, were analyzed independently by the two techniques using similar GC columns and conditions. Table 4 presents a list of the compounds used to make up the synthetic mixture as well as the results of a computer search of the digitized spectral libraries loaded on the FTIR data system (11). The results show that every compound for which there was a corresponding library entry was identified correctly.

Commercially available digitized FTIR libraries are continually growing and are a significant contributor to the success of GC/FTIR as a means of identifying unknowns in a chromatographic separation. For example, the Sadtler Starter Library of digitized FTIR spectra currently contains over 11,000 entries. Many specialized libraries are available for searching diverse classes of compounds. This broad data base significantly contributes to the rapid identification of unknowns in a mixture. In addition, FTIR spectrometer manufacturers are developing more rapid search algorithms that contribute to the efficiency with which unknowns are identified.

Table 2. Infrared Band Positions of Matrix-Isolated Heteroaromatics

Compound	Band Maxima (cm^{-1})[a]
Nitrogen-containing aromatics	
Pyrrole	1050.5, 1015, 866.5, *730*, 520
Carbazole	1496, 1464.5, 1454, 1397.5, *1338*, 1328, 1241, 754, 729.5, 618, 569
13*H*-Dibenzo[*a,i*]carbazole	1391, 1310, 885, *802*, 779, 740, 725
Indole	1457, 1413, 1354, 1333, 766, (*747, 744, 741*)
2-Methylindole	*1457*, 1416, 1287, 782, 748, 735
3-Methylindole	*1458*, 1420, 1249, 741
Pyridine	1582, 1483.5, 1440, 1031, 991, 745, (*703.5, 701.5*), 602
Quinoline	*1504*, 1316, 1119.5, 941, (*812, 809, 806*), 787, 611
Isoquinoline	1631, (1503, 1500), 1385, 1275, 946, (*833, 830, 828*), (*747.5, 745, 742*), (641, 639, 637)
Lepidine (4-methylquinoline)	1574, 1511, 1459, 1394, 1308, 1246, 1161, 1139, 859.5, 843, 814, (*762, 760, 758*), 705.5, 563.5
7-Methylquinoline	1632, 1504, (1323, 1321), 1145, 889, 835, *831*, 806, 785, 639
2,4,6-Trimethylpyridine	1616, 1578, *1463*, 1378, 1220.5, 929.5, (842, 840, 839)
Acridine	1516, (1370, 1367), 1233, (*743, 739*), 601
Phenazine	1516, 1435, 1139, 1112, (821, 818), (*757, 754*), 598
1,2-7,8-Dibenzacridine	1381, (*842, 840*), 754
4-Azafluorine	1458, 1417, 1406, 1175, 779, (*749, 747, 746*)
3,4,7,8-Tetramethyl-1, 10-phenanthroline	1511, 1465, 1443, 1424, 864, 852, 815, 732
Quinaldine (2-methylquinoline)	1509, 1423, 1374, 1223, 1115, (*824, 821, 819*), 786, (*751, 747*), 619
6-Methylquinoline	1503, 1449, 1376, 1326, 1120, 894, 880, (*836, 832.5*), 799, 765, 615
8-Methylquinoline	1501, 1394, 1375, 1317, 1073, 870, (825.5, 822.5), *793.5*, 568
3-Methylisoquinoline	(1638, 1636), 1498, 1275, 950, (877, 873.5), (*752, 749*), (*722, 720*)
2,7-Dimethylquinoline	1442, 1307, 1147, 990, (892, 888), 845, (*839, 838*), 783
2,6-Dimethylquinoline	1442, 1396, 1375, (1372, 1370), 1120, 1095, 882, (*837, 834*)
2,4-Dimethylquinoline	1455, 1412, 1377, 1372, 1333, 1192, 862, (*763, 760*)

Table 2. (*Continued*)

Compound	Band Maxima (cm^{-1})^a
4,6-Dimethylquinoline	1440, 1380, 1247, 858, (*849, 847, 844*)
Aniline	*1507*, 1177, *753*, 692
1-Naphthylamine	1620, 1518, 1410, 1290, (*776, 773, 772*)
2-Naphthylamine	*1519*, 1340, 1227, 1184, (842, 840, 829), (814, 810), (*748, 746, 744*)
Oxygen-containing aromatics	
Phenol	1609, *1502*, 1473, 1189, 1072, 812, (*756.5, 754.5, 753.5*), (691.5, 689)
o-Cresol	(*1470, 1468*), 1328, 1261, 1169, 1103, 843, (*756, 754.5*), (751.5, 750.5)
m-Cresol	1494, *1286*, 1158, (932, 927), multiplet near 776, 688
p-Cresol	*1518*, 1260, 1172, 1105, 841, (824, *820*, 818, 812.5, 808), 739.5
1-Naphthol	*1391*, 1369, 1285, 1249, 1085, 1045, 1017, 793, (*777, 775, 773*)
2-Naphthol	*1518*, 1472, 1264, (1224, 1220), (1190, 1184, 1179), 1123, *814*, 750, 718, 622
2,3-Dihydroxynaphthalene	1530, 1492, 1378, *1274*, 1160, *844*, 748

^{a} Resolution = 1 cm^{-1}; parentheses indicate multiplets; strongest bands are in italics.

As noted earlier, one of the advantages which IR detection has to offer to chromatographic separation is the ability to distinguish between isomeric compounds, which may not be done with mass spectral data alone. This is illustrated in Figure 1, which shows the mass spectra and IR spectra of 2- and 1-methylnaphthalenes (11). The mass spectra clearly do not provide enough information to distinguish between these two isomers, whereas the IR spectra are sufficiently different to distinguish the two compounds. Often, compounds that cannot be differentiated based on the mass spectral data may be distinguished by significant differences in their chromatographic retention behavior. However, this requires that standard compounds be available so that their retention behavior can be studied on a particular chromatographic system. When dealing with a complex environmental sample, it may be unrealistic to expect to have standards for all of the eluted compounds. In these cases, the IR spectra of the components are invaluable as aids in the unambiguous identification. Furthermore, isomers may not always be well resolved by a particular chromatographic column, in which case retention data may not provide a good basis for identification even when standards are

Table 3. Infrared Band Positions for Nitro-PAH compounds in a Nitrogen Matrix

Compound	Frequencies (cm^{-1})^{a,b}
1-Nitronaphthalene	*1537s*, 1515w, 1367w, 1355w, 1348m, 1263w, 875w, 811w, 791w, 774m, 658w, 506w
1,2-Dinitronaphthalene	*1537s*, 1513m, 1391w, 1386w, 1381w, 1366w, 1359w, 1350m, 1276w, 868w, 816m, 796m, 783w, (747, 745)w, 737w, 663w, 646w
1-4-Dinitronaphthalene	*1549s*, 1540w, 1433w, 1349s, 836m, (775, 772)w
1,8-Dinitronaphthalene	*1553*, 1416, 1361s, 1349w, 838w, 806m, (762, 759)w, 541m
2,3,5-Trinitronaphthalene	*1557s*, 1546s, 1416w, 1363m, 1349m, 854w, 803w, 509w
9-Nitroanthracene	*1544s*, 1533m, 1451m, 1380w, 1349w, 1325m, 1281m, 869m, 789w, 777m, 740m, 598w, 537w
2-Nitrofluorene	1546w, 1534m, 1475w, 1455w, *1348s*, 1076w, 807w, 746m
2,7-Dinitrofluorene	1543m, 1535m, *1349s*, 1262m, 1079w, 1037w, 1029m, 1016w, (819, 816)m, 807m, 744w
1-Nitropyrene	1562w, *1528s*, 1358s, 1349s, 1343s, 1317m, 1239w, 1195w, 1184w, 886m, 855m, 847w, 833w, 800m, 758w, 711m, 531w
1,3-Dinitropyrene	1570w, (1544, 1540)m, 1529m, 1463w, 1424w, (1382, 1380)w, *1360s*, 1347s, 1316w, 1310m, 1255w, 1225m, 1195w, 900m, 883w, 850m, 837w, 799m, 711w, 501w
1,6-Dinitropyrene	1540m, 1527w, 1383w, 1357m, *1350s*, 1316m, 1198w, 1184w, 855w, 850m, (818, 815)w
1,8-Dinitropyrene	1570w, 1539s, 1529m, 1460w, 1355s, *1348s*, 1320w, 1198w, 1029w, 902w, 868m, 863w, 835w, (820, 817)w, 801w, 774w

^{a} Abbreviations: s, strong; m, medium; w, weak.
^{b} Parentheses indicate multiplets; strongest bands are in italics.

available. This case is illustrated in Figure 2, in which the GC/MS total-ion plot for the synthetic mixture is shown (11). The unresolved peaks labeled 9 and 10 are due to two isomers of ethylnaphthalene. These peaks could not be assigned correctly using only the retention and mass spectral data, but they were identified correctly as corresponding to 2- and 1-ethylnaphthalene by FTIR detection.

Table 4. Components of the Known Mixture and Results of the IR Computer Search

Peak	Compound	Results of IR Computer Search
1	N-Methylaniline	N-Methylaniline
2	2,3-Dimethylphenol[a]	5-Ethyl-m-cresol
3	2,6-Dimethylaniline	2,6-Dimethylaniline
4	Naphthalene	Naphthalene
5	Isoquinoline	Isoquinoline
6	2-Methylnaphthalene	2-Methylnaphthalene
7	2-Methylquinoline	2-Methylquinoline
8	1-Methylnaphthalene	1-Methylnaphthalene
9	2-Ethylnaphthalene	2-Ethylnaphthalene
10	1-Ethylnaphthalene	1-Ethylnaphthalene
11	2,6-Dimethylnaphthalene	2,6-Dimethylnaphthalene
12	1,4-Dimethylnaphthalene	1,4-Dimethylnaphthalene
13	2,3-Dimethylnaphthalene[a]	1,4-Dimethylnaphthalene
14	1-Cyanonaphthalene	1-Cyanonaphthalene
15	Dibenzofuran	Dibenzofuran
16	2-Nitronaphthalene	2-Nitronaphthalene
17	9-Fluorenone[a]	1-Indanone
18	Dibenzothiophene	Dibenzothiophene
19	Phenanthrene	Phenanthrene
20	Carbazole	Carbazole

[a] Spectrum not in library.

Further examples of the complementary nature of the techniques of GC/MS and GC/FTIR were presented by Chiu et al. in conjunction with the analysis of a complex environmental sample (11). A limitation of the analysis of PACs by capillary column GC/FTIR has been raised by several workers (12, 13). Gurka et al. concluded that GC/FTIR is 40–440 times less sensitive than GC/MS using similar chromatographic conditions (12). Furthermore, it was shown that some PACs show a very weak IR response when compared with 15 typical environmental contaminants. In this study, the minimum identifiable amount of phenanthrene for GC/FTIR was 2.2 μg. In another paper, Gurka et al. reported that the sensitivity order of several types of environmental contaminants studied by GC/FTIR is: organophosphorus and thio-organophosphorus compounds > aliphatic organochlorides > polynuclear aromatics (13).

The above discussion pertains to "normal" GC/FTIR, in which the GC effluent, or some split portion of the effluent, is passed through a gold-coated light pipe with IR-transparent windows while interferograms are continu-

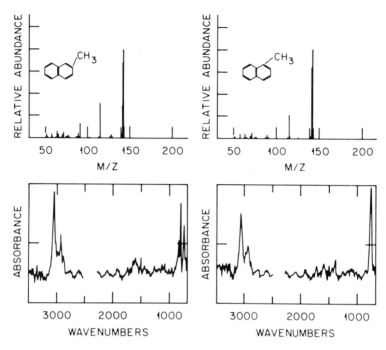

Figure 1. Mass spectra and IR spectra of 2- and 1-methylnaphthalenes. The spectral region from 2450 to 2300 cm^{-1} in the IR was not reported because of CO_2 background.

ously collected. Interferograms collected while a component from the mixture passes through the light pipe are co-added and Fourier transformed to produce the IR spectrum of the component.

Another GC/FTIR detection scheme for PACs has been described by Hembree et al. (14). This scheme utilizes the technique of matrix isolation (see below). This detection system provides for the trapping of eluted species in a matrix of frozen, IR-transparent gas on a rotating gold-plated disk. Schematic diagrams of the detection apparatus and of the GC interface to the detection apparatus are shown in Figures 3 and 4, respectively (14). Reflectance–absorbance spectra of the trapped eluted components were recorded after the GC run was completed. Characterizations of synthetic and complex real samples were made. The detection limit of the technique for one compound, pyrene, was 710 ng. This technique provides for a way to increase the signal/noise ratio of IR spectra of components in a GC effluent. Interferograms are collected after the GC run is finished, allowing for any number of co-additions. Also, because of the nature of MI, IR bands are narrowed, thereby reducing the possibility of overlap of bands from different com-

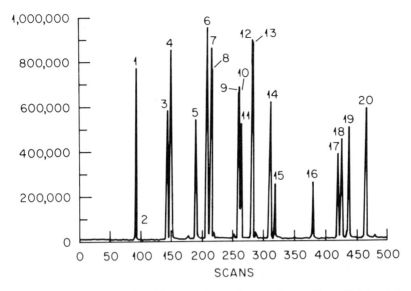

Figure 2. GC/MS total-ion plot of known mixture. GC conditions: 15-m DB-5 fused silica capillary column, temperature programmed from 45°C to 250°C at 5°C/min.

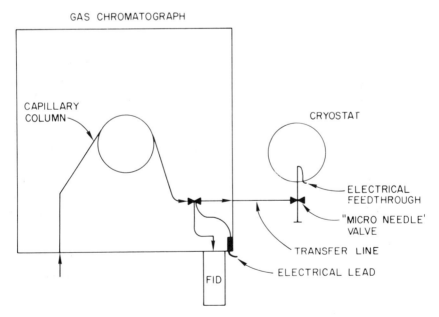

Figure 3. Schematic diagram of experimental system for gas chromatographic detection by matrix isolation FTIR spectrometry. FID stands for flame ionization detector.

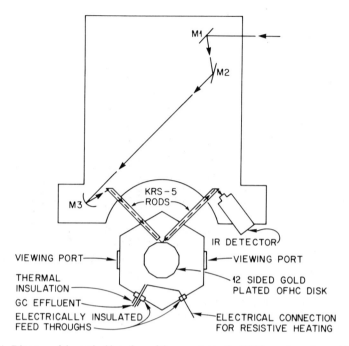

Figure 4. Diagram of the optical interface of the cryostat to the FTIR spectrometer and fittings in the vacuum shroud. M1 and M2 are planar mirrors; M3 is a spherical mirror. Details of the vacuum seals in the shroud are not shown.

pounds. Since MI IR spectra closely resemble gas-phase spectra, normal gas-phase spectral libraries may be searched for possible identification of components. However, analysis of a real sample showed that poor chromatographic resolution reduced the probability of correct identification based on searched libraries. This means that identification has a greater dependence on comparison to reference spectra previously collected using the same system. This problem could be solved by utilizing a higher-resolution chromatographic system such as a capillary column rather than a wide-bore column, which was used for this study. Long analysis times are virtually guaranteed with this technique. The authors reported a data collection time of 37–90 min for each spectrum after the chromatographic run. Also, since there were only 12 sides to the deposition surface, the number of eluted bands which can be spectroscopically studied was limited. Other approaches to this problem have been recently reviewed (5).

Several papers have appeared in which an FTIR spectrometer was used as the detector for PACs in liquid chromatographic effluent. Of course, one of the largest problems associated with IR detection of LC effluent is the strong absorption of the solvent in regions of interest. This encourages the use of solvent systems that provide an IR window in the mobile phase such that the analyte species can be detected without spectral overlap. For example, Johnson and Taylor have used Freon-113 as the mobile phase in LC separations of solutions containing aliphatic, monoaromatic, and polyaromatic hydrocarbons with FTIR detection (15). The IR windows of Freon-113 are shown in Figure 5 (15). These workers used 1- and 0.2-mm flow cells in the sample beam of an FTIR spectrometer to detect the chromatographic effluent of real and model solutions containing a number of PACs. Separations were achieved using semipreparative, analytical, and microbore columns using silica gel as the stationary phase. Real samples of the nonpolar fraction of solvent-refined coal and jet fuel were also separated using these systems. In each case, a class separation was observed; thus, the aliphatics eluted first, followed by the monoaromatics and then the polyaromatics.

A novel method for recording IR spectra of components separated by HPLC has been reported by Gagel and Biemann (16). This method involves continuous deposition of the effluent on a rotating reflecting surface. Spectra

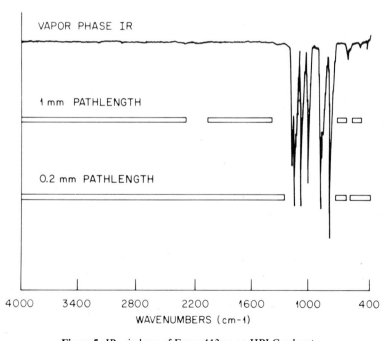

Figure 5. IR windows of Freon-113 as an HPLC solvent.

of the deposited effluent are subsequently collected with a standard reflectance accessory incorporating beam condensing optics. A schematic of the deposition apparatus is shown in Figure 6 (16). The effluent from the LC column, after passing through a standard UV detector, is passed through a nebulizer and is then directed to the reflective surface for deposition through a syringe tip. The nebulized effluent emerges as a fine spray, and nonvolatile components are deposited on the continuously rotating reflecting surface. Traces of solvent are removed by passing a gentle stream of nitrogen over the deposit. The deposition disk is attached to a reflectance accessory with beam condensing optics for collection of IR reflection absorbance spectra of the deposition.

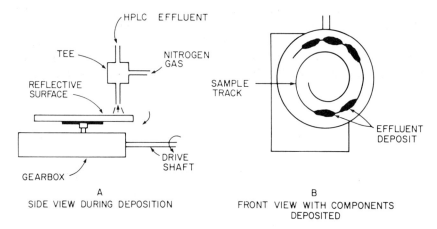

A
SIDE VIEW DURING DEPOSITION

B
FRONT VIEW WITH COMPONENTS DEPOSITED

Figure 6. Diagrams of the continuous collection device. (*A*) Deposition of HPLC effluent. (*B*) View of deposited components.

This technique was demonstrated with a five-component mixture of PACs. It was demonstrated that a 40-min chromatogram could be collected on the rotating disk. Less than 125 ng of anthracene injected on the column could be detected using this technique.

Several advantages of this technique should be pointed out. First of all, the solvent used is of no consequence to the collected spectra since the solvent is evaporated prior to data collection. This removes the handicap of insufficient IR windows in LC mobile-phase solvents. Also, since FTIR data are not collected until after the chromatographic separation is complete, there are no constraints on data collection time. Thus, large numbers of scans may be co-added to improve S/N.

4. SPECIALIZED TECHNIQUES

4.1. Photoacoustic Spectroscopy

In photoacoustic spectroscopy (PAS), IR energy is absorbed by the sample in an acoustically sealed cell, resulting in vibrational excitation of molecules in the surface of the sample. Some of this vibrational energy is converted to thermal energy by nonradiative decay. This thermal energy is transmitted to the gas surrounding the sample in the form of pressure variations. These pressure variations, or acoustic waves, are detected by a microphone and are transduced to an electrical signal corresponding to the IR absorbance signal. Rosencwaig has treated the theory of PAS in detail (17).

PAS in the IR region (IR/PAS) has been studied as a technique for the analysis of PACs adsorbed on particulate surfaces (18, 19). PACs were adsorbed on the surface of alumina and silica gel by evaporation of a solution of the PACs dissolved in appropriate solvent. The coverage of the particle surface by the PACs was estimated to be 2.2 monolayers and 0.2 monolayers of 9-nitroanthracene on alumina and silica gel, respectively (18). It was estimated that the minimum detectable amount of 9-nitroanthracene in the sample cell is approximately 65 μg (19). It was found that the PAC signal/noise ratio is very dependent on the surface area exposed to the incident radiation (19). For a given monolayer coverage, the greater the surface area, the greater the PAC signal. This principle was demonstrated by taking the IR/PAS spectrum of a whole KBr disk containing 5% nitroanthracene and comparing it with an IR/PAS spectrum of the ground disk. The spectrum of the ground disk shows a much higher signal/noise ratio than the unground disk, as can be seen in Figure 7 (19).

The analytical utility of the IR/PAS technique is questionable because of the poor sensitivity, which requires relatively large amounts of analyte to observe spectral features. Determining quantities of adsorbed species on particulate matter is better handled by extraction techniques to remove the adsorbed species for subsequent analysis. The real utility of the IR/PAS technique arises from the fact that it allows for nondestructive observation of the adsorbed species and its interactions with the substrate.

Cabaniss et al. have studied the interaction of 9-nitroanthracene with the alumina substrate on which it is adsorbed (18). Comparison of literature values for vibrational modes of the compound with bands observed in the spectrum of the species adsorbed on alumina indicated good agreement for all bands except one. The asymetrical N–O stretch band was shifted to a higher energy by 12 cm^{-1}. The authors suggest that this perturbation is caused by interaction of the alumina substrate with one of the oxygens in the nitro

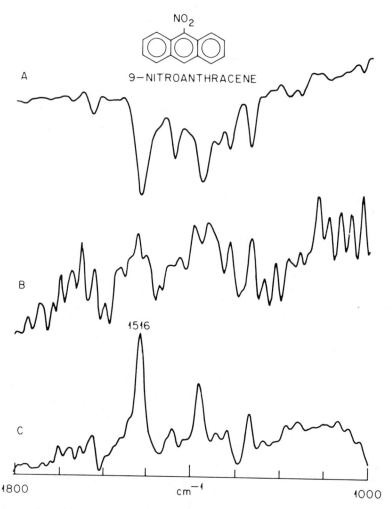

Figure 7. Transmission and photoacoustic spectra of 9-nitroanthracene in KBr. (*A*) Transmission spectrum of pressed pellet. (*B*) Photoacoustic spectrum of pressed pellet. (*C*) Photoacoustic spectrum of ground pellet.

group. Chemisorption of the PACs by way of hydrogen bonding to surface water was favored.

Photochemical reactions of adsorbed species have also been studied by Cabaniss et al. (18) and Saucy et al. (19) using IR/PAS. For these experiments, about one monolayer of 9-nitroanthracene was solution deposited on the surface of silica gel. The mixture was then dispersed on a glass slide and exposed to UV radiation for 11 h. Spectra of the sample before and after irradiation are shown in Figure 8 (19). The nitroanthracene was clearly diminished in the irradiated sample, and bands corresponding to anthraquinone are present. The presence of anthraquinone and the disappearance of nitroanthracene were confirmed by GC analysis of the extract. IR/PAS has thus been shown to be useful for irradiation studies of PACs adsorbed on particulate substrates. However, relatively large amounts of PACs are required. The main advantage is that IR/PAS is a nondestructive technique, and the adsorbed species need not be extracted to be studied.

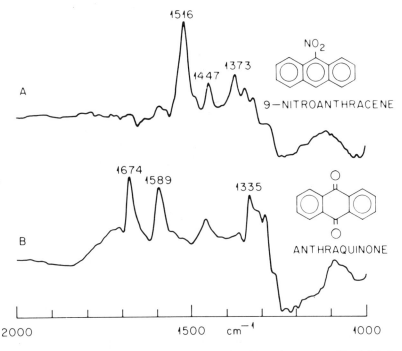

Figure 8. Photoacoustic spectra of ~ 1 monolayer of 9-nitroanthracene-on-silica (*A*) before irradiation and (*B*) after 11 h of UV irradiation.

4.2. Triplet-State Spectroscopy

Excited triplet-state IR spectrometry (ETSIR) has been suggested as an analytical technique (20). In this technique the FTIR spectrum is simplified by reducing the number of bands in a spectrum and thereby reducing the possibility of overlap of bands in a mixture. This is accomplished by creating a steady-state population of molecules in their triplet state by UV excitation of a frozen, dilute solution.

Bugay and Leenstra (20) produced excited triplet-state spectra of a mixture of naphthalene and naphthalene-d_8. A 0.1 M solution of the mixture in Nujol was pipetted into a KBr IR cell with a path length of 0.2 mm and was frozen at about 90 K using a liquid-nitrogen cold finger. The sample was excited with a UV arc lamp simultaneous to data collection. A schematic of the experiment is shown in Figure 9 (20).

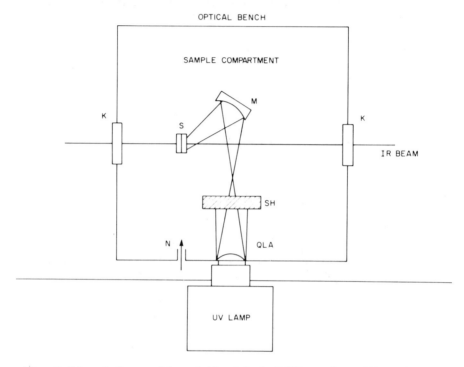

Figure 9. Schematic diagram of the optical bench for the ETSIR experiment. Abbreviations: M, reflective mirror; K, potassium bromide optical windows; S, sample cell; SH, Shutter; QLA, quartz lens assembly; N, nitrogen inlet.

Spectra were collected with the UV source "on" as well as "off" so that both ground-state and excited-state spectra were obtained. The spectrum taken with the UV source "on" contained features due to the ground-state absorptions as well as to the excited triplet-state absorptions. To produce a spectrum containing features due only to the triplet-state absorptions, a ratio of the triplet-state spectrum to the ground-state spectrum was made. This spectrum exhibits both positive-going and negative-going bands, as can be seen in Figure 10 (20). The positive-going bands are a result of excited triplet-state

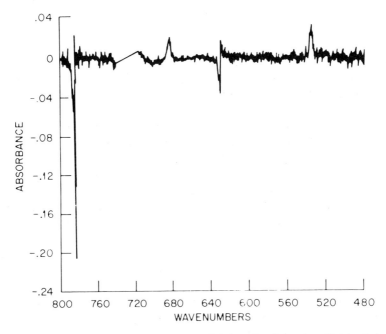

Figure 10. ETSIR spectrum of a naphthalene/naphthalene-d_8 mixture in a Nujol solution at 0.5-cm^{-1} resolution. The 740–715-cm^{-1} region is blanked because of excessive absorption of Nujol.

absorptions. The negative-going bands are caused by ground-state bands that have different populations depending on whether the sample is being irradiated. Thus, the positive- and negative-going bands should always appear in pairs, since the vibrational mode in the ground state that has been depopulated should show an absorption in the populated excited state. The relative concentration of the triplet-state can be determined from the following equation (20):

$$\% \text{ triplet concentration} = \frac{\text{area } T_1 \text{ band}}{\text{area } S_0 \text{ band}} \times 100 \qquad (1)$$

with the S_0 band measured when the excitation source is "off". Bugay and Leenstra reported a triplet-state concentration of about 2% using this method (20).

The same type of ETSIR spectrum may be obtained by taking a difference between excited- and ground-state spectra as by taking a ratio of these two spectra. The difference method has been demonstrated by Baiardo et al. for triphenylene (21). The method used was similar to that of Bugay and Leenstra, with the exception that Baiardo et al. used a sample frozen in a nitrogen matrix at 15 K.

Mitchell et al. have obtained triplet-state IR spectra of acridine and phenazine isolated in an argon matrix at 13 K (22). For these experiments, an argon-ion laser was used as the excitation source. The excitation source was modulated, and IR data were collected continuously. A lock-in amplifier was used.

The simplified spectra produced by the ETSIR method could conceivably be used to identify specific compounds in a mixture of unknown composition. A library of ETSIR spectra would be necessary. No data on the sensitivity of the technique have been reported, although the sensitivity is presumed to be poor because it depends on the population of the excited state. Since the population of the excited state will be much less than the population of the ground state, the utility of this technique seems limited to mixtures with fairly high concentrations of analyte species.

4.3. Other Techniques

Several other papers have appeared which pertain to the analysis of PACs by an IR technique but which do not fit into the categories previously discussed. These will be mentioned here.

Mielczarski et al. have studied the application of attenuated total reflectance (ATR) IR spectroscopy to the characterization of coal (23). These workers studied the ATR spectra of solid coal ground to a particle size of 50–60 μm and less than 50 μm. The theory of ATR is treated exhaustively in the monograph by Harrick (24). Spectra of the coal were collected after varying times of low-temperature oxidation. Correlations of the intensity of the band at 1600 cm^{-1} to the rank of the coal were made. An inverse relationship between the intensity of this band and the degree of condensation of aromatic rings on the coal was suggested. Lower-rank coals exhibit a lower degree of condensation of aromatic rings than do higher-rank coals, and so the suggestion that the intensity of this band may be correlated to rank may be valid. There are other possible explanations for the increase in the intensity of the band at 1600 cm^{-1}, such as the presence of phenolic groups, which must be considered.

The IR spectra of 11 PACs at several pressures up to 45 kbar were studied by Hamann (25). Spectra of the PACs at various pressures were collected using a diamond anvil cell. Seven of the 11 PACs examined showed evidence of having undergone phase transformations in the pressure range 15–40 kbar at 25°C. Phase transformations were indicated in the spectra by a reduction in the number of IR bands observed, indicating a change to a more symmetrical arrangement. The molecules in their normal crystal structures are tilted with respect to each other and to the crystal axis. The high pressure apparently causes a compression that results in a simple two-dimensional close-packed arrangement in stacked parallel layers.

5. CONCLUSION

The analysis of PACs by IR spectroscopy has been made possible to a large degree by the widespread availability of FTIR spectrometers. Real samples generally contain very low levels of PACs, which require highly sensitive techniques for detection after either optical or physical separation of the PACs from a matrix environment. Future improvements in detection limits of PACs by IR spectroscopy will probably come as a result of improvements in FTIR-based chromatographic detectors. Although FTIR spectroscopy will probably never be as sensitive as mass spectrometry, the complementary information obtained will be an impetus for future improvement in sensitivity.

ACKNOWLEDGMENTS

We would like to thank E. L. Wehry, Arlene Garrison, and Don Dunstan for useful comments.

REFERENCES

1. National Academy of Sciences, *Particulate Polycyclic Organic Matter*, Washington, D.C. (1972).
2. R. M. Silverstein and G. C. Bassler, *Spectroscopic Identification of Organic Compounds*, Wiley p. 99 (1967).
3. E. Whittle, D. A. Dows, and G. C. Pimentel, *J. Chem. Phys.*, **22**, 1943 (1954).
4. G. Mamantov, A. A. Garrison, and E. L. Wehry, *Appl. Spectrosc.*, **36**, 339 (1982).
5. E. L. Wehry and G. Mamantov, *Progr. Anal. Spectrosc.*, **10**, 507 (1987).
6. A. A. Garrison, G. Mamantov, and E. E. Wehry, *Appl. Spectrosc.*, **36**, 348 (1982).
7. P. J. Stout and G. Mamantov, *Appl. Spectrosc.*, **41**, 1048 (1987).
8. S. S. Hecht, W. E. Bondinell, and D. Hoffman, *J. Nat. Cancer Inst.*, **53**, 1121 (1974).
9. S. S. Hecht, M. Loy, and D. Hoffman, in *Carcinogenesis—A Comprehensive Survey, Vol. 1*, R. I. Freudenthal and P. W. Jones, Eds., Raven Press, New York, p. 325 (1976).
10. W. H. Griest, B. A. Tomkins, J. L. Epler, and T. K. Roa, in *Polynuclear Aromatic Hydrocarbons*, P. W. Jones and P. Leber, Eds., Ann Arbor Science Publishers, Ann Arbor, Mich. (1979).
11. K. S. Chiu, K. Biemann, K. Krishnan, and S. L. Hill, *Anal. Chem.*, **56**, 1610 (1984).
12. D. F. Gurka, M. Hiatt, and R. Titus, *Anal. Chem.*, **56**, 1102 (1984).

13. D. F. Gurka, M. Umana, E. D. Pellizzari, A. Moseley, and J. de Haseth, *Appl. Spectrosc.*, **39,** 297 (1985).
14. D. M. Hembree, A. A. Garrison, R. A. Crocombe, R. A. Yokley, E. L. Wehry, and G. Mamantov, *Anal. Chem.*, **53,** 1783 (1981).
15. C. C. Johnson and L. T. Taylor, *Anal. Chem.*, **55,** 436 (1983).
16. J. J Gagel and K. Biemann, *Anal. Chem.*, **58,** 2184 (1986).
17. A. Rosencwaig, *Photoacoustics and Photoacoustic Spectroscopy*, Wiley, New York (1980).
18. G. E. Cabaniss, D. A. Saucy, and R. W. Linton, in *Polycyclic Aromatic Hydrocarbons: Formation Metabolism and Measurement*, 7th edition, M. Cooke and A. J. Dennis, Eds., p. 243 (1982).
19. D. A. Saucy, G. E. Cabaniss, and R. W. Linton, *Anal. Chem.*, **57,** 876 (1985).
20. D. E. Bugay and W. R. Leenstra, *Anal. Chem.*, **58,** 2335 (1986).
21. J. Baiardo, R. Mukherjee, and M. Vala, *J. Mol. Struct.* **80,** 109 (1982).
22. M. B. Mitchell, G. R. Smith, and W. A. Guillory, *J. Chem. Phys.*, **75,** 44 (1981).
23. J. A. Mielczarski, A. Denca, and J. W. Strojek, *Appl. Spectrosc.*, **40,** 998 (1986).
24. N. J. Harrick, *Internal Reflection Spectroscopy*, Interscience, New York (1967).
25. S. D. Hamann, *High Temp. High Pressures*, **10,** 503 (1978).

CHAPTER

14

RAMAN SPECTROSCOPY

D. L. GERRARD AND H. J. BOWLEY

BP Research Centre
Sunbury-on-Thames
Middlesex, England

1. INTRODUCTION

Since Raman spectroscopy is still very rarely used as a broad-based analytical technique it is worth while, at the outset, to consider very briefly the nature of the Raman effect, because this is what gives the technique some distinct advantages over other spectroscopic methods. If a molecule is irradiated with photons of a single fixed frequency, v_0 (energy $= hv_0$), then apart from the well-known effects of reflection and absorption, two molecular scattering processes can also occur. The first of these is elastic scattering—that is, there is no change in energy. This is known as the *Rayleigh scatter* and gives rise to photons in the scattered beam with energy hv_0. Other collisions between the incident photons and the molecule may, however, be inelastic. These collisions may cause the molecule to undergo a quantum transition to a higher energy level, with the result that the photon giving rise to this transition loses energy and is scattered with a lower frequency, v_R. In this case, $\Delta v = v_0 - v_R$ and has a positive value. If the molecule undergoing the interaction is already in an energy level above its lowest one, then the collision with a photon may cause a transition to a lower energy, in which case the scattered photon will have a higher frequency, $v_{R'}$. In this case $\Delta v = v_0 - v_{R'}$ has a negative value. Thus the observed Raman shifts, $v_0 - v_R$ and $v_0 - v_{R'}$, are equivalent to the energy changes involved in transitions of the scattering species and are therefore characteristic of it. Also, since under normal conditions of temperature and pressure the population of a higher energy level is less than that of a lower level and falls off exponentially with the energy, the Raman shifts with positive values of Δv are significantly more intense than those with negative Δv, and for the latter the intensity falls off rapidly as $|\Delta v|$ increases.

The range of frequencies in the scattered beam (positive and negative Δv) constitute the Raman spectrum of the molecule being irradiated. Frequency shifts with positive Δv are known as *Stokes lines*, and those with negative Δv

are *anti-Stokes lines* (or *bands*). Hence, the overall Raman spectrum comprises a range of frequency shifts on either side of the Rayleigh line, normally expressed in units of Δ cm^{-1}, which are mirror images of each other in terms of Δ cm^{-1} (the Stokes and anti-Stokes spectra), and this is illustrated in Figure 1. This shows the overall Raman spectrum for a relatively simple molecule, carbon tetrachloride. The intense band at $\Delta v = 0$ cm^{-1} is the Rayleigh line, and on either side are the Stokes (positive Δv) and anti-Stokes (negative Δv) Raman lines. The figures associated with the peaks are the photon counts recorded on the detector for the appropriate band. It can be seen from these values that the Stokes spectrum is about 20 times stronger than the anti-Stokes spectrum and that the intensity of the anti-Stokes spectrum relative to the Stokes spectrum falls off with increasing Δv, although this latter effect is not particularly strong because of the relatively short range of Δv covered in this spectrum.

These observed Raman shifts correspond to vibrational (or, in the case of gases, rotational) transitions of the scattering molecule. Such frequencies, when observed by absorption techniques, occur in the infrared region of the spectrum. In Raman spectroscopy, since v_0 is normally in the UV/visible region, the observed Raman bands are also in this region. For example, if the value of v_0 is 20,000 cm^{-1} (i.e., $\lambda = 500$ nm), then the C–H stretching mode of a methyl group will come at a Δv of abour 2950 cm^{-1}, causing a shift in the scattered photons to a new frequency of $20{,}000 - 2950 = 17{,}050$ cm^{-1} or $\lambda = 586.5$ nm (Stokes). It is worth noting at this point that for most practical purposes it is only the Stokes spectrum that is used in conventional Raman spectroscopy. Much more highly specialized nonlinear Raman effects may use the anti-Stokes spectrum, but these techniques lie outside the area of reference of this chapter, and the reader is referred to the excellent book by Harvey (1) for a detailed account of the various types of nonlinear Raman spectroscopy. The area of interest to the vibrational spectroscopist is normally in the region 0–4,000 Δ cm^{-1} and so for most Raman work this will be within about 150 nm of the wavelength used to obtain the spectrum (known as the *exciting wavelength*, λ_e). This gives several advantages in terms of detection and sample handling. Very high efficiency detectors are available in the UV/visible region of the spectrum, notably photomultiplier tubes, diode arrays, and, more recently, charge coupled devices (2,3). Sample handling is simple because the fact that the Raman effect relies on a scattering rather than an absorption process means that solids can be examined directly by supporting them in the laser beam, and size, shape, and transparency are not limitations. Liquids and gases can be contained in glass or quartz vessels (usually depending on the value of λ_e), neither of which give stronger Raman spectra.

Although the Raman technique observes vibrational transitions, as does infrared absorption spectroscopy, it is a mistake to see Raman spectroscopy

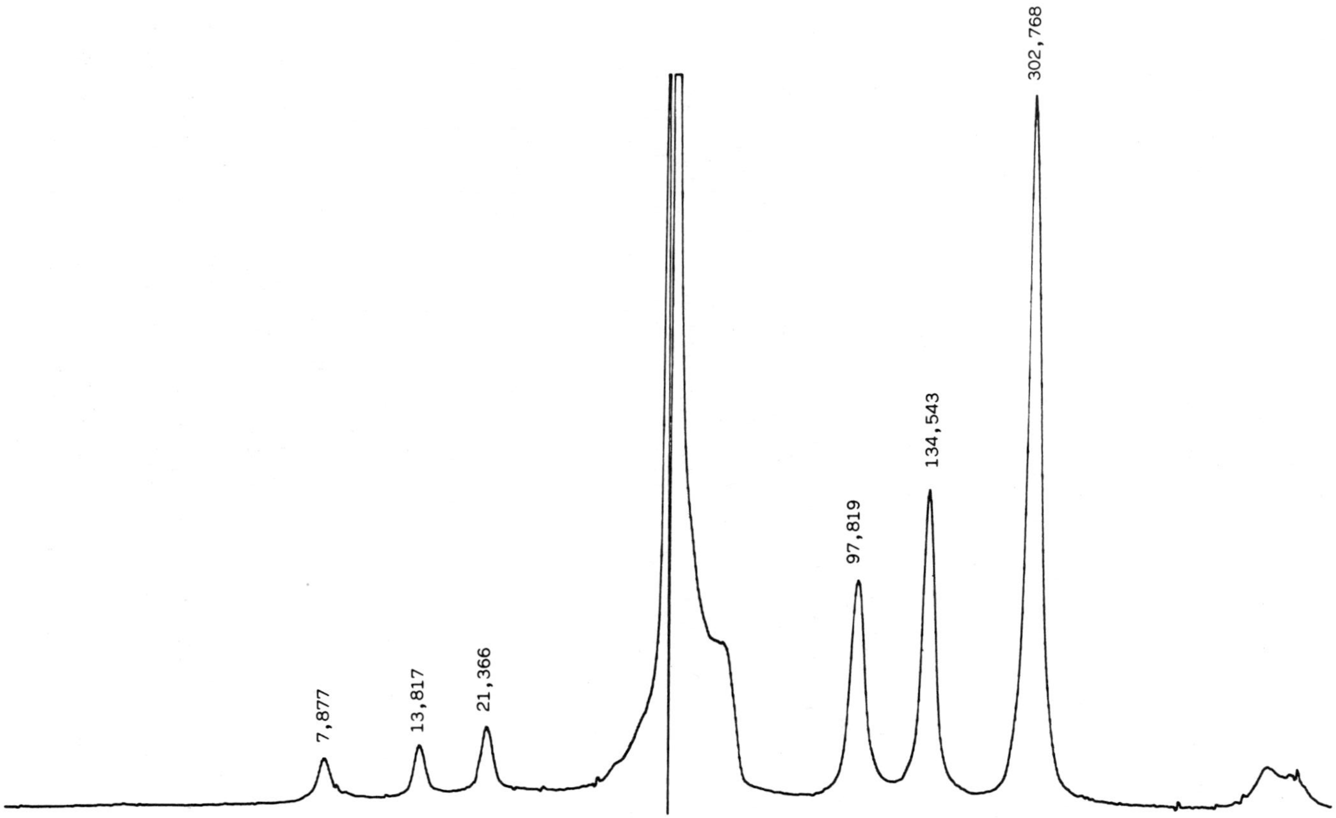

Figure 1. Raman spectrum showing a range of frequency shifts on each side of the Rayleigh line.

302,768

134,543

97,819

21,366

13,817

7,877

435

as just an alternative way of obtaining results that could be obtained by the infrared method. The two techniques differ in their mechanism, the Raman effect being a scattering phenomenon and the infrared being an absorption effect. Hence the data obtainable by the two methods are not identical. Frequencies permitted in the infrared may be forbidden in the Raman and vice versa. The two techniques are therefore complementary in nature, and data from both are required to give a full vibrational picture of any molecule. It is often the case that vibrational modes that are strong in the infrared are weak in the Raman and vice versa, and this often means—especially when dealing with mixture of compounds—that one technique will be much more effective than the other from a qualititative and /or quantitative point of view.

Since it deals with vibrational transitions, the Raman technique, like the infrared, can be used for characterization and identification of unknown compounds, as of course can other spectroscopic techniques such as nuclear magnetic resonance and mass spectrometry. The great disadvantage with the Raman effect for this purpose is its very weak nature—typically only 1 in 10^7, or less, of the incident photons exhibit the effect. It was not, therefore, until the development of continuous-wave (CW) gas ion lasers in the late 1960s that the technique began to realize some of its latent potential as an analytical method. By this time, other spectroscopic techniques were well established and had such large data bases that for routine characterization and identification of unknown samples conventional Raman spectroscopy is not the preferred method. However, as indicated above, it does have some advantages in terms of sample handling:

1. Glass or quartz (both very weak Raman scatterers) can be used to contain the sample. This gives a great deal of scope for presenting the sample to be analyzed to the spectrometer. It also means that samples that may be affected by exposure to air can be readily studied in sealed glass–quartz containers.

2. Water (also a weak Raman scatterer) is an ideal solvent for Raman work. In particular, this can be of value in biological and pollution studies.

3. Low-frequency vibrations (e.g., in macromolecules) are easily obtained in the Raman spectrum without the need for any ancillary equipment.

4. The use of visible light means that it is a simple procedure to use a conventional visible microscope attached to a spectrometer for Raman studies. This enables samples down to about 1 μm in size to be examined routinely.

5. At the other end of the scale the scattering nature of the Raman effect means that virtually any size or shape of sample can be handled.

The hardware necessary for Raman studies is essentially very simple and, once it is working, has a relatively long lifetime and very little downtime. It comprises:

1. Some means of holding/containing the sample.
2. A monochromatic light source (laser).
3. A collection optic to collect the Raman scattered photons coming off the sample. This is usually a camera lens (visible), a Cassegrain (UV/visible), or a parabolic mirror (UV).
4. A monochromator. This may be single, double, triple, or single plus double.
5. A photon detector. This can be single channel (photomultiplier tube) which is normally used with a double monochromator, the combination giving a high degree of resolution. This can be used in either the visible UV but is normally best with CW lasers. Alternatively, multichannel devices can be used—typically diode arrays or charge-coupled devices. Such detectors are of greatest value when used with pulsed lasers and combined double–single monochromators. Resolution is normally inferior to the photomultiplier-tube–double-monochromator system, but speed is considerably greater.
6. A data handling system. Commercial systems are not yet comparable with those available for infrared data handling, but transfer of data from a Raman spectrometer to an infrared instrument for the purposes of spectral enhancement is quite feasible.

In principle, then, the technique of Raman spectroscopy is very simple and versatile, but, for the reasons outlined above, its strength does not normally lie in the routine qualitative or quantitative analysis of unknown samples. The reason for this is that better established methods are available which are often more sensitive, more rapid, or more generally applicable because of their better established and larger data bases. So, unless one of the specific sampling advantages which the nature of the Raman effect confers can be used effectively, then infrared absorption spectroscopy, NMR, or mass spectrometry will normally be preferred.

In addition to the disadvantages of Raman spectroscopy which have already been mentioned, it has another major drawback, which is often encountered when dealing with polycyclic aromatic compounds. This is the fact that the laser light being used to produce the Raman signal may well produce, in addition, a broad fluorescence spectrum in the same spectral region as the Raman signal. It is often the case that, when this occurs, the fluorescence is so intense that it can obscure the Raman signal. In such cases, it may be possible to extract the Raman signal by computer subtraction of a simulated fluorescence background. This is illustrated in Figure 2, which shows two spectra of pentacene: (a) the untreated data obtained using $\lambda_e = 514.5$ nm and (b) the same data after computer subtraction of a simulated fluorescence background.

An alternative method is to use a value of λ_e which produces either no fluorescence or a very much reduced fluorescence. This usually involves either using a laser line in the far-red region of the spectrum [typically the 752.5- or 799.3-nm lines of a krypton ion laser (4)] or, as a result of recent work by Chase and co-workers (5,6), using a near-infrared laser in conjunction with an FTIR spectrometer with optics and detector suitable for operation in the near-infrared region. The use of the 1.064-μm line of a CW neodymium/YAG laser with an FTIR spectrometer is illustrated in Figure 3. This shows the spectrum of rubrene obtained (a) with $\lambda_e = 514.5$ nm, (b) with $\lambda_e = 632.8$ nm, (c) with $\lambda_e = 752.5$ nm, and (d) with $\lambda_e = 1064$ nm. This shows the value of near-infrared lasers for the analysis of fluorescent materials.

In the specific case of polycylic aromatic compounds, where does the value of Raman spectroscopy lie? The Raman spectrum will give additional vibrational information not available in the infrared spectrum which will be of value for detailed structural analysis. On the whole, however, for conventional Raman studies, it is only in those cases where use can be made of the handling advantages mentioned above that Raman spectroscopy will be the preferred technique for the analyst. It is largely for this reason that the majority of the reported Raman work on polycyclic aromatic systems to date has related to biological and biochemical applications where the presence of large amounts of water is a key factor and where the spatial resolution of Raman microscopy can be used to advantage.

There is, however, one other feature of the Raman effect which, when it can be used to advantage, opens up considerable possibilities for the analysis of unsaturated systems in general and polycyclic aromatic (PCA) systems in particular—this feature is the *resonance Raman effect*.

2. THE RESONANCE RAMAN EFFECT

When the laser wavelength being used to obtain the Raman spectrum approaches or enters a region of electronic (vibronic) absorption of the molecule being irradiated, then the observed spectrum may exhibit resonance enhancement. Resonance excitation raises the scattering power of many substances, including PCAs, often by several orders of magnitude. In many cases, this enables spectra to be obtained for such compounds at extremely

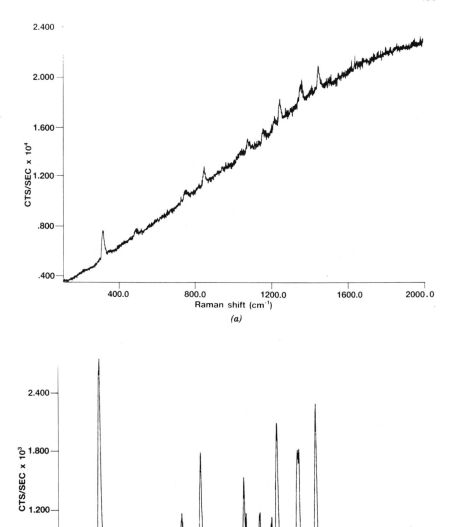

Figure 2. Two Raman spectra of pentacene. (*a*) Untreated data obtained using $\lambda_e = 514.5$ nm. (*b*) The same data after computer subtraction of a simulated fluoroscence background.

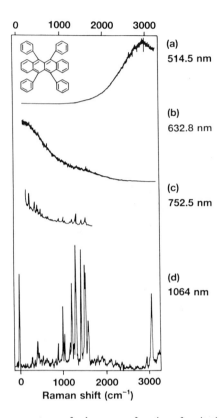

Figure 3. Raman spectrum of rubrene as a function of excitation wavelength.

low levels, and it is in this respect that the greatest value of Raman spectroscopy lies for the analysis of PCAs. For a detailed account of the theory of the resonance effect, the reader is referred to the article by Behringer (7). It is worth noting with respect to the resonance spectrum that only vibrations associated with the chromophore producing the electronic absorption exhibit resonance enhancement. Another important point to be aware of is that the closer the exciting line is to the electronic transition (i.e., the closer λ_e to λ_{max}, the absorption maximum), the greater the resonance enhancement.

The CW gas ion lasers that are normally used for Raman spectroscopy are the argon and krypton ion lasers, which give a range of wavelengths shown in Table 1. Apart from a small number of lines in the near-UV, the majority of these output wavelengths lie in the visible region of the spectrum. Another useful laser wavelength is the 632.8 nm availabe from a helium–neon laser. Since it is usually desirable to use a value of λ_e as close to λ_{max} as possible, a

Table 1. Wavelengths Available from Argon and Krypton Ion Lasers (nm)

Ar$^+$	Kr$^+$	Ar$^+$	Kr$^+$
528.7	799.3	454.5	476.2
514.5	793.1	379.5	468.0
501.7	752.5	363.8	415.4
496.5	676.5	351.4	413.1
488.0	647.1	351.1	406.7
476.5	568.2	335.9	356.4
472.7	530.9	334.5	350.7
465.8	520.8	333.6	337.5
457.9	482.5		323.9

much greater degree of wavelength tunability is required than can be achieved by using the range of wavelengths available from Ar$^+$ and Kr$^+$ lasers. This can be accomplished by using a dye laser. A dye laser comprises a narrow jet of a solution of an appropriate dye through which is passed a "pump" laser beam. This produces a coherent, broad-band output, which is tunable by means of an optical wedge. In this way, complete tunability can be achieved, using a high-power argon ion laser and a range of different dyes (8), over the range 400–900 nm. This gives the ability to tune the laser wavelength to coincide with the electronic absorption maximum, and thus the greatest degree of resonance enhancement can be achieved. Such a system is only useful, however, for the study of PCAs that have an absorption maximum in this spectral range (visible/near-infrared). Most PCAs exhibit strong absorptions in the UV and relatively few in the visible region, and so, before Raman spectroscopy could become a technique that is widely applicable to the analysis of PCAs, a laser source that is tunable in the UV region needed to be produced.

This occurred with the development of highly sophisticated and expensive laser systems based on dye lasers pumped by either neodymium/YAG or excimer lasers. In both cases the pulsed output from the pump laser is used in conjunction with the dye laser in much the same way as the CW laser system. The laser pulses are of very short duration (typically 5–10 ns) at a repetition rate of between 10 and 300 Hz, depending on the particular system. Although each pulse is only of low energy (usually <1 mJ), the short duration means that they are of very high power (in the kilowatt or even megawatt range). This high power produces very high efficiency in the dye laser, the output from which can be passed through a frequency-doubling crystal to produce an output in the UV. By using a range of dyes in conjunction with different types of frequency-doubling crystals and by using frequency tripling and mixing, such laser systems can give tunability over the range 190 nm to 1 μm, which covers the range where most organic compounds, including PCAs, exhibit electronic absorption. The use of such a system was first described by Asher et al. (9), and this group is still the most significant contributor in this area. The experimental expertise required for this type of study is considerable, and because of the low repetition rate of pulsed lasers, gated diode array detectors have to be used, which collect a significant portion of the total spectrum simultaneously. Typically, a diode array will collect about a 1000-cm^{-1} portion of the spectrum without the need to scan, which is necessary if a photomultiplier tube is used.

3. RAMAN SPECTROSCOPY OF POLYCYCLIC AROMATIC COMPOUNDS

3.1. Nonresonance Raman Spectroscopy

As mentioned above, the Raman spectrum is rich in vibrational information, and this can be used, as with infrared spectroscopy, for the characterization and identification of PCAs. For example, a considerable amount of work has been carried out on the Raman spectrum of naphthalene, and vibrational assignments have been made for the spectra of naphthalene itself and for several of the deuterated forms (10–12). Spectra have also been recorded for a range of α- and β-monosubstituted naphthalenes (13). The Raman spectra of these compounds show bands that can be assigned to α- and β-substitution for a variety of substituents, including OH, CH_3, CN, SH, Cl, and Br. A series of monoalkyl naphthalenes has also been studied (14), as have several dimethyl- and trimethyl-substituted naphthalenes (15). The various dimethyl compounds can be distinguished by characteristic Raman bands in the 1640–1570 and 800–550 cm^{-1} ranges.

Anthracene has been studied in solution (16, 17) and in the solid state (18) as have various mono- and disubstituted compounds (19–21). Again, the various substitution patterns give characteristic Raman bands. Other polycyclic aromatic hydrocarbons for which Raman spectra are reported include acenaphthene (22), phenanthrene (23), pyrene (24), and triphenylene (25). On the whole, the use of nonresonance Raman spectroscopy for identification of specific PCAs has been very limited because in most cases this can be done by other, more readily available, spectroscopic methods. As detailed above, it is a technique that would normally only be used analytically in those cases where one of the specific advantages conferred by the Raman effect can be used to advantage. Otherwise it is only of value in the assignment of vibrational modes to specific observed frequencies.

3.2. Resonance-Enhanced Raman Spectroscopy

This is undoubtedly the most significant area of application of the Raman technique to the analysis of PCAs, and it will be considered in some detail.

3.2.1.   Resonance Raman Spectroscopy Using CW Lasers

The resonance effect enables those compounds that exhibit it to be identified at very low concentrations. Since many polycyclic aromatic compounds have significance in the areas of biology, biochemistry, medicine, and pollution, it is often a requirement to identify and quantify them at extremely low levels. It may well be a further requirement to carry out this analysis in aqueous systems—ideal for Raman spectroscopy. The early work on resonance Raman spectroscopy was limited by the relatively small number of wavelengths available from argon and krypton ion CW lasers (~ 350–800 nm). This imposed a considerable limitation on the use of the technique for PCAs, the majority of which exhibit absorbances at much shorter wavelengths, well into the UV region. Characteristic resonance Raman spectra have, however, been obtained for a limited number of compounds which exhibit absorptions in the visible region, notably perylene, anthranthrene, 1,12-benzperylene, and naphthacene (26). The problem with these compounds is that even at very low concentrations they can exhibit such severe fluorescence that even the resonance enhanced Raman signal may be obscured. In such cases it may be possible to reduce or even eliminate this fluorescence by using a fluorescence quenching agent such as butane 2,3-dione or nitrobenzene (22). In most cases a large excess of such a compound needs to be used, and the conventional Raman spectrum of the quenching agent may obscure some of the resonance-enhanced bands of the PCA being studied. Of course, all organic dyes exhibit absorptions in the visible region and, as such, lend themselves very well to examination by Raman spectroscopy, although with such compounds the problem of fluorescence is often severe. An example of a PCA dye that has been characterized by this method is indigo carmine (27).

Another group of PCAs which has been studied by resonance Raman spectroscopy in the visible region is a series of flavin derivatives (28). Compounds containing chemical modifications in positions 7 and 8 of the isoalloxazine ring can be distinguished by their resonance-enhanced spectra. The compounds used in this work were riboflavin, 7,8-dichlororiboflavin, 8-chlororiboflavin, 8-bromoriboflavin, 8-(methylmercapto)riboflavin, 7-chlorolumiflavin, 8-norlumiflavin, and 3-CH_2COOH lumiflavin. Spectra were measured of the complexes of these compounds with riboflavin-binding protein, which also acted as a fluorescence quenching agent.

The range of wavelengths available from CW lasers in the visible region can be extended, as mentioned above, by the use of a dye laser to give complete tunability over the whole of the visible range. Even so, the vast majority of organic compounds, including PCAs, are precluded from resonance Raman studies because their electronic transitions lie in the UV. One way around this problem, which has been used with limited success, is to derivatize molecules that absorb only in the UV region to form chromophores that absorb in the visible region. This method has been used for the detection of biogenic amines, such as adrenaline, which can be readily oxidized to the corresponding aminochrome with a λ_{max} in the visible and can be detected at levels of 10^{-5} M using 488-nm excitation (29). This approach is limited to a relatively small number of compounds and may well reduce the specificity of the technique.

An alternative approach is to convert the PCA to be analyzed to a charge-transfer complex. Most PCAs are strong electron donors and, if mixed with an appropriate electron acceptor, will form charge-transfer complexes that show a strong, broad absorption that is considerably red-shifted from the electronic absorption of either the PCA or the electron acceptor. Typical electron acceptors used for this purpose are tetracyanoethylene (TCNE), tetracyanobenzene (TCNB), and tetracyanoquinodimethane (TCNQ). Bandrauk et al. (30) have studied the perylene–TCNQ system in the solid state, Umemura et al. (31) have examined the anthracene– and perdeuteroanthracene–TCNB complexes, and Kuok and Tang (32) have assigned vibrational modes to the anthracene–tetrachlorophthalic anhydride charge-transfer complex. Although the use of charge-transfer complexes in this way provides a method for studying PCAs at low concentrations, the interaction with the electron acceptor and the resonance enhancement of some of the vibrational bands of the acceptor can often produce spectra that are difficult to interpret, and a comprehensive data base needs to be established for the donor–acceptor systems of interest if the technique is to be used for identification purposes.

3.2.2. Resonance Raman Spectroscopy Using Pulsed UV Lasers

Although this is by far the most important application of Raman spectroscopy to the study of PCAs, it must be emphasized that it is a technique that is still in its early stages of development. It is extremely demanding experimentally and is expensive. The experimental difficulties and cost need to be weighed against the value of the results likely to be obtained. There is no doubt that UV resonance Raman spectroscopy provides an extremely sensitive and specific method for the identification and quantification of PCAs, and if there is a need for this type of work within any particular analytical group, then it is a method that should be seriously considered. At the present time, there is only a handful of groups worldwide undertaking this type of work, and most of these are involved with biological systems where the ability to use a tunable laser to probe chromophores within complex organic molecules provides workers in this field with a unique capability. For detailed accounts of the Raman spectroscopy of biological systems, the reader is referred to the books by Carey (33) and Tu (34).

Asher et al. (9) at Pittsburgh have been the pioneer group in the broad-based analytical application of a pulsed UV laser in conjunction with a Raman spectrometer. The system they have described still remains the best for this type of work. One of their first published spectra was that of adenine in water (9). This exhibited Raman bands at 1583, 1486, 1340, and 1249 cm^{-1} and gave a recognizable spectrum at concentrations as low as 4×10^{-5} M (~ 6 ppm).

In this case the exciting line used was 252 nm, and the time taken to obtain the spectrum was 20 min. Asher's group has also studied a limited range of polycyclic aromatic hydrocarbons using the pulsed UV laser (35). Compounds for which spectra have been obtained include naphthalene, anthracene, phenanthrene, and pyrene. In each case the compound was excited at its absorption maximum as follows:

Naphthalene	220 nm
Anthracene	252 nm
Phenanthrene	240 nm
Pyrene	240 nm

The results obtained for these compounds are given in Table 2.

Table 2. Observed Resonance Raman Bands for Some Polycyclic Aromatic Hydrocarbons (cm^{-1})

Naphthalene	Anthracene	Phenanthrene	Pyrene
766	399	386	592
	756	745	1393
	1407	1386	1622

The vibrations that exhibit the greatest resonance enhancement with UV excitation are those that provide the most significant contribution to the electronic transition producing the absorption spectrum. The spectra were obtained as solutions in acetonitrile, and, in the case of pyrene, spectra were also obtained for very dilute solutions in water (10^{-7} M). Although the limit of detection of polycyclic aromatic hydrocarbons has not yet been realistically established for a wide range of such compounds, it seems from the limited number of results so far available that detection at the parts per billion level is feasible. The technique is not only useful for the detection and quantification of such compounds, but it is also reliable for the structural differentiations of very similar species. This is illustrated by a comparison of the spectra of anthracene, 9-methyl, 2-methyl, 9-phenyl, and 9,10-diphenyl anthracenes (35), given in Table 3. Furthermore, by changing the exciting line, different species will be excited into resonance at different values of λ_e, and so it is possible to selectively excite individual classes of PCAs in mixtures. This selectivity and sensitivity could well have considerable significance in the analysis of petroleum fractions or in the study of environmental samples such as polluted water. A similar study on naphthalene, 1-hydroxynaphthalene, and 2-hydroxynaphthalene is summarized in Table 4.

Table 3. Observed Resonance Raman Bands for Some Substituted Anthracenes

Compound	Raman Bands (cm^{-1})
Anthracene	399, 756, 1407
9-Methyl anthracene	387, 752, 1395
2-Methyl anthracene	371, 1398
9-Phenyl anthracene	388, 1391
9,10-Diphenyl anthracene	398, 1299, 1393

Table 4. Observed Resonance Raman Bands for Naphthalene and Hydroxynapthalenes ($\lambda_e = 235$ nm)

Compound	Raman Bands (cm^{-1})
Naphthalene	508, 762, 1031, 1145, 1240, 1374, 1456, 1579, 1626
1-Hydroxynaphthalene	713, 1013, 1187, 1282, 1380, 1424, 1635, 1728, 1902
2-Hydroxynaphthalene	716, 768, 1294, 1393, 1465, 1588, 1623, 1855

Another advantage of the use of UV excitation is a consequence of the fact that few compounds that have their first excited singlet states below 250 nm have significant fluorescence quantum yields. Hence, excitation below 250 nm gives rise to few fluorescence problems, especially in those cases where the Raman spectrum is resonance enhanced. This has again been shown by Asher and co-workers, who used UV resonance Raman spectroscopy to characterize polycyclic aromatic hydrocarbons in a distillate (340–510°C) from hydrogenated coal (36). The coal liquid was dissolved in acetonitrile or methanol, and spectra were obtained using 256-, 234-, 230-, and 220-nm excitation. By comparison of the spectra obtained with those of standard polycyclic aromatic hydrocarbons such as naphthalene, fluorene, anthracene, phenanthrene, triphenylene, and pyrene, indications of the presence of compounds containing all of these ring structures were found. This technique has great potential for the analysis of PCAs in such samples because of its speed, sensitivity, and selectivity.

The technique has also been applied to the analysis of eluant from high-performance liquid chromatography columns. Conventional detection of PCAs is by a fluorescence detector that provides a sensitive method for indicating the presence of an eluant but does not spectroscopically identify it. Asher and co-workers (29) have applied UV resonance Raman spectroscopy to a mixture of five polycyclic aromatic hydrocarbons (naphthalene, fluorene, anthracene, pyrene, and chrysene). The mixture containing each component at a concentration of 10^{-3} M was separated on an HPLC column, and spectra were obtained for each component as it eluted using 230-nm excitation and a spectral accumulation time of 1–2 min. The spectra obtained were sufficient to identify the hydrocarbon in each case. This type of analysis could be of great value in the assessment of complex mixtures of polycyclic aromatic hydrocarbons for the presence of known carcinogens or of compounds that poison catalysts.

Another hydrocarbon that has been examined with UV excitation is azulene. Cable and Albrecht (37) obtained an excitation profile for azulene using a range of different excitation wavelengths in the UV region and used the results to confirm symmetry assignments for all of the symmetric ring vibrations.

The application of UV resonance Raman spectroscopy to the study of polycyclic aromatic hydrocarbons, particularly in systems of commercial importance, is still in its infancy, but its use in the area of biological and biochemical studies is much better established. Aromatic amino acids have been a popular subject for investigation. Tryptophan, for example, has been examined using 200-, 218-, 240-, and 266-nm excitation in dilute aqueous solution (38). Significant changes in the resonance enhancement pattern were observed for the different values of λ_e as the different chromophores that make up the overall UV absorption spectrum were excited into resonance.

The nonresonance Raman spectra of polynucleotides and DNA or RNA are complex and contain a number of bands, most of which are attributable to the nucleic acid base modes, although some features originating from the phosphate groups have been identified and they are useful probes of backbone conformation. The bases have intense Π–Π^* electronic transitions in the 260–280-nm region and thus lend themselves to study by UV resonance Raman spectroscopy. This technique gives much better sensitivity (down to

$10 \, \mu g \, ml^{-1}$), and the selectivity of the resonance effect allows the selective enhancement of a molecular species, for example, in a ligand–base complex. Thus resonance Raman spectroscopy can be used to selectively probe different aspects of nucleic acid structures (39).

4. QUANTITATIVE STUDIES

The Raman effect has an intrinsic advantage over infrared absorption spectroscopy from the point of view of quantitative analysis. This advantage is that the intensity of a Raman band is directly proportional to the concentration of the species that produces it. In the infrared case the relationship is logarithmic. This means that Raman spectroscopy has good accuracy over a very wide range of concentrations and can be used to quantify accurately two different components present in the same solution at widely different concentrations. The great drawback of the Raman technique, however, is that, because it relies on a scattering effect, the observed intensity will be a function of several parameters, including (a) sample orientation relative to the laser, (b) the collection optic, (c) laser intensity, (d) laser wavelength, (e) slit width, (f) detector, (g) spectrometer efficiency, etc. In fact, the intensity depends on every variable associated with obtaining the spectrum, since there is no constant that is equivalent to the extinction coefficient used in absorption spectroscopy. However, given a little care, very good quantitative results can be obtained.

The normal technique employed is to use an internal standard. Calibration of the compound whose concentration is to be measured against this standard is also necessary. Such a standard, whether in the solid or liquid phase, needs to fulfil certain criteria. It must be inert with respect to the system being examined. It must give a band that does not overlap with any of the bands of the compounds present in the system. It needs to give a band fairly close to the one being measured for the unknown compound to avoid problems of variable detector response. At first, these problems might appear daunting, but with a little care an appropriate material can normally be found. In the case of solution studies, a band due to the solvent is often appropriate. Quantitative studies using the resonance-enhanced spectrum of indigo carmine have been reported (27).

5. CONCLUSIONS

The value of Raman spectroscopy in the study of PCAs can be summarized as follows:

(a) Nonresonance Raman spectra will supply additional information on the vibrational structure of molecules that is not available from the infrared spectra alone. On the whole, such information is of most interest in the assignment of vibrational modes and in biological studies.

(b) From an analytical point of view, the use of nonresonance Raman spectroscopy is normally limited to situations that do not lend themselves easily to studies by other spectroscopic techniques. These include aqueous systems (particularly biological), micro- and macrosamples, *in situ* studies (particularly *in vivo*) where the use of fibre optics to transmit the laser light and collect the Raman scatter is gradually becoming popular, and the study of samples that may be sensitive to exposure to the atmosphere.

(c) The main application of Raman spectroscopy to the study of PCAs lies in the use of resonance-enhanced Raman spectroscopy and, in particular, the use of tunable UV lasers to obtain the resonance spectra. Although they are expensive and experimentally demanding, they can achieve a degree of sensitivity and specificity not available using other techniques. The spectra can be obtained sufficiently rapidly to allow the use of a Raman spectrometer in conjunction with a separation system such as HPLC. The value of complete tunability means that the technique can be used as a probe for complex mixtures and for examining different functionalities in complex molecules.

The potential of resonance Raman spectroscopy in the study of complex systems, notably oil-based and biological samples, is considerable, and although it has taken a long time to get this type of study under way there are already signs that it will be a technique of major importance in the next few years in the study of PCAs.

ACKNOWLEDGMENT

Permission to publish this work has been given by the British Petroleum Company.

REFERENCES

1. A. B. Harvey, Ed., *Chemical Applications of Non-Linear Raman Spectroscopy*, Academic Press, New York (1982).
2. J. Janesick and M. Blouke, *Sky and Telescope*, 238 (1987).
3. C. A. Murray and S. B. Dierker, *J. Opt. Soc. Am. A*, **3**, 2151 (1986).
4. D. L. Gerrard and K. P. J. Williams, *Opt. Laser Technol.*, 245 (1985).
5. D. B. Chase, *J. Am. Chem. Soc.*, **108**, 7485 (1986).
6. T. Hirschfeld and D. B. Chase, *Appl. Spectrosc.*, **40**, 133 (1986).

7. J. Behringer, in *Raman Spectroscopy, Theory and Practice*, H. A. Szymanski, Ed., Plenum Press, New York (1967).
8. M. Maeda, *Laser Dyes*, Academic Press, London (1984).
9. S. A. Asher, C. R. Johnson and J. Murtagh, *Rev. Sci. Instrum.* **54**, 1657 (1983).
10. S. S. Mitra and H. J. Bernstein, *Can. J. Chem.*, **37**, 553 (1959).
11. A. Bree and R. A. Kydd, *Spectrochim. Acta*, **26A**, 1791 (1970).
12. P. Dizabo, H. E. Gatica and N. LeCalve, *J. Chim. Phys.*, **66**, 1497 (1969).
13. H. Gockel, *Z. Phys. Chem.*, **29B**, 79 (1935).
14. H. Luther, *Z. Electrochem.*, **52**, 210 (1948).
15. H. Luther and C. Reichel, *Z. Phys. Chem.*, **195**, 103 (1950).
16. N. Abasbegovic, N. Vakotic and L. Colombo, *J. Chem. Phys.*, **41**, 2575 (1964).
17. M. Brigodiot and J. M. Lebas, *C. R. Acad. Sci. Ser. B*, **268**, 51 (1969).
18. L. Colombo and J. P. Mathieu, *Bull. Soc. Fr. Mineral Cryst.*, **83**, 250 (1960).
19. R. Manzoni-Ansidei, *Gazz. Chim. Ital.*, **67**, 790 (1937).
20. M. Brigodiot and J. M. Lebas, *Spectrochim. Acta.* **27A**, 1315 (1971).
21. M. Brigodiot and J. M. Lebas, *Spectrochim. Acta*, **27A**, 1325 (1971).
22. A. Bree, R. A. Kydd and T. N. Misra, *Spectrochim. Acta*, **25A**, 1815 (1969).
23. V. Schettino, *J. Chem. Phys.*, **46**, 302 (1967).
24. N. Neto and C. DiLauro, *Spectrochim. Acta*, **26A**, 1175 (1970).
25. V. Schettino, *J. Mol. Spectrosc.*, **34**, 78 (1970).
26. D. L. Gerrard and W. F. Maddams, *Appl. Spectrosc.*, **30**, 554 (1976).
27. H. Baranska and A. Labudzinska, *Chem. Anal.*, **25**, 607 (1980).
28. L. M. Schopfer and M. D. Morris, *Biochemistry*, **19**, 4932 (1980).
29. C. M. Jones, T. A. Naim, M. Ludwig, J. Murtaugh, P. L. Flaugh, J. M. Dudik, C. R. Johnson and S. A. Asher, *Trends Anal. Chem.* **4**, 75 (1985).
30. A. D. Bandrauk, K. D. Truong and C. Carlone, *Can. J. Chem.*, **60**, 588 (1982).
31. J. Umemura, L. V. Haley, D. G. Cameron, W. F. Murphy, C. F. Ingold and D. F. Williams, *Spectrochim. Acta*, **37A**, 835 (1981).
32. M. H. Kuok and S. H. Tang, *Spectrochim. Acta*, **42A**, 515 (1986).
33. P. R. Carey, *Biochemical Applications of Raman and Resonance Raman Spectroscopies*, Academic Press, New York (1982).
34. A. T. Tu, *Raman Spectroscopy in Biology*, Wiley, New York (1982).
35. S. A. Asher, *Anal. Chem.*, **56**, 720 (1984).
36. C. R. Johnson and S. A. Asher, *Anal. Chem.* **56**, 2261 (1984).
37. J. R. Cable and A. C. Albrecht, *J. Chem. Phys.*, **84**, 1969 (1986).
38. R. P. Rava and T. G. Spiro, *J. Phys. Chem.*, **89**, 1856 (1985).
39. P. Y. Turpin, L. Chinsky, B. Jolles and A. Laigle, *Spectrosc. Biol. Mol. Proc. Eur. Conf. 1st* (1985).

CHAPTER

15

SURFACE-ENHANCED RAMAN SPECTROSCOPY

TUAN VO-DINH

Advanced Monitoring Development Group
Health and Safety Research Division
Oak Ridge National Laboratory
Oak Ridge, Tennessee

1. INTRODUCTION

An important technique for chemical and biological analysis is Raman spectroscopy because of its excellent specificity for chemical group identification. Recently, the Raman technique has enjoyed an increased interest among analytical chemists following observations of enormous Raman enhancement for molecules adsorbed on special metal surfaces (1–3). In 1974, Fleischmann et al. reported strongly enhanced Raman scattering from pyridine molecules adsorbed on silver electrode surfaces that had been roughened electrochemically by oxidation–reduction cycles (1). This increase in Raman signal, originally attributed to a high surface density provided by the roughened electrodes (1), was later identified by Jeanmaire and Van Duyne (2) and Albrecht and Creighton (3) as coming from a surface enhancement process, hence the term *surface-enhanced Raman scattering* (SERS) *effect*. The observed Raman scattering signals for the adsorbed molecules were found to be more than a million times larger than those expected from gas-phase molecules or from nonadsorbed compounds. These enormous enhancement factors, which help overcome the normally weak Raman scattering process, open new horizons to the Raman technique for ultratrace analysis.

The discovery of the SERS effect has stimulated an immense interest in fundamental research on the SERS effect, as evidenced by a number of review articles and monographs dealing with various photochemical and physical processes related to SERS (4–9). However, this early enthusiasm for widespread and practical applications of SERS quickly waned because the Raman enhancement effect has been observed for only a limited number of molecules. Most fundamental studies mainly used samples at concentrations between 10^{-1} and 10^{-3} M, which are well above the useful concentration ranges for trace analysis. Most fundamental studies were also limited to a few highly polarizable small molecules, such as pyridine, benzoic acid and its derivatives, or some ionic species such as the cyanide radical CN^- and the anion of dithiozone (10). Information about the general applicability and reproducibility of the SERS method was not available. The lack of practical SERS-active substrate materials or sample media was the major barrier for analytical applications.

In spite of extensive basic research in the 1970s, only recently has SERS been developed into a useful and practical analytical technique. The general applicability of SERS as an analytical technique has been reported for a variety of chemicals, including several homocyclic and heterocyclic polyaromatic compounds on cellulose-based substrates covered with silver-coated microspheres (11). The method consists of first coating various surfaces such as glass plates or filter paper with submicron spheres and then depositing a layer of silver to produce a uniformly rough metal surface (11–15). Another approach involving etching a crystalline SiO_2 surface to produce prolate posts (16,17) was investigated and evaluated for use as substrates for SERS analyses (18–20). More recent research has focused on the development of other experimental approaches for the production of SERS-active substrates that are practical to use and yet yield results with relatively good sensitivity and reproducibility. One simple experimental method involving the use of special cellulose substrates coated with silver metal has been reported (21). Pemberton and Buck reported trace detection of the anion diphenylthiocarbazone adsorbed on a silver electrode (10). Several alternative approaches have used silver colloids on cellulose substrates and in solutions. Tran reported subnanogram detection of dyes spotted on filter paper and on chromatographic paper that had been treated with colloidal silver sols (22,23). Fan Ni and Cotton described the deposition of silver on frosted glass for use as an SERS-active substrate (24). The use of silver hydrosols has also been investigated by Morris and co-workers (25) and Winefordner and co-workers (26–28). Recently, SERS was studied by Ahern and Garrell, who used an *in situ* laser-induced photoreduction method to produce "photocolloids" in silver nitrate solutions (29). Other applications of SERS dealt with the detection of copper and zinc phthalocyanine complexes on silver island films (30) and tributylphosphate after adsorption on silver hydrosols (31). The SERS effect has also been reported in studies using a variety of media such as electrolyte interfaces, mechanically polished polycrystalline silver, matrix-isolated metal clusters, holographic gratings, silver island films, silver and gold colloids, and tunnel junctions with a silver electrode on a rough surface (8).

2. PRINCIPLE OF THE SERS EFFECT

The important discovery of SERS has stimulated a great deal of research activity over the past few years. Extensive research efforts have been devoted to the determination and understanding of the sources of enhancement. To date, there is no single theoretical model capable of accounting for all the experimental observations related to SERS, and the origin of the enormous Raman enhancement is believed to come from the results of several mechanisms. There are at least two major types of mechanisms that contribute to the SERS effect: (a) an electromagnetic effect associated with large local fields caused by electromagnetic resonances occurring near metal surface structures and (b) a chemical effect involving a scattering process associated with chemical interactions between the molecule and the metal surface. Some aspects of SERS, such as the contribution of electromagnetic interactions, have been extensively investigated and are reasonably well understood. Other aspects, such as the contribution of chemical effects, are less known and are currently topics of extensive research. Some of the main theoretical models for SERS are briefly summarized here. The reader is referred to a number of excellent reviews for further details (4–9).

2.1. Electromagnetic Model for SERS

Electromagnetic interactions between the molecule and the substrate are believed to play a major role in the SERS process. These electromagnetic interactions are divided into two major classes: (a) interactions that occur only in the presence of a radiation field and (b) interactions that occur even without a radiation field.

The first class of interactions has been extensively studied and can be further divided into several enhancement processes. A major contribution to electromagnetic enhancement is due to surface plasmons. Surface plasmons are associated with collective excitations of surface conduction electrons in metal particles (32). Raman enhancements result from excitation of these surface plasmons by the incident radiation. At the plasmon frequency, the metal becomes highly polarizable, resulting in large field-induced polarizations and thus large local fields on the surface. These local fields increase the Raman emission intensity, which is proportional to the square of the applied field at the molecule. Another enhancement is due to excitation of surface plasmons by the Raman emission radiation of the molecule.

A commonly investigated electromagnetic model of SERS processes involves a single metal spheroid onto which are adsorbed the molecules of interest. The size of the spheroids is small compared with the wavelength of light (8). Interactions between the adsorbed molecules are ignored, as are the interactions between the molecules and the metal. The problem consists of solving Maxwell's equations to calculate the electric field experienced by each molecule when the whole system is irradiated by a laser radiation field. The importance of multiple plasmon resonances, the interaction of light with clusters, and random distributions of metal hemispheroids were also investigated (33). Theoretical models of concentric spheres comprising a metallic and dielectric region have also been investigated (34).

Electromagnetic enhancement mechanisms are characterized by the following features: (a) The effects are long range in nature since the dipole fields induced in polarizable metal particles vary as the inverse cube of the distance to the center of the particle; (b) unlike chemical interactions, electromagnetic effects are generally independent of the adsorbed molecule; and (c) the enhancements depend on the electronic structure of the substrate and the roughness of the surface since the frequency of the surface plasmon resonances depends on these factors.

Theoretical calculations were presented that show that the enhancement predicted by the particle plasmon model of SERS is limited by radiation damping (35). The damping process becomes more severe as particle size increases, whereas the enhancement produced by small particles is limited by surface scattering. The effects of radiation damping and dynamic depolarization have been analyzed further by various investigators (7, 36, 37).

Surface plasmons are not the only sources of enhanced local electromagnetic fields. Other types of electromagnetic enhancement mechanisms are due to: (a) concentration of electromagnetic field lines near high-curvature points on the surface, that is, the "lightning-rod" effect (38, 39); (b) polarization of the surface by dipole-induced fields in adsorbed molecules, that is, the image effect (39–41); and (c) Fresnel reflection effects.

Gersten and Nitzan have discussed the electromagnetic model in detail (39). They considered the problem of Raman scattering from a metallic ellipsoid with a molecule adsorbed at the tip, and they calculated the enhancement in terms of Legendre polynomials of the first and second kind. Three sources of enhancement were discussed, including the image dipole effect, the increase in local field by the lightning-rod effect, and the resonant particle plasmon effect. Zeman and Schatz recently described an accurate but simple method for determining electromagnetic fields near the surfaces of small metal spheroidal particles that is used to calculate Raman enhancements for 10 metals (42).

The mechanisms discussed above occur in the presence of a radiation field. There are also electromagnetic interactions that occur even in the absence of the radiation field. One such mechanism is associated with the modulation of the metallic reflectance due to adsorbate vibrations. This effect is due to the interaction of the oscillation dipole and/or quadrupole of the vibrating adsorbate with surface electrons, resulting in Raman-shifted reflection of light from the surface (39–42).

2.2. The Chemical Effect in SERS

The chemical effect is associated with the overlap of metal and adsorbate electronic wavefunctions, which leads to ground-state and light-induced charge-transfer processes (8, 42–47).

In the charge-transfer model, an electron of the metal, excited by the incident photon, tunnels into a charge-transfer excited state of the adsorbed molecule. The resulting negative ion (adsorbate molecule–electron) has a different equilibrium geometry than the original neutral adsorbate molecule. Therefore, the charge-transfer process induces a nuclear relaxation in the adsorbate molecule which, after the return of the electron to the metal, leads to a vibrationally excited neutral molecule and to emission of a Raman-shifted photon.

The "adatom model" also suggests additional Raman enhancement for adsorbates at special active sites of atomic-scale roughness, which may facilitate charge-transfer enhancement mechanisms (42, 44).

In general, the chemical effect contribution to SERS is necessarily short-ranged (0.1–0.5 nm). This mechanism depends on the adsorption site, the geometry of bonding, and the energy levels of the adsorbate molecule. For specific adsorbate–surface systems, enhancements may be large. The contribution of charge–transfer processes to SERS has been estimated to be approximately $10–10^3$ (45–47). Chemical enhancement can provide useful information on chemisorption interactions between metal and adsorbate. However, this enhancement is not a general mechanism and is restricted by its chemical specificity.

2.3. Characteristics of SERS

The SERS phenomenon has excitation characteristics that are similar to those observed with normal Raman scattering (NRS). The intensity of the scattered light is linear with that of the incident light. The scattered light is depolarized even with molecules such as pyridine, which exhibits highly polarized NRS.

The SERS effect appears to occur under specific experimental conditions based on (a) special requirements on the dielectric constant and (b) the morphology of the surface under study. Selection criteria of the type of substrates and media that are SERS active are based on several consider-ations.

The type of metal on the surfaces is an important factor. The SERS phenomenon occurs mostly on specific metallic surfaces. Silver exhibits the strongest enhancement effects followed by copper and gold. Certain transition metals that have been shown to be SERS active include Pt and Ni (8). Other materials, such as Li (48), Na (49), Cd (50), and Al (51) have also been investigated for SERS.

The metal alone is not sufficient to induce the SERS phenomenon since Raman enhancement was not observed on monocrystalline silver surfaces (52). When appropriately roughened, a silver surface can induce an SERS enhancement factor of 10^6, whereas a smooth silver surface produces only 400-fold Raman enhancement (53). The dependence of the SERS enhancement on surface roughness exhibits different excitation profiles for different surface preparations. For silver colloids and arrays of posts, maximum enhancements were reported at a frequency that depends on the shape of the metal particles in the colloid solutions or on the posts (54,55). An optimal range of roughness for maximum SERS can be estimated for many types of substrates. For example, diameter sizes in the range of 10–100 nm appear to be optimal for spheroidal silver particles. Further studies are required in order to determine the optimal sizes of roughness for each specific type of surface structure.

3. INSTRUMENTATION

The instrumentation required for SERS is relatively simple and is essentially similar to that used for NRS. The basic components include (a) a monochromatic excitation source (e.g., a laser), (b) a high-resolution dispersive element, such as a monochromator or a polychromator, and (c) an appropriate detector. Since the SERS signal is much stronger than the NRS signal and is often higher than the laser-scattered stray light, which is a major problem in conventional Raman spectrometry, the monochromators and related optical instrumentation for SERS do not need to be sophisticated and costly.

Figure 1 schematically shows a typical arrangement of an SERS experiment. The excitation light source is usually a laser. The laser radiation is generally passed through optical filters to reject unwanted plasma lines and is directed to the sample. Selection of the laser polarization may affect the intensity of the observed SERS signal. The SERS emission from the sample is collected through appropriate optics and is focused onto the entrance slit of a dispersive element. There are two basic classes of Raman spectrometers: single-channel instruments and multichannel instruments. The first type of instruments generally uses a double-grating monochromator coupled to a photomultiplier, whereas the latter uses a polychromator interfaced to a multichannel detector, such as a vidicon, a photodiode array, or a charge-coupled device.

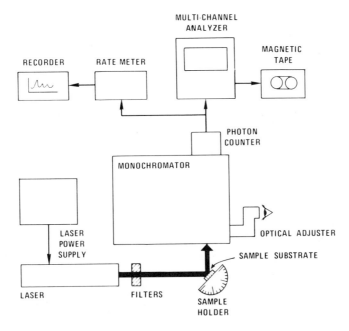

Figure 1. Typical experimental setup for SERS measurements. Reproduced from ref. 11.

There are several manufacturers of spectrometers, each offering a variety of models with different performance characteristics and each offering different options. Basic instrument components are also commercially available. An investigator may assemble off-the-shelf components for his or her particular applications.

Selection of laser excitation sources is determined by the frequencies that could be matched to the surface plasmon frequencies in order to take advantage of maximum enhancement. If time-resolved measurements are performed, the pulse width of the laser is an important factor to consider. A number of devices, such as a spike filter or a single monochromator, may be used to reject the Rayleigh scattered light. The recent development of notch filters, which consist of crystalline arrays of polystyrene spheres, gives rise to filters with very high rejection efficiency of laser lines (56).

A photomultiplier coupled with a double-grating monochromator is the most commonly used Raman spectrometer. The spectral resolution and the sensitivity of the spectrometer are strongly dependent on the characteristics of the monochromator and the photomultiplier tube. Modern holographic gratings generally offer excellent stray-light rejection. The single-photon counting method is often used for detection. Unlike the more conventional analog detection method, the single-photon counting signal output is digital in nature, producing discrete pulses of charge. The digital technique has proved to have several advantages over the analog method, especially for low-level signal detection (57–59).

Photomultipliers are widely used for their high spectral sensitivity, wide operating range, low cost, and relatively simple electronics. Since low-intensity signals are frequently observed in SERS measurements, it is desirable to improve the sensitivity of detection by recording the entire spectrum simultaneously. Interferometric methods (e.g., Fourier transform techniques) can enhance the detection sensitivity only if the noise of the measured signal is detector limited. The major limiting noise sources in SERS are generally the scattered light or the sample background fluorescence rather than the detector noise. Therefore, neither the throughput (or "Jacquinot") advantage nor the "Fellgett" (or Fourier) advantage of an interferometric instrument can be used for SERS. The use of monochromators helps minimize the stray light often encountered in SERS measurements. However, with monochromators, only one spectral resolution element, or channel, can be monitored at a time. Detectors that permit the recording of the entire spectrum simultaneously, thus providing the multiplex advantage, are known as *multichannel detectors*. A well-known example of a multichannel detector is the photographic plate. More modern multichannel detectors comprise the vidicon, the diode array, or the charge-coupled device (60,61). In multichannel spectrometers, the detector is placed at the focal plane of a polychromator, which is a monochromator with the exit slit removed. As a result, the entire emission dispersed at all wavelengths within the polychromator is detected simultaneously. The simultaneous detection of all the dispersed emission using n spectral resolution elements reduces the measurement time by a factor of n in the case of an S/N (signal-to-noise)-limited measurement, or it improves the S/N ratio by a factor of $\sqrt{n}$ in the case of a time-limited measurement.

Several types of multichannel detectors are currently available. One device is the vidicon, which is essentially a television-type device consisting of an array of microscopic photosensitive diode junctions that are grown upon a single silicon-crystal wafer. These diode junctions provide individual microelements that are used to detect simultaneously the radiation dispersed by the polychromator. Prior to detection, each diode is charged to a preset reversed-bias potential by a fast scanning electron beam. The electromagnetic radiation that strikes the diode causes depletion of the charge in each diode. A current proportional to the incident radiation recharges the diode to the preset potential when the electron beam scans again, thus providing information on the intensity of the incident light that has struck the diode. Consequently, an almost instantaneous recording of the spectral intensity at each diode is generated in the computer memory of the vidicon detector. Vidicons are capable of integrating radiation intensity over multiple scanning cycles because of their charge storage capabilities. Sensitivity can be further enhanced by incorporating an image-intensification section in front of the vidicon to yield a silicon-intensified target (SIT) vidicon.

Another multichannel detector is the diode array, which also utilizes photodiodes as detection elements. Recording of the signal is performed with direct on-chip circuitry rather than with a scanning electron beam. Signal amplification is achieved by a microchannel plate image intensifier. Gated detection down to 5-ns time resolution can be performed with the intensified diode array. Intensified diode arrays are currently the most widely used multichannel detectors. Recently, other types of devices, such as charge-coupled devices (CCD), have received increasing use because of their high quantum yield, two-dimensional imaging capability, and very low dark current.

4. SERS TECHNIQUES

One of the major difficulties in the development of the SERS technique for analytical applications is the production of surfaces or media that have an easily controlled protrusion size and reproducible structures. Roughened metal electrodes, metal island films, and metal colloids are among the first SERS-active media that have been used.

4.1. Metal Electrodes

Following the discovery of the SERS effect on silver electrodes (2, 3), many studies have been conducted on compounds adsorbed at the surface of metal electrodes (8). Electrochemical cells for SERS studies generally employed silver electrodes, although SERS studies have also been conducted using other metal electrodes (62–64). The working electrode is generally placed in a position such that the laser excitation can be focused onto its surface, and the Raman scattered light can be efficiently collected by appropriate optics. Strong SERS signals appear only after an electrochemical oxidation–reduction cycle, often referred to as "activation cycle," is performed on the metal electrode. During the first half of the cycle, silver at the electrode is oxidized by the reaction $Ag \rightarrow Ag^+ + e^-$. During the reduction half cycle, a roughened silver surface is reproduced by the reaction $Ag^+ + e^- \rightarrow Ag$. This oxidation–reduction procedure generally produces surface protrusions in the size range of 25–500 nm on the electrode surface.

4.2. Colloid Hydrosols

Metal colloid hydrosols are often used to produce SERS-active media in solutions. There are several reasons for using colloid hydrosols, including ease of colloid formation and straightforward characterization of the colloid solutions by simple UV absorption. Silver colloids are generally prepared by rapidly mixing a solution of $AgNO_3$ with ice-cold $NaBH_4$.

Hildebrandt and Stockburger investigated SERS of rhodamine 6G, a polycyclic aromatic dye, using silver colloids as the SERS-active medium (65). Rhodamine 6G is a strongly fluorescent xanthene derivative that shows a molecular resonance Raman effect when excited into its visible absorption band. The combined effect of resonance Raman and SERS can result in very high sensitivity, thus allowing investigation of submonolayers of adsorbates where there are no interactions between the dye molecules. The results of this study indicate that two different mechanisms, which occur at different adsorption sites, can contribute to Raman enhancement. One mechanism is already effective when rhodamine 6G molecules are spontaneously adsorbed on the silver colloids, whereas the second mechanism requires activation of the colloidal solution by a supporting electrolyte. The first type of enhancement is rather unspecific and shows a high surface coverage. The second type of enhancement is only observed in the presence of anions (Cl^-, I^-, Br^-, F^-, SO_4^{2-}), with specific sites being formed at an extremely low surface coverage. It was concluded from the adsorption isotherm study that at such sites the molecules are chemisorbed. The two types of adsorption sites could be distinguished by characteristic Raman features. Overall Raman enhancement factors of up to 10^6 were found for dye molecules at anion-activated sites. Figure 2 shows an example of the SERS spectrum of a 10^{-9} M solution of rhodamine 6G.

A disadvantage of the colloid systems is its tendency to coagulate. Stabilizers such as poly(vinylalcohol), poly(vinylpyrrolidone), and sodium dodecyl sulfate have been used to minimize this coagulation problem (66–68). However, the use of such stabilizers could produce interferences (68). Tran has demonstrated that silver colloids stabilized by filter paper supports (cellulose, glass, and quartz fibers) could enhance the Raman emission of various dyes adsorbed onto them (22–23). The enhancement was found to depend on the difference in the average distance between the dye and silver particles induced by the paper fibers and, therefore, on the specific type of filter paper that was used. Figure 3 shows the SERS spectrum of 10 ng of crystal violet adsorbed on 22.5 μl of silver hydrosols on Whatman glass fiber GF/F support (22). With this simple technique, detection limits in the subnanogram range were achieved for various azo dyes, triphenylmethane, and cyanine using only a small 3-mW helium–neon laser.

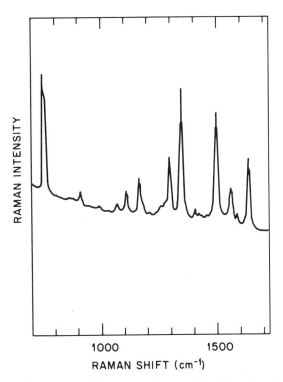

Figure 2. SERS spectrum of rhodamine 6G in a silver colloid solution. Reproduced from ref. 65.

Recently, certain aspects of the coagulation problem have been overcome, and silver colloid hydrosols were prepared at room temperature (27, 28). Measurements with the colloid hydrosols indicated that for carefully prepared samples, the unaggregated colloids were stable for periods of more than 3 weeks (69). Figure 4 shows the UV absorption spectra of a silver colloid hydrosol solution measured at different time periods. No significant change was observed in the absorption spectrum of the silver sols on aging.

Torres and Winefordner reported trace determination of nitrogen-containing PACs, such as 2-aminofluorene, by SERS on silver colloids (26). Figure 5 shows the SERS spectra of 2-aminofluorene and a silver hydrosol blank solution (26). The analytical calibration curve of 2-aminofluorene yielded a straight line over nearly two orders of magnitude with a slope of the log–log plot close to unity. The detection limit of 2-aminofluorene was 7 ng.

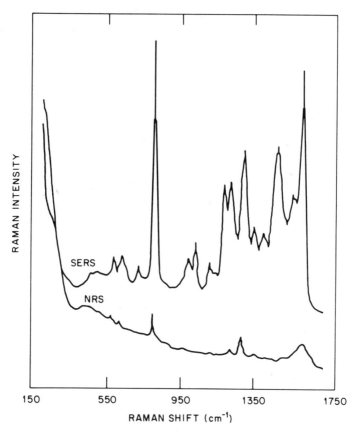

Figure 3. SERS spectrum of crystal violet in a silver hydrosol solution. Reproduced from ref. 22.

4.3. Metal Island Films

SERS from copper and zinc phthalocyanine complexes from silver and indium island films were reported (70, 71). The silver and indium films were vacuum evaporated ($p < 10^{-6}$ torr) onto tin oxide glass slides and were then coated with copper and zinc phthalocyanine complexes in a vacuum system at a base pressure of 5×10^{-7} torr. Metal thickness was about 7.5 nm on the substrates, and the analyte coatings ranged from 7.5 to 200 nm. Figure 6 shows the SERS spectrum of zinc phthalocyanine on a 7.5-nm-thick silver island film (70). The spectra were obtained using 150 mW of the 514-nm line of an argon ion laser. The optimal experimental geometry was obtained when the film and the incident laser line were forming a 30° angle. The SERS

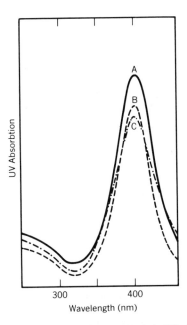

Figure 4. Stability of silver colloid solutions measured by their UV absorption spectra: curve A, freshly prepared solution; curve B, after 1 week; curve C, after 2 weeks.

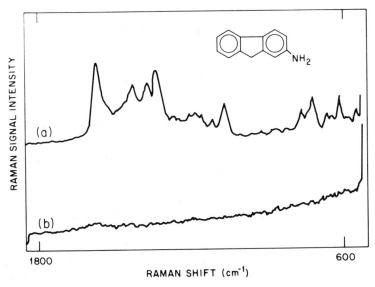

Figure 5. SERS spectra of 2-aminofluorene (curve a) and of the silver hydrosol blank solution (curve b). Reproduced from ref. 26.

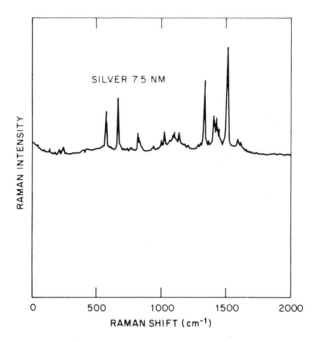

Figure 6. SERS spectrum of zinc phthalocyanine on silver island film. Reproduced from ref. 70.

spectrum corresponds to a 25-nm film of zinc phthalocyanine. It was observed that the SERS signal from a 7.5-nm phthalocyanine film on silver under identical experimental conditions was comparable to a 25-nm film on silver, thus indicating that the SERS effect occurred with the first monolayer of adsorbate molecules. The vibrational structures of the NRS and SERS spectra were found to be similar. Only minor differences in relative intensities were noticeable, and no significant frequency shift could be observed. The enhancement factor was determined to be approximately 500. Samples left exposed to air for 30–60 days continued to show SERS effects, although a decrease in the signal was observed.

4.4.   Silver-Coated Microsphere Substrates

In general, a 50-μl volume of a suspension of latex or Teflon submicron spheres was applied to the surface of the substrate. The different types of substrates investigated include filter paper, cellulosic membranes, glass plates, or quartz materials (11–15). The substrate was then placed on a high-speed spinning device and spun at 800–2000 rpm for about 20 s. The silver was deposited on the microsphere-coated substrate in a vacuum evaporator at a deposition rate of 2 nm/s. The thickness of the silver layer deposited was generally 50–100 nm. Figure 7 shows the SERS spectrum of benzo[a]pyrene (BP) (4-μl sample of 10^{-4} M in ethanol) adsorbed on a silver-coated microsphere substrate (14). The SERS-active substrate used in this study consisted of a microscope glass slide covered with polystyrene microspheres having a 364-nm diameter and covered by a 75-nm-thick layer of silver.

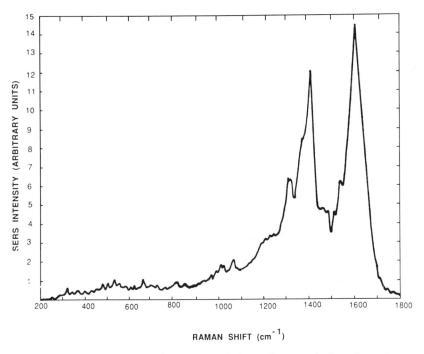

Figure 7. SERS spectrum of benzo[a]pyrene adsorbed on a silver-coated microsphere substrate. Reproduced from ref. 14.

4.5. Substrates with Silver-Coated Titanium Dioxide Particles

Titanium dioxide materials were first deposited on glass and cellulose substrates and were then coated with a 50–100-nm layer of silver by thermal evaporation as described previously. Prior to deposition, titanium dioxide was prepared as a suspension in water (10% concentration by weight). The silver-coated titanium dioxide substrates obtained by this method appear to provide a new type of efficient SERS-active substrate.

4.6.   Silver-Coated Fumed Silica Substrates

Another type of substrate that is quite SERS active and easy to prepare is the fumed silica-based substrate. Fumed silica has been used as a thickening agent in various industrial processes, including coating and cosmetics preparations. In the preparation of SERS materials, the selection of the appropriate types of fumed silica is important. Fumed silica is manufactured in different grades, which vary with respect to surface area, particle diameter, and degree of compression. The fumed silica particles were suspended in a 10% water solution and were then coated onto a glass plate or filter paper. The substrate was then coated with a 50–100-nm layer of silver by thermal evaporation. With these types of substrates, the fumed silica materials, which have submicron-size structures, provide the rough-surface effect for the SERS process. Figure 8 shows an example of the SERS spectrum of a 3-μl (10^{-3} M) sample of phthalic acid using the fumed silica-based substrate.

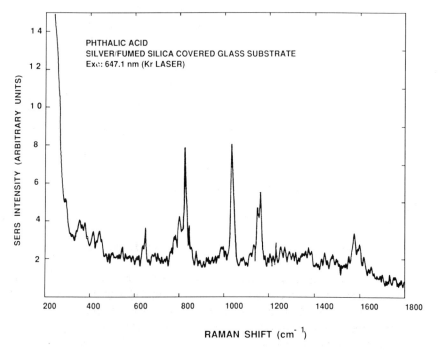

Figure 8. SERS spectrum of phthalic acid on silver-coated fumed silica substrate.

4.7. Silver-Coated Quartz Posts

The preparation of SiO$_2$ prolate posts is a multistep operation that involves plasma etching of SiO$_2$ with a silver island film as an etch mask (16–20). Since fused quartz etches much more slowly than thermally deposited quartz, a 500-nm layer of SiO$_2$ was first thermally evaporated onto fused quartz at a rate of 0.1–0.2 nm/s. The resulting crystalline quartz was annealed to the fused quartz for 45 min at approximately 950°C. A 5-nm silver layer was then evaporated onto the thermal SiO$_2$ layer, and the substrate was flash-heated for 20 s at 500°C. This heating causes the thin silver layer to bead up into small globules, which act as etch masks. The substrate was then etched for 30–60 min in a CHF$_3$ plasma to produce submicron prolate SiO$_2$ posts, which were then coated with an 80-nm layer of silver at normal evaporation angle.

Figure 9 shows the SERS spectrum of 74 ng of 1-nitropyrene adsorbed on a silver-coated quartz post substrate (19). Figure 10 shows the SERS spectrum of another important PAC, pyrene, using the 647.1-nm krypton ion laser excitation line. Pyrene is an important environmental chemical commonly

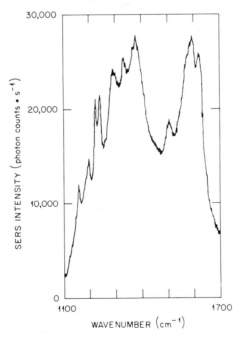

Figure 9. SERS spectrum of 1-nitropyrene on a quartz post substrate coated with a 80-nm silver layer. Reproduced from ref. 19.

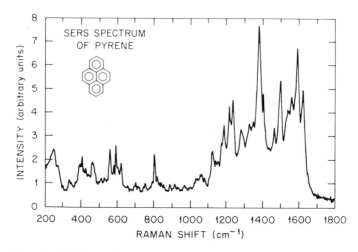

Figure 10. SERS spectrum of pyrene on a silver-coated quartz post substrate.

found in polluted atmospheres, automobile emissions, and combustion emissions of organic products. The SERS spectrum of pyrene was obtained using the prolate post substrate. The signal shown in Figure 10 corresponds to a 7-ng sample of pyrene. The SERS spectrum of pyrene is similar to the NRS spectrum and exhibits a series of sharp peaks. The major peaks at 385, 785, 1038, 1180, 1220, 1235, 1380, 1490, 1540, and 1618 cm^{-1} have frequencies that are close to a_g symmetry bands observed in the Raman spectrum of pyrene in solution at 408, 802, 1040, 1192, 1233, 1242, 1395, 1504, 1553, and 1632 cm^{-1}, respectively. The SERS spectrum also exhibits a peak at 1582 cm^{-1}, which has a frequency close to the b_{3g} symmetry band at 1597 cm^{-1} observed in the conventional Raman spectrum of pyrene in solution. It is noteworthy that some changes are expected in SERS spectra relative to conventional Raman spectra. This indicates that the vibrations responsible for the shifted peaks are affected by adsorption onto the metal. Spectral shifts are greater for chemisorption than for simple adsorption of the analyte molecule on the metal surface.

4.8. Metal-Coated Cellulose Substrates

Certain types of micropore filter papers coated with a thin layer of evaporated silver appear to provide efficient SERS-active substrates. Scanning electron micrographs of these cellulosic materials showed that these surfaces consist of fibrous 10-μm strands with numerous tendrils that provide the necessary microprotrusions required for the SERS enhancement. The simple technique employing silver-coated cellulose substrate was used to analyze benzo[a]pyrene–DNA adducts (21). The carcinogenic activity of benzo[a]-pyrene (BP) in animals and its metabolic activation to the ultimate carcinogenic metabolite, BP-7,8-diol-9,10 epoxide (BPDE) have been well established (72, 73). Figure 11 shows the SERS spectra of hydrolyzed products of BPDE–DNA adducts (8×10^{-6} M in 0.1 N HCl), BP–tetrol (10^{-4} M in methanol), and DNA (calf thymus DNA; 1.2 mg/ml in H_2O). The spectra were obtained using the krypton laser line at 647.1 nm (40 mW) as the excitation (21). A number of Raman peaks in the BPDE–DNA adduct spectrum (Figure 11, part a) can be attributed to DNA. The strong SERS band at 737 cm^{-1}, prominent in both spectra, is characteristic of DNA Raman emission. This Raman peak has been attributed to the 740-cm^{-1} ring-breathing mode of adenine. The results indicated that, whereas many spectral features from DNA could be found in the BPDE–DNA adduct spectrum, several peaks at 620, 1310, 1390, and 1655 cm^{-1} were unique to BPDE–DNA adducts. Comparison of the SERS spectra in parts a and b of Figure 12 clearly indicates that an intense and sharp peak at 1240–1245 cm^{-1}, which is characteristic of BP-tetrol, is also observed in the BPDE–DNA spectrum. The results of SERS measurements shown in Figure 12 demonstrate the capability of detecting the BP-tetrol portion of the intact (nonhydrolyzed) BPDE–DNA adduct. This feature is an important advantage in DNA adduct analysis since it avoids the acid hydrolysis procedure (21).

4.9. Silver Membranes

One of the simpler types of solid substrates we have investigated is the silver membrane used for air particulate sampling. The filter already has micropores that provide the microstructure required to induce SERS. These substrates consist of silver membranes and can, therefore, be used directly as SERS-active substrates without requiring silver coating. Figure 13 shows the SERS spectrum of 1-nitropyrene adsorbed directly on a silver membrane. The silver membrane substrate was used directly as SERS substrate without requiring any chemical treatment or metal evaporation.

To our knowledge, this is the first reported case where SERS is observed directly on a simple type of substrate readily available. One should mention that optimal conditions may not be achieved because these silver membranes were off-the-shelf and were not freshly produced in the laboratory as were other substrates. The metal surface of these silver membranes might be oxidized or contaminated by polluted atmospheres. The SERS spectrum of 1-nitropyrene obtained with the silver membrane is similar to that observed with the microsphere-coated substrates or the prolate quartz post substrates reported previously (19). Although these substrates are simple to use and are

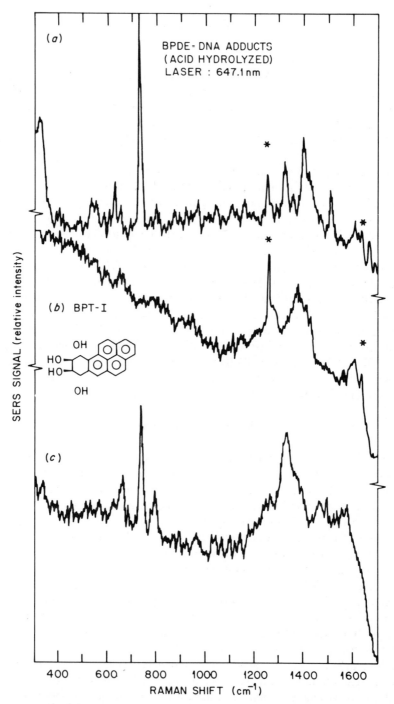

Figure 11. SERS spectra using silver-coated cellulose substrate: (*a*) hydrolyzed benzo[*a*]pyrene-7,8-diol-9,10-epoxide (BPDE)–DNA adducts; (*b*) benzo[*a*]pyrene–tetrol-1; (*c*) calf thymus DNA. Reproduced from ref. 21.

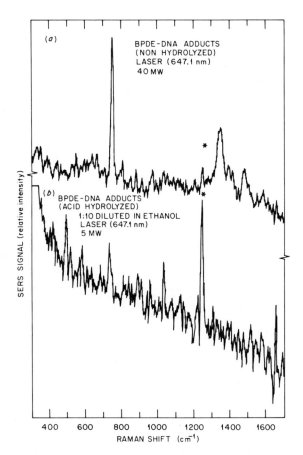

Figure 12. SERS spectra using silver-coated cellulose substrate: (*a*) BPDE–DNA adducts (8×10^{-6} M in H_2O); (*b*) hydrolyzed BPDE–DNA adducts (1:10 dilution in H_2O of an 8×10^{-6} M solution). Reproduced from ref. 21.

commercially obtained, they are quite fragile and usually not as efficient as the above-mentioned substrates. Also, very often the silver surface of the membrane has to be freshly recoated in order to improve the SERS effect.

4.10. Chemically Etched Metal Surfaces

Miller et al. described two simple etching procedures used to produce SERS-active copper surfaces (74). In the first procedure, copper foil was etched for 40 min in $2 \, mol \, cm^{-3}$ nitric acid at room temperature. The second procedure consisted of sandblasting copper foil with Al_2O_3 at 4 bar pressure and

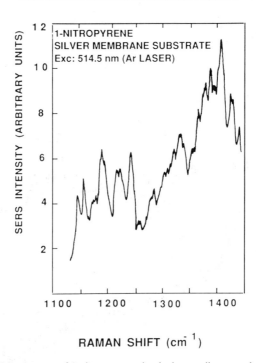

Figure 13. SERS spectrum of 1-nitropyrene adsorbed on a silver membrane substrate.

subsequently etching for 2 min. An electron scanning microscope picture of the metal surfaces indicated that both etching procedures could produce surface roughness on the 10–100-nm scale. The etched copper surfaces were used to investigate the SERS emission of a polycyclic aromatic dye, Nile blue. The resonance Raman signals of Nile blue were enhanced by 10^3–10^4 relative to the molecule in solution. This enhancement was only a factor of 2 smaller than the corresponding value for a silver-particle surface, when also excited at 662 nm. Figure 14 shows an SERS spectrum of Nile blue adsorbed on a copper surface using laser excitation at 662 nm. The spectrum was obtained using a substrate produced by a combination of sandblasting and a 70-s chemical etch (74).

Figure 15 shows scanning electron micrographs of some types of practical SERS-active substrates. The SERS-active surface consists of randomly distributed surface agglomerates and protrusions in the 10–100-nm range. These structures produce large electromagnetic fields on the surface when the incident photon energy is in resonance with the localized surface plasmons.

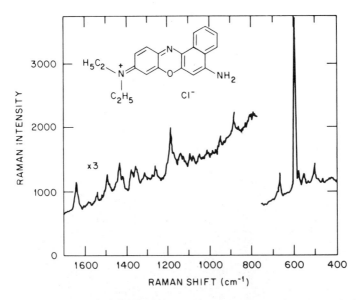

Figure 14. SERS spectrum of Nile blue on a sandblasted copper substrate etched for 70 s (approximate coverage of 2.5×10^{13} molecules·cm^2). Reproduced from ref. 74.

Another demonstration of the advantage in sensitivity and selectivity of the SERS technique is illustrated in Figure 16. This figure depicts the SERS spectrum of anthracene, an important environmental PAC that is on the EPA Priority Pollutants List. The substrate was a silver-coated quartz post substrate, and excitation at 514.5 nm was from an argon ion laser. A noteworthy observation is that most of the SERS band positions are essentially similar to those obtained in conventional Raman. The largest frequency shift is less than 20 cm^{-1}. Therefore, the previously assigned bands of the normal Raman spectra can be compared with the SERS spectra. Several major SERS peaks observed at 1275, 1412, and 1564 cm^{-1} correspond to the normal Raman bands observed in polycrystalline anthracene at 1258, 1403, and 1556 cm^{-1}. These three Raman peaks were assigned to a_g symmetry, with the peaks at 1403 and 1556 cm^{-1} attributed to totally symmetric modes. Several nontotally symmetric modes were observed in the SERS spectrum at 405, 1190, and 1477 cm^{-1}. These SERS peaks could be assigned to the b_{3g} Raman bands observed in anthracene at 394, 1185, and 1477 cm^{-1}. It is important to mention that for large molecules, one should not expect the SERS spectra to be exactly the same as the normal Raman spectra. With small molecules (one benzene ring or smaller), most of the vibrations of the molecule may be enhanced. With large molecules, only chemical groups that are closed to, or adsorbed on, the SERS-active surface will experience Raman enhancement.

5. ANALYSIS OF COMPLEX MIXTURES

The spectral selectivity of the technique is illustrated in Figure 17, which shows the SERS spectrum of a synthetic mixture containing three chemicals of environmental and biological interest, namely, benzo[a]pyrene, 1-nitropyr-

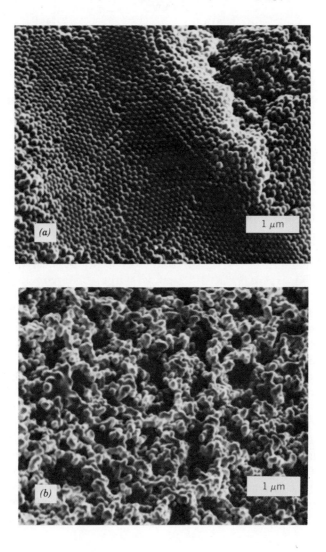

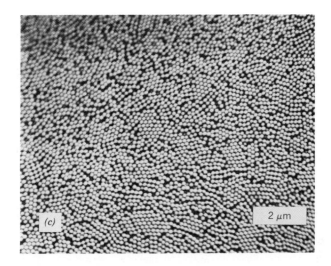

Figure 15. Scanning electron micrographs of some SERS-active solid substrates. (a) Whatman paper covered with 0.495-μm spheres and coated with a 200-nm layer of silver. (b) Millipore paper covered with 0.495-μm spheres (without spinning) and coated with a 200-nm layer of silver. (c) Silica gel plate coated with 0.365-μm spheres and covered with a 150-nm layer of silver. (d) Fumed silica coated on a glass plate and covered with a 75-nm layer of silver.

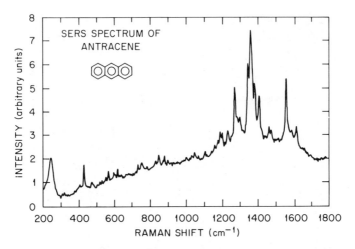

Figure 16. SERS spectrum of anthracene.

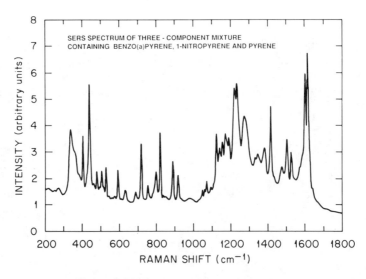

Figure 17. SERS spectrum of a synthetic mixture.

ene, and pyrene. 1-Nitropyrene is one of the family of nitro-PNA compounds that have received considerable interest. These species are often produced in atmospheric reactions of PACs with NO_x or during combustion processes in motor vehicle engines (75). Nitro-PAC compounds are also detected in polluted air particulates, diesel exhaust emissions, and xerographic toners.

Studies have indicated that nitro-PAC compounds induce cancer in rats. Extensive investigations are currently being conducted for the detection of 1-nitropyrene, which is a potent direct-acting bacterial mutagen. In this example we illustrate the use of SERS to detect 1-nitropyrene in the presence of other PAC species. Figure 17 illustrates that the SERS spectrum of the three-component mixture exhibits many sharp Raman peaks that can be assigned to 1-nitropyrene, benzo[a]pyrene, and pyrene. Table 1 summarizes the spectral characterization of the mixture. The results show that 1-nitropyrene can be easily differentiated from its parent compound, pyrene.

Another example of multicomponent analysis by SERS is illustrated in Figure 18, which shows the SERS spectrum of a mixture containing 7,8-benzoquinoline, phenanthrene, and coronene (18). The remarkable spectral selectivity of SERS was demonstrated by the fact that most of the 30 Raman peaks of the mixture's spectrum can be assigned to the individual component.

The selectivity of the technique for real-life sample analysis is demonstrated in Figure 19, which shows the SERS spectrum of a diesel particulate sample. The sample, received from the National Bureau of Standards, was directly diluted in ethanol and spotted on a quartz post substrate. No chemical extraction of the PACs from the sample was performed. The 647.1-nm laser line was used for excitation at 50-mW power. Six compounds, including anthracene, phenanthrene, fluoranthrene, pyrene, benzo[a]pyrene, and 1-nitropyrene, could be identified. The results demonstrated that SERS is a useful technique for trace analysis of complex mixtures.

6. ANALYTICAL CONSIDERATIONS

Raman spectroscopy has a number of important advantages for chemical analysis. The technique is well known for its high spectral selectivity of vibrational structures to provide fingerprints for molecules. This technique, however, is limited by the inherently weak cross section of Raman scattering. The SERS technique, which can detect the Raman signal amplified by factors from 3 to 7 orders of magnitude, will therefore provide a technique with sensitivity similar to that of luminescence techniques with the added merit of spectral specificity. Detection limits of SERS are in the picogram range level, and the calibration plots exhibit linear responses over several orders of magnitude. However, for most compounds, the actual enhancement factor is difficult to determine since the NRS signal could not be detected under similar experimental conditions.

Techniques that employ metal electrodes and colloidal solutions have been used first for SERS and will remain widely used. Among the more recent techniques based on solid substrates, the methods using simple submicron

Table 1. Spectral Characterization of a Three-Component Mixture

SERS Peaks of Mixture (cm^{-1})	SERS Peaks of Individual Compound (cm^{-1})		
	Pyrene	Benzo[a]pyrene	1-Nitropyrene
342	337		339
409	407		406
444			441[a]
465			462[a]
481			479[a]
507			505[a]
530			529[a]
565	560[a]		
594	591		590
635			632[a]
720			719[a]
801	804		800
824			820[a]
893	890		890
921	917		916
1050		1045	1046
1074	1081[a]		
1127	1125	1127	1124
1176		1177[a]	
1194	1189		1190
1224	1220		1224
1240	1239	1248	1235
1277	1280		1278
1335	1328		
1350	1351	1348	
1385	1383	1382	1380
1419			1414[a]
1506	1501	1504	1500
1527			1525[a]
1606			1601[a]
1622	1625	1620	1616

[a] Spectral position unique to one component.

materials, such as fumed silica and Teflon or latex spheres, appear to be the simplest to prepare. Teflon and latex spheres are commercially available in a wide variety of sizes. The shapes of these materials are very regular, and their size can be selected for optimal enhancement. The effect of the sphere size and metal layer thickness upon the SERS effect can be easily investigated. The

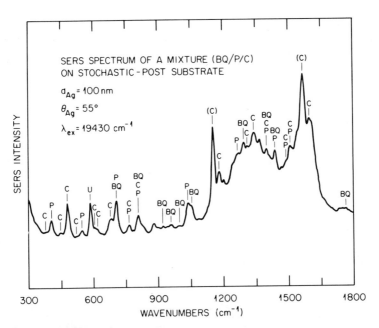

Figure 18. SERS spectrum of a three-component mixture containing 7,8-benzoquinoline (BQ), phenanthrene (P), and coronene (C). Reproduced from ref. 18.

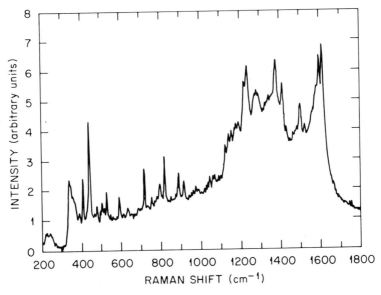

Figure 19. SERS spectrum of a diesel emission particulate sample.

results indicated that, for each sphere size, there is an optimum silver layer thickness for which the maximum SERS signal is observed (14). The effect of the sphere size and the silver thickness on the SERS intensity is illustrated in Figures 20 and 21 (14).

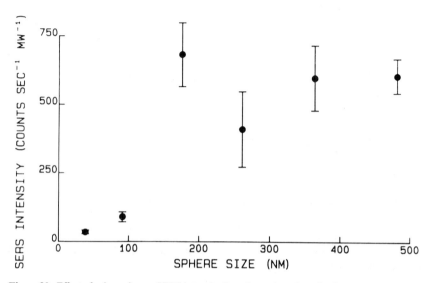

Figure 20. Effect of sphere size on SERS intensity for microsphere-based substrates. Reproduced from ref. 14.

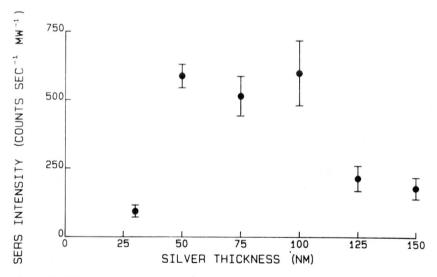

Figure 21. Effect of silver thickness on SERS intensity for microsphere-based substrates. Reproduced from ref. 14.

The method using prolate quartz posts requires several operations, including SiO$_2$ evaporation, silver island mask deposition, plasma etching, and chemical cleaning prior to silver coating. All these operations have to be performed in vacuum chambers. Once these quartz posts are made, they can be reused. Comparative studies indicate that the simple fumed silica or microsphere-based substrates are equally good or more efficient than the quartz post substrate. The fumed silica and microsphere-based substrates are simpler to prepare than the quartz posts since they do not require silver mask deposition and plasma etching operations. Fumed silica is also very simple to handle.

Fumed silica appears to be one of the most efficient materials for the production of SERS-active substrates. One important advantage of fumed silica over Teflon or latex microspheres is its very low cost. This important feature makes fumed silica a promising material for SERS-active substrate development and makes it most suitable for practical applications.

Substrates using titanium dioxide coated with silver are alternate materials that can be used for SERS. This technique was found to produce efficient Raman enhancement. The procedures to prepare substrates are simple. The titanium dioxide materials are also commercially available but cost more than fumed silica.

Special filter papers coated with silver could provide useful substrates. We have evaluated a large variety of cellulosic materials, and only a very few of them are efficient. Because of proprietary reasons the exact nature and composition of these cellulosic materials are not made available to us by the manufacturers. For this reason, it is sometimes difficult to investigate the physicochemical factors that could produce the optimal SERS conditions.

Colloid silver hydrosols provide efficient media for SERS measurements. An attribute of the technique using silver hydrosols is that it does not require the use of evaporation vacuum chambers. The main advantage of the colloid SERS technique is the relatively simple experimental procedures for preparation of the hydrosol solutions. Compared with the method using roughened electrodes, the molecular structure of the sample is not influenced through the oxidation–reduction cycle during pretreatment. On the other hand, unlike microstructures on solid substrates, the sizes of the hydrosols are not uniform from batch to batch and are unstable over long time periods. Colloidal stability has been a major problem that has been extensively investigated. Recently, Winefordner and co-workers have investigated the possibility of developing the hydrosol technique using reproducible flow injection operations for flow-through systems (27, 28). Another disadvantage of colloid hydrosols is their tendency to flocculate at the bottom of spectroscopic cells. Tran reported that this limitation could be avoided by application of the silver colloidal hydrosols on filter paper (22). This method has been further developed for direct *in situ* analysis of subnanogram levels of nucleic purine derivatives on thin-layer chromatograms (76). Recently, Fately and co-workers have developed the SERS colloid technique for high-performance liquid chromatography detection (77).

The versatility of the SERS technique is demonstrated by the variety of methods using different types of solid and liquid media. Very often, one approach should not be directly compared with another because each uses a specific sample medium appropriate for certain applications. Although its potential as an analytical tool is now firmly established, the SERS technique is still in the burgeoning stage, and many analytical figures of merit remain to be discovered.

7. CONCLUSION

With the renewed interest in the SERS effect, it is expected that the SERS technique will soon become a useful and powerful tool for spectrochemical analysis. Vibrational information obtained from adsorbed molecules that exhibit SERS provides high-resolution data on their identity, orientation, and bond configuration. For instance, SERS provides important information related to surface adsorption processes. Because the scattering cross section of normal Raman is small, it has been difficult to measure Raman signals of monolayers of molecules without, at the same time, using high laser power levels at which the molecules became thermally desorbed or photodecom-posed. The SERS technique allows investigation of new problems, such as characterization and dynamics of adsorbed layers at electrodes (78), confor-mation changes (79) and charge-transfer kinetics (80) upon adsorption of organic molecules on colloidal metal particles, and conformation changes in adsorbates upon wetting (81).

ACKNOWLEDGMENTS

This research was sponsored by the Office of Health and Environmental Research, U.S. Department of Energy, under contract DE-AC05-84OR21400 with Martin Marietta Energy Systems, Inc. It is a pleasure to acknowledge the contribution and careful thoughts of many of my students, research associates, and colleagues, Y. K. Hiromoto, R. L. Moody, M. A. Morrison, Drs. A. Alak, P. Enlow, J. Bello, and Professor W. D. Fletcher.

REFERENCES

1. M. Fleischmann, P. J. Hendra, and A. J. McQuillan, *Chem. Phys. Lett.*, **26**, 163 (1974).

2. D. J. Jeanmaire and R. P. Van Duyne, *J. Electroanal. Chem.*, **84**, 1 (1977).
3. M. G. Albrecht and J. A. Creighton, *J. Am. Chem. Soc.*, **99**, 5215 (1977).
4. M. Moskovits, *Rev. Mod. Phys.*, **57**, 783 (1985).
5. A. Wokaun, in *Solid State Physics*, Vol. 38, H. Ehrenreich, F. Seitz, and D. Turnbull, Eds. Academic Press, New York, p. 223 (1984).
6. G. C. Schatz, *Acc. Chem. Res.*, **17**, 370 (1984).
7. M. Kerker, *Acc. Chem. Res.*, **17**, 271 (1984).
8. R. K. Chang and T. E. Furtak (Eds.), *Surface-Enhanced Raman Scattering*, Plenum Press, New York (1982).
9. I. Pockrand, *Surface-Enhanced Raman Vibrational Studies at Solid/Gas Interfaces*, Springer, Berlin (1984).
10. J. E. Pemberton, and R. P. Buck, *Anal. Chem.*, **53**, 2263 (1981).
11. T. Vo-Dinh, M. Y. K. Hiromoto, G. M. Begun, and R. L. Moody, *Anal. Chem.*, **56**, 1667 (1984).
12. J. P. Goudonnet, G. M. Begun, and E. T. Arakawa, *Chem. Phys. Lett.*, **92**, 197 (1982).
13. A. Alak, and T. Vo-Dinh, *Anal. Chem.*, **59**, 2149 (1987).
14. R. L. Moody, T. Vo-Dinh, and W. H. Fletcher, *Appl. Spectrosc.*, **41**, 966 (1987).
15. A. Alak and T. Vo-Dinh, *Anal. Chim. Acta*, **206**, 333 (1988).
16. P. F. Liao, in *Surface Enhanced Raman Scattering*, R. K. Chang and T. E. Furtak, Eds., Plenum Press, New York, p. 379 (1982).
17. M. C. Buncick, R. J. Warmack, J. W. Little, and T. L. Ferrell, *Bull. Am. Phys. Soc.*, **29**, 129 (1984).
18. T. Vo-Dinh, M. Meier, and A. Wokaun, *Anal. Chim. Acta*, **181**, 139 (1986).
19. P. D. Enlow, M. C. Buncick, R. J. Warmack, and T. Vo-Dinh, *Anal. Chem.*, **58**, 1119 (1986).
20. M. Meier, A. Wokaun, and T. Vo-Dinh, *J. Phys. Chem.*, **89**, 1843 (1985).
21. T. Vo-Dinh, M. Uziel, and A. Morrison, *Appl. Spectrosc.* **41**, 605 (1987).
22. C. D. Tran, *Anal. Chem.*, **56**, 824 (1984).
23. C. D. Tran, *J. Chromatog.*, **292**, 432 (1984).
24. Fan Ni and T. M. Cotton, *Anal. Chem.*, **58**, 3159 (1986).
25. R.-S. Sheng, L. Zhu, and M. D. Morris, *Anal. Chem.*, **58**, 1116 (1986).
26. E. L. Torres and J. D. Winefordner, *Anal. Chem.*, **59**, 1626 (1987).
27. J. J. Laserna, A. Berthod, and J. D. Winefordner, *Appl. Spectrosc.*, **41**, 1137 (1987).
28. A. Berthod, J. J. Laserna, and J. D. Winefordner, *Talanta*, **34**, 745 (1987).
29. A. M. Ahern and R. L. Garrell, *Anal. Chem.*, **59**, 2816 (1987).
30. C. Jennings, R. Aroca, A. M. Hor, and R. O. Loutfy, *Anal. Chem.*, **56**, 2033 (1984).
31. E. Gautner, D. Steiner, and J. Reinhardt, *Anal. Chem.*, **57**, 1658 (1985).
32. R. H. Ritchie, *Phys. Rev.*, **106**, 879 (1957).
33. U. Laor and G. C. Schatz, *Chem. Phys. Lett.*, **82**, 566 (1981).
34. M. Kerker and C. G. Blatchford, *Phys. Rev. B*, **26**, 4052 (1982).
35. A. Wokaun, J. P. Gordon, and P. F. Liao, *Phys. Rev. Lett.*, **48**, 957 (1982).
36. P. W. Barber, R. K. Chang, and H. Massoudi, *Phys. Rev. Lett.*, **50**, 997 (1983).
37. M. Meier and A. Wokaun, *Opt. Lett.*, **8**, 581 (1983).
38. J. I. Gersten, *J. Chem. Phys.*, **72**, 5779 (1980).
39. J. Gerstan and A. Nitzan, *J. Chem. Phys.*, **73**, 3023 (1980).

40. G. C. Schatz and R. P. Van Duyne, *Surf. Sci.* **101,** 425 (1980).
41. A. Otto, *Surf. Sci.* **75,** 1392 (1978).
42. E. J. Zeman and G. C. Schatz, *J. Phys. Chem.,* **91,** 634 (1987).
43. E. Burstein, Y. J. Chen, C. Y. Chen, S. Lundquist, and E. Tosath, *Solid State Commun.,* **29,** 567 (1979).
44. T. E. Furtak, in *Advances in Laser Spectroscopy,* Vol. 2, B. A. Garet and J. R. Lombaradi, Ed., Wiley, New York (1983), p. 175.
45. P. K. Pandey, and G. C. Schatz, *J. Chem. Phys.,* **80,** 2959 (1984).
46. F. J. Adrian, *J. Chem. Phys,* **77,** 5302 (1982).
47. B. N. J. Persson, *Chem. Phys. Lett.,* **82,** 561 (1981).
48. M. Moskovits and D. P. DiLella, in *Surface-Enhanced Raman Scattering,* R. K. Chang and T. E. Furtak, Eds, Plenum, New York (1982), p. 243.
49. P. A. Lund, R. R. Smardzewski, and D. E. Terault, *Chem. Phys. Lett.,* **89,** 508 (1982).
50. B. H. Loo, *J. Chem. Phys.,* **75,** 5955 (1981).
51. T. H. Wood and M. V. Klein, *Solid State Commun.,* **35,** 263 (1980).
52. T. E. Furtak, *Surf. Sci.,* **101,** 1 (1981).
53. M. Udagawa, C. Chou, J. Hemminger, and S. Ushioda, *Phys. Rev.,* **B23,** 6843 (1981).
54. J. A. Creighton, C. B. Blatchford, and M. C. Albrecht, *J. Chem. Soc. Faraday Trans.,* **2**(75), 790 (1979).
55. P. F. Liao and M. B. Stern, *Opt. Lett.,* **7,** 483 (1982).
56. S. A. Asher, P. Flaugh, and G. Washinger, *Spectroscopy,* **1,** 26 (1986).
57. H. V. Malmstadt, M. L. Franklin, and G. Horlick, *Anal. Chem.,* **44,** 63A (1972).
58. T. Vo-Dinh and U. P. Wild, *Appl. Opt.,* **12,** 1286 (1973).
59. T. Vo-Dinh and U. P. Wild, *Appl. Opt.,* **13,** 2899 (1974).
60. Y. Talmi, *Anal. Chem.,* **47,** 658A (1975).
61. A. Campion and W. H. Woodruff, *Anal. Chem.,* **59,** 1299A (1987).
62. B. Pettinger, U. Wenneng, and H. Wetzel, *Surf. Sci.,* **101,** 409 (1980).
63. M. Fleischmann, P. R. Graves, and J. Robinson, *J. Electroanal. Chem.,* **182,** 87 (1985).
64. B. H. Loo, *J. Phys. Chem.,* **87,** 3003 (1983).
65. P. Hildebrandt and M. Stockburger, *J. Phys. Chem.,* **88,** 5935 (1984).
66. O. Siiman, L. A. Bumm, R. Callaghan, C. G. Blatchford, and M. Kerker, *J. Phys. Chem.,* **87,** 1014 (1983).
67. S. M. Heard, F. Grieser, and C. G. Barracloough, *Chem. Phys. Lett.,* **95,** 154 (1983).
68. P. C. Lee and D. Meisel, *Chem. Phys. Lett.,* **99,** 262 (1983).
69. T. Vo-Dinh, A. Alak, and R. L. Moody, *Spectrochim. Acta,* **4/5,** 605 (1988).
70. C. Jennings, R. Aroca, A. Hor, and R. O. Loufty, *Anal. Chem.,* **56,** 2033 (1984).
71. R. Aroca and F. Martin, *J. Raman Spectrosc.,* **17,** 243 (1986).
72. G. Grimmer, Ed. *Environmental Carcinogens: Polycyclic Aromatic Hydrocarbons,* CRC Press, Boca Raton, Fla. (1983).
73. H. V. Gelboin and O. P. Tso, *Polycyclic Hydrocarbons and Cancer,* Academic Press, New York (1978).
74. S. K. Miller, A. Baiker, M. Meier, and A. Wokaun, *J. Chem. Soc. Farad. Trans. 1,* **80,** 1305 (1984).

75. C. M. White, Ed. *Nitrated Polycyclic Aromatic Hydrocarbons*, Huething Publishers, New York (1985).
76. J. M. L. Sequaris and E. Koglin, *Anal. Chem.*, **59,** 525 (1987).
77. R. D. Freeman, R. M. Hammaker, C. E. Meloan, and W. G. Fately, *Appl. Spectrosc.* **42,** 456 (1988).
78. R. K. Chang and B. L. Laube, *CRC Crit. Rev. Solid State Mater. Sci.*, **12,** 1 (1984).
79. E. Koglin and J. M. Sequaris, *J. Phys. Colloq.*, **44,** C10–487 (1983).
80. C. J. Sandroff, D. A. Weitz, J. C. Chung, and D. R. Herschbach, *J. Phys. Chem.*, **87,** 2127 (1983).
81. C. J. Sandroff, S. Garoff, and K. P. Leung, *Chem. Phys. Lett.*, **96,** 547 (1983).

INDEX

Absorbance, 80, 85, 88
  detectors, 85
Aceanthrylene, 231
Acenaphthalene, 300
Acenaphthene, 158, 235, 351, 413
Acenaphthylene, 234, 351
Acephenanthrylene, 231
2-Acetylaminofluorene, 20
2-Acetylaminophenanthrene, 20
Acridine, 178, 179, 224, 416
Adatom model, 455
Aerodynamic mean diameter, 34
Aflatoxin B_1 (AFL-B_1), 406
  DNA adduct, 406
Airborne particulate, 2
Air pollution, 16
Alanine, 300
Alkylated PAHs, 220
Alkylation, 75
Alkylbenzenes, 162
Alkylcarbazoles, 135
Alkylfluorenes, 6
Alkylnapthalenes, 91, 251
Alkylphenanthrenes, 6
Alkylpyrenes, 91
Alkylsilane, 77
3-Amino-N-methylphthalimide, 207
Amperometric electrochemical detection, 166
Amplitude modulators, 205
Analine, 274
Aniline, 417
Anthanthrene, 50, 443
1-Anthracenamine, 223
Anthracene, 2, 33, 65, 158, 164, 174, 189,
  208, 213, 229, 232, 235, 300, 318,
  351, 413, 424, 445, 446, 476
  decomposition, 42
  derivatives, 220
  half-life, 45

hydroxy, 446
  in vivo metabolism, 2
  substituted, 446
  vapor pressure, 33
9-Anthracene carbonitrile, 224
Anthracenocarbazole, 175
Anthraquinone, 37, 427
Antibody, 22
  development, 391–393
  monoclonal, 24, 392, 398
  polyclonal, 392
Anti-Stokes lines, 434
Aoki's resonance count, 85
Aromatic amines, 15, 230
Autoionization, 287, 304
Autoionize, 266, 302
Aza-arenes, 15, 230
4-Aza-fluorene, 41
Azulene, 235, 298

BaP, 4, 6, 8, 9, 16, 17, 24, 34, 46, 50, 154.
  See also Benz(a)pyrene
  half-lives, 45
  in mussels, 9
  quinones, 38
  1,6-quinones, 38
  3,6-quinones, 38
  6,12-quinones, 38
BaP-7,8-diol, 18
BaP-7,8-diol-9,10-oxide, 17
BaP-tetrol, 25, 470
"Bay region" theory, 2
Benz[a]anthracene, 2, 6, 37, 43, 45, 50, 66,
  185, 232, 235, 352, 413
  quinone, 43
  reactions with N_2O_5, 49
1H-Benz[de]anthracene-7-one, 247
Benzene, 164, 267, 286, 291, 293–298, 302,
  307–315

(*continued from front*)